Building Construction and Architecture

Building Construction and Architecture

Edited by **Seth Royal**

Willford Press

New York

Published by Willford Press,
118-35 Queens Blvd., Suite 400,
Forest Hills, NY 11375, USA
www.willfordpress.com

Building Construction and Architecture
Edited by Seth Royal

International Standard Book Number: 978-1-68285-306-1 (Hardback)

Printed in China.

Contents

Preface

The purpose of the book is to provide a glimpse into the dynamics and to present opinions and studies of some of the scientists engaged in the development of new ideas in the field from very different standpoints. This book will prove useful to students and researchers owing to its high content quality.

The principles and techniques of building construction and architecture have constantly evolved in the recent past. This comprehensive book presents the latest and most innovative models of building construction through lucid descriptions of topics such as building material, building architecture and engineering, etc. It aims to expand the scope of research in this discipline and aid its advancement. Students, researchers, experts and all associated with the field of construction and architecture will benefit alike from this book.

At the end, I would like to appreciate all the efforts made by the authors in completing their chapters professionally. I express my deepest gratitude to all of them for contributing to this book by sharing their valuable works. A special thanks to my family and friends for their constant support in this journey.

Editor

Performance Evaluation of Different Masonry Infill Walls with Structural Fuse Elements Based on In-Plane Cyclic Load Testing

Andrew Kauffman [1] and Ali M. Memari [2,*]

[1] Barnhart Crane and Rigging Company, Monroe, MI 48161, USA;
E-Mail: kauffman205@gmail.com

[2] Department of Architectural Engineering and Department of Civil and Environmental Engineering, The Pennsylvania State University, University Park, PA 16802, USA

* Author to whom correspondence should be addressed; E-Mail: memari@engr.psu.edu

External editor: Chimay J. Anumba

Abstract: This paper discusses the performance of a structural fuse concept developed for use as a seismic isolation system in the design and retrofit of masonry infill walls. An experimental program was developed and executed to study the behavior of the structural fuse system under cyclic loads, and to evaluate the performance of the system with various masonry materials. Cyclic tests were performed by applying displacement controlled loads at the first, second, and third stories of a two-bay, three-story steel test frame with brick infill walls; using a quasi-static loading protocol to create a first mode response in the structural system. A parametric study was also completed by replacing the brick infill panels with infill walls constructed of concrete masonry units and autoclaved aerated concrete blocks, and applying monotonically increasing, displacement controlled loads at the top story of the test frame.

Keywords: masonry infill walls; seismic isolation; cyclic lateral load testing; structural fuse

1. Introduction

Masonry infill walls are a common building element found throughout the world. Infill walls constructed of various masonry materials are often used in both concrete and steel structures to infill the frame openings [1]. This type of construction is particularly common in developing countries where masonry materials such as clay bricks, concrete masonry units, and hollow clay tiles are readily available [2]. In many cases, infill walls are treated as architectural elements and their influence on the behavior of the structure is not considered. This design philosophy can lead to uneconomical design as well as unexpected behavior and even catastrophic collapse.

It has been widely documented by many researchers that masonry infill walls significantly influence the in-plane behavior and response of structural frames [3–18]. Masonry infill walls increase the stiffness of structural frames, and in general help to limit building deflection under lateral loads. Although this increase in stiffness is beneficial for limiting building drift during wind storms and minor to moderate earthquakes, it can have a negative impact on the performance of structures during major seismic events. A comprehensive literature review on these issues is presented by Aliaari [19].

The structural properties of masonry infill walls are often overlooked by designers who do not consider the increase in stiffness and potential decrease in ductility introduced into structural frames by the addition of unreinforced masonry infill materials. Typically, concrete and steel building frames are designed to resist all of the gravity and lateral loads, including wind and seismic forces. Infill walls are often treated as nonstructural elements even though they have a significant influence on the in-plane behavior and seismic performance of the structure [20].

Ignoring the contribution of masonry infill walls to the strength and stiffness of building frames can lead to damage in the masonry walls as well as resulting in an inefficient use of materials and uneconomical design [21]. According to Colombo *et al.* [22], neglecting the effects of the infill panels, as suggested by various building codes, does not lead to a safe seismic design. The assumption that masonry infill walls in concrete and steel frames will only increase the lateral load capacity of these structures is a common misconception [23]. This oversight can result in severe structural damage and collapse in buildings where the ability of the frame to safely dissipate seismic input energy has been significantly overestimated. Overstressing of the masonry walls and the formation of a collapse mechanism in the structural frame can occur if the composite interaction between masonry infill walls and bounding frames is not accounted for during design.

Two common design philosophies have been developed which consider the influence of infill walls on the response of structural frames to lateral loads [24]. One approach is to isolate the masonry infill walls from the bounding frame with a physical gap [25]. This allows the frame to act independently of the masonry walls and to be designed without consideration of the interaction between the frame elements and the infill panels. Isolating the infill panels from the frame can prevent severe cracking and damage to the masonry materials. This is an important consideration since falling debris from damaged and collapsed infill walls is a major life safety issue for buildings in seismic areas. By isolating the infill walls from the concrete or steel frame, the structure can dissipate the seismic input energy in a predictable and safe manner.

A second common design method for infilled structural frames is to recognize the contribution of the infill panels to the strength and stiffness of the building frame and design the masonry walls as structural elements [26]. In many cases, the size of the structural members in a building frame is dictated by building drift limits or other serviceability criteria rather than strength considerations. By taking into account the structural properties of the masonry infill walls in the calculation of building drift, the framing members can be reduced in size, resulting in a more economical design. This approach requires careful consideration of the composite interaction between the structural frame and the infill panels so that both can be designed to safely resist lateral loads. A disadvantage of this method is that the structural frame cannot act independently of the infill walls when subjected to large seismic forces. The added stiffness of the infill panels decreases the natural period of the structure, which results in higher seismic loads [7]. Severe cracking of the masonry infill walls and shedding of debris can take place if the infill walls are not properly designed and detailed.

A new concept in the performance and design of masonry infill walls is the idea of a structural fuse system [19,25–27]. The structural fuse concept combines the two common design approaches by allowing masonry infill walls to be engaged with the bounding frame up to a predetermined level of lateral load. Brittle failure of the infill walls or frame elements is prevented by the introduction of a fuse mechanism, which isolates the infill material from the frame under higher loads. For lower levels of load, the strength and stiffness of the masonry material work compositely with the structural frame to limit lateral deflections. Under higher lateral loads, the infill panels are disengaged from the structure using the fuse mechanism, which prevents damage to the masonry walls and the formation of a frame failure mechanism. With this system, the structural frame can be designed to resist high lateral forces without the influence of the masonry material.

The fuse element is the key component of the structural fuse system. The purpose of the fuse is to serve as a link between the structural frame and the masonry infill walls and prevent damage to the infill material. An extensive analytical and experimental study has been carried out on fuse materials and systems [19,25,28]. Under typical loading conditions, the fuse mechanism transfers the story shear forces from the structural frame to the masonry infill panels, which help to resist the in-plane lateral loads and limit frame deflections. If the story shear forces are sufficiently high to cause inelastic behavior in the masonry panels, the fuse element is designed to "break" and disengage the infill wall from the frame before damage occurs to the masonry material.

An initial experimental study [25,28] showed that the structural fuse concept works well as a seismic isolation system. In that study, monotonically increasing, displacement controlled loads were applied at the top story of a two-bay, three-story steel test frame with brick infill walls and lumber disk fuse elements. According to the test results, the brick infill walls made a significant contribution to the in-plane stiffness of the test frame, up to the point where the fuse elements failed. The fuse mechanism successfully isolated the infill panels from the test frame, preventing damage to the brick masonry material. Since the initial study [25,28] concentrated on concept development and proof-of-concepts based on monotonic tests, a follow-up experimental program was developed to study the response of the structural fuse system under a cyclic loading history, and to evaluate the performance of the system with various masonry infill materials.

2. Experimental Program

This paper discusses the details of the follow-up testing program [29]. In the test program discussed here, two series of tests were performed in the Building Components and Envelopes Research Laboratory (BCERL) at the Pennsylvania State University, on the same test frame used by Aliaari [19]. For the first test series, quasi-static, cyclic in-plane loads were applied to the test frame with brick infill wall panels in place. In the second series of tests, a parametric study was completed by replacing the brick infill panels with infill walls constructed of concrete masonry units (CMU) and autoclaved aerated concrete blocks (AAC). For these tests, displacement-controlled monotonic loads were applied in only one direction to match the loading protocol used by Aliaari [19].

Both series of tests were performed on the 1/4 scale, two-bay, three-story steel test frame shown in Figures 1 and 2, respectively. The test frame shown in Figure 1 consists of HSS 127 mm × 127 mm × 9.5 mm (5 × 5 × 3/8 in.) hollow structural steel members pin connected using 38.1 mm diameter steel bars to allow for adequate floor movements during racking and to enable the study of the system in a braced (stiff) and partially braced (flexible) configuration. The masonry infill wall specimens were supported by the bottom beam of each bay of the test frame with a 152.5 mm gap between the sides of the walls and the HSS columns, and a gap of between 38.1 mm and 60.3 mm between the top of the walls and the upper beam. Steel angles were welded to the top of the support beams on either side of the infill panels to prevent the base of the walls from sliding in the in-plane direction. Out-of-plane movement at the bottom of the infill panels was prevented by steel angles welded to the sides of the HSS beams. In order to prevent overturning of the infill panels, steel plates were secured to the top of the walls using 12.7 mm diameter threaded rods bolted to the outstanding legs of these angles.

Figure 1. 1/4 scale, two-bay, three-story steel test frame.

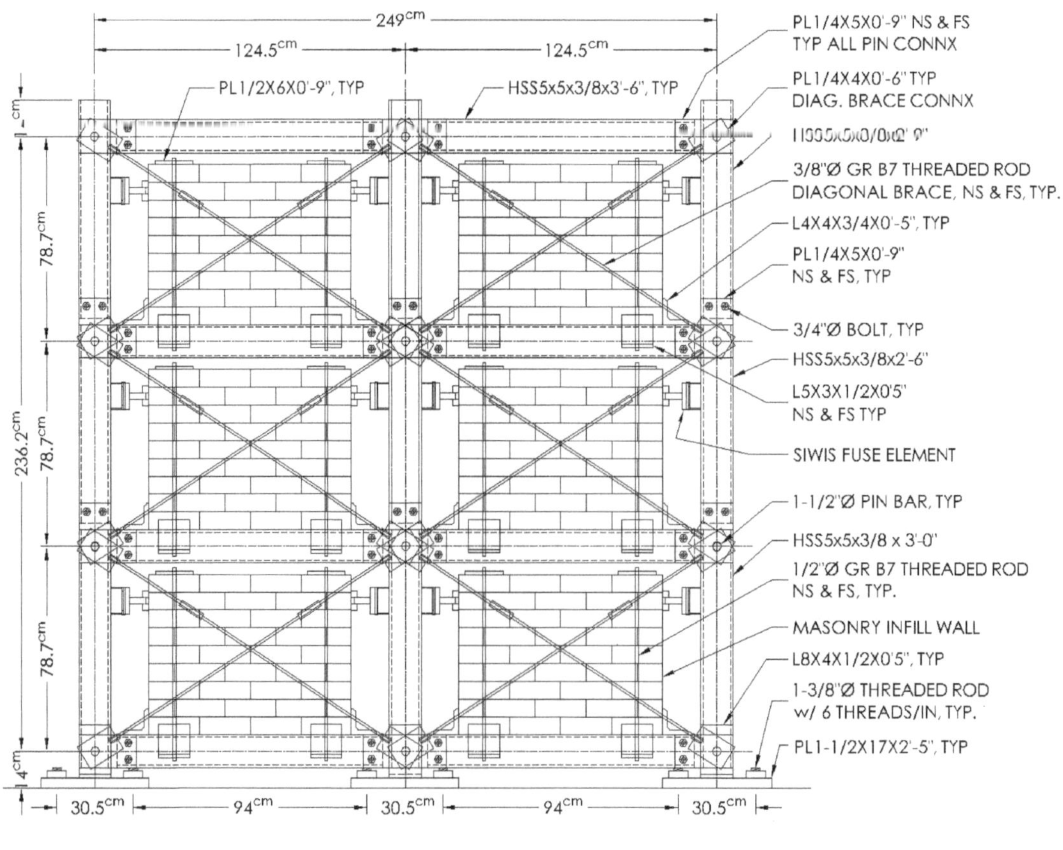

Figure 2. Concrete test platform and steel reaction frame.

In-plane, lateral stability of the test frame was provided by diagonal steel bracing rods. In the fully braced configuration, a total of four diagonal braces were used in each bay, with two bracing rods on each side of the infill panels. For the half-brace configuration, bracing rods were used in only one bay of each story. The diagonal bracing system was designed so that in the fully braced configuration, the pin-jointed steel frame had the same in-plane stiffness as an equivalent steel frame with moment resisting connections. Each diagonal brace member consisted of two 9.5 mm diameter, Grade B7, threaded rods with opposite hand threads joined together using a special coupling nut with right-hand threads in one side and left-hand threads in the other. The bracing rods were attached to the test frame using steel connection plates that fit over the 38.1 mm diameter steel pin bars. The rods were attached to the connection plates using welded coupling nuts.

The masonry infill walls used in the cyclic and parametric tests were constructed in the BCERL test facility by professional masons. Destructive shear tests were performed to determine the in-plane shear strength of each type of masonry wall panel. The brick, CMU, AAC, and hollow clay tile (HCT) infill wall specimens used for this testing program are shown in Figure 3. The brick wall panels were used for the cyclic tests, while the CMU and AAC specimens were used for the parametric study. The HCT infill wall specimens were included in the destructive shear tests, but not used in the structural fuse system tests due to their low in-plane shear capacity. The set-up for the destructive shear tests is shown in Figure 4. The results of these tests are given in Table 1.

During the structural fuse system tests, lateral loads were applied to the test frame at each story level using hydraulic cylinders attached to the steel reaction frame shown in Figure 2. The in-plane lateral loads were transferred from the steel test frame to the masonry infill walls through the fuse mechanism shown in Figure 5. As shown in this figure, the 100 mm diameter, lumber disk fuse elements were held in place by a steel seat disk and pipe assembly welded to the HSS columns near the top of the infill panels. The fuse punching mechanism consisted of a 22.4 mm diameter, half-threaded steel rod attached to a steel bearing plate which was epoxied to the infill panels. The lumber disk fuse elements used in the cyclic and parametric testing programs were cut from hard maple boards. Lumber disk puncture tests were performed to evaluate the strength of the fuse elements and determine the appropriate disk thickness to use with each type of masonry infill material [29]. The results of the initial series of lumber disk puncture tests performed as part of this testing program are provided in Table 2. The structural fuse test presented here builds on the results obtained from tests presented in an earlier study [28] where different materials including concrete and wood disks were tested. The study also provided load-displacement diagrams for typical fuse elements and indicated that wood disks have better deformation capacity and provide desirable ductility for the infill wall system with fuse elements. Calibration issues related to the fuse elements as needed for the test programs may be found in the documents developed by Aliaari [19] and Kauffman [29].

Figure 3. Masonry infill wall specimens.

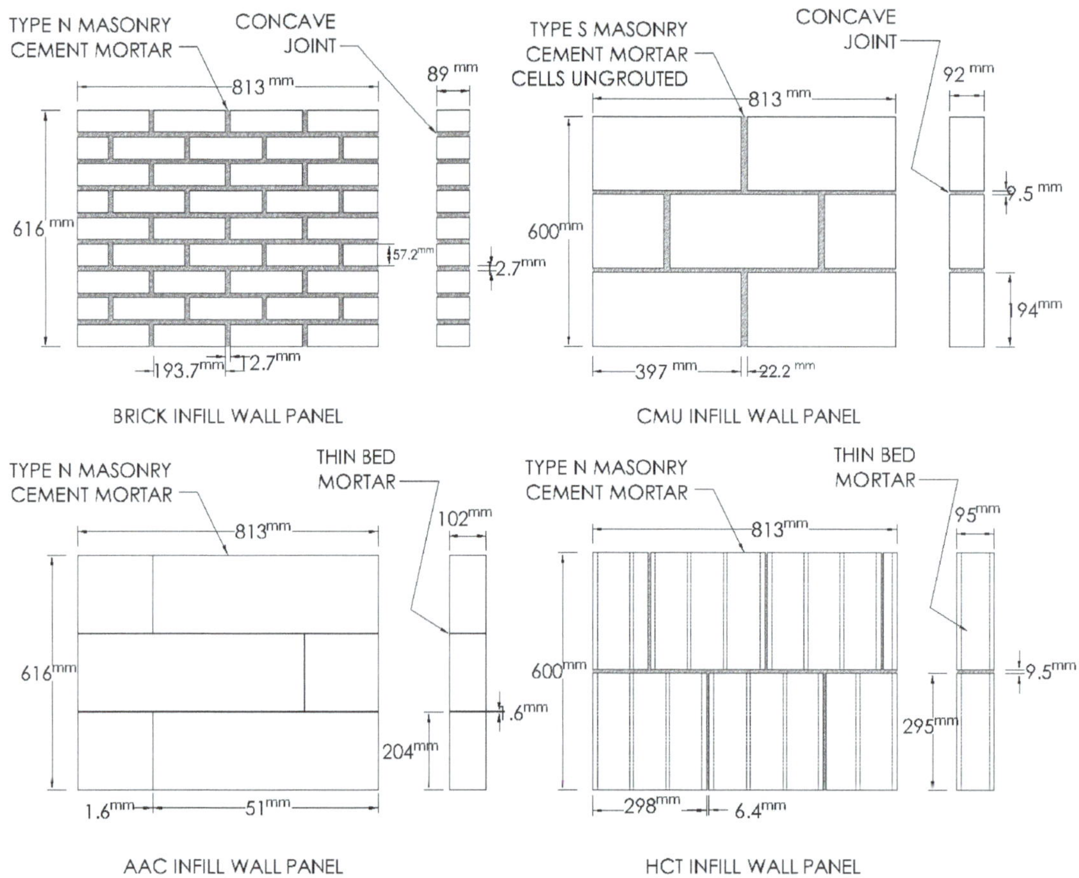

Figure 4. In-plane, destructive shear test set-up.

Table 1. Masonry wall panel shear tests results.

Wall Specimen	Masonry Material	Peak Capacity kN	Average kN	Standard Deviation kN	Coefficient of Variation CV%	Stiffness kN/mm	Average kN/mm	Standard Deviation kN/mm	Coefficient of Variation CV%
W1	AAC	61.47	56.8	6.2	11.3	5.62	7.0	1.5	21.3
W2	AAC	52.09				6.37			
W3	AAC	61.16				8.27			
W4	AAC	52.44				7.76			
W5	CMU	114.50	97.7	20.0	20.5	13.38	15.5	2.3	14.7
W6	CMU	108.49				16.41			
W7	CMU	87.36				17.64			
W8	CMU	80.51				14.73			
W9	HCT	9.70	18.6	6.2	78.4	1.66	3.7	1.5	80.5
W10	HCT	27.49				5.66			

Figure 5. Fuse mechanism detail.

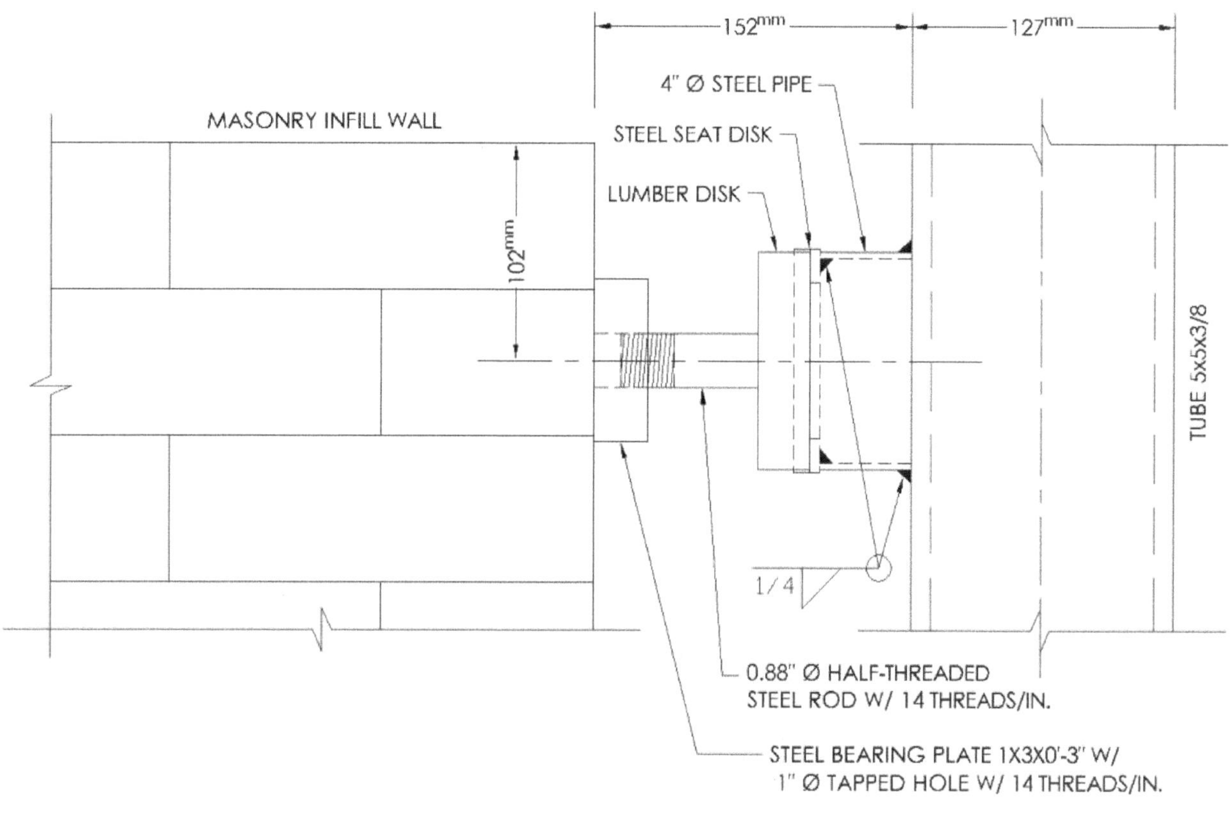

For the cyclic testing program, quasi-static, displacement-controlled loads were applied simultaneously at the first, second and third story levels using three hydraulic jacks. Two different bracing configurations were tested, including full-brace and half-brace, in order to observe the performance of the structural fuse system for frames with different in-plane stiffness properties. The fuse element thicknesses were also varied. For some of the tests, the same fuse element thickness was used at each story level. For other tests, a different fuse element thickness was used at each story level. A test matrix of the cyclic testing program is given in Table 3. As the table indicates, the test specimen included bare frame, full brace, and half brace configurations. Diagrams of each test configuration are shown in

Figure 6. Figure 7 shows the location of the fuse elements on the steel test frame during the cyclic tests. The data collected during the cyclic tests included the load applied to the test frame by the hydraulic cylinders, the in-plane deflection of the test frame at each story level, and the shear forces transferred from the test frame to the brick infill walls through the structural fuse elements.

Table 2. Lumber disk puncture test results.

Disk Specification	Thickness	Weight	Volume	Density	Average	Standard Deviation	Coefficient of Variation	Capacity	Average	Standard Deviation	Coefficient of Variation
	mm	gr	cm³	gr/cm³	gr/cm³	gr/cm³	CV%	kN	kN	kN	CV%
D1	12.9	74.9	100.9	0.74				8.47			
D2	12.9	74.0	100.9	0.73	0.72	0.0249	3.3	11.21	10.66	1.97	18.5
D3	12.9	70.4	100.9	0.70				12.29			
D4	19.1	106.7	149.1	0.72				18.72			
D5	19.3	108.5	151.1	0.72	0.71	0.0111	1.7	19.16	19.11	0.37	1.9
D6	19.3	105.1	151.1	0.70				19.46			
D7	25.4	139.4	197.9	0.70				22.46			
D8	25.6	142.3	199.8	0.71	0.70	0.00830	1.2	19.94	21.99	1.87	8.5
D9	25.4	137.6	197.8	0.70				23.59			
D10	38.3	243.1	301.7	0.81				38.31			
D11	38.1	236.1	298.2	0.79	0.77	0.0498	6.4	35.53	37.05	1.41	3.8
D12	38.1	215.2	301.2	0.71				37.32			

Table 3. Cyclic testing program test matrix.

Test	Type	Description	Fuse Element Thickness (mm)		
			1st Story	2nd Story	3rd Story
BFMT1	Bare Frame Monotonic Test	Fully braced, no fuse elements	NA	NA	NA
BFMT2	Bare Frame Monotonic Test	Fully braced, no fuse elements	NA	NA	NA
FBMT1	Full Brace Monotonic Test	Fully braced, with fuse elements	25.4	19.1	12.7
FBMT2	Full Brace Monotonic Test	Fully braced, with fuse elements	0.5	12.7	12.7
BFCT1	Bare Frame Cyclic Test	Fully braced, no fuse elements	NA	NA	NA
BFCT2	Bare Frame Cyclic Test	Fully braced, no fuse elements	NA	NA	NA
FBCT1	Full Brace Cyclic Test	Fully braced, with fuse elements	25.4	19.1	12.7
FBCT2	Full Brace Cyclic Test	Fully braced, with fuse elements	25.4	19.0	12.7
FBCT3	Full Brace Cyclic Test	Fully braced, with fuse elements	12.7	12.7	12.7
FBCT4	Full Brace Cyclic Test	Fully braced, with fuse elements	12.7	12.7	12.7
HBCT1	Half Brace Cyclic Test	Half braced, with fuse elements	25.4	19.1	12.7
HBCT2	Half Brace Cyclic Test	Half braced, with fuse elements	25.4	19.1	12.7

Figure 6. Test frame configurations for cyclic test program.

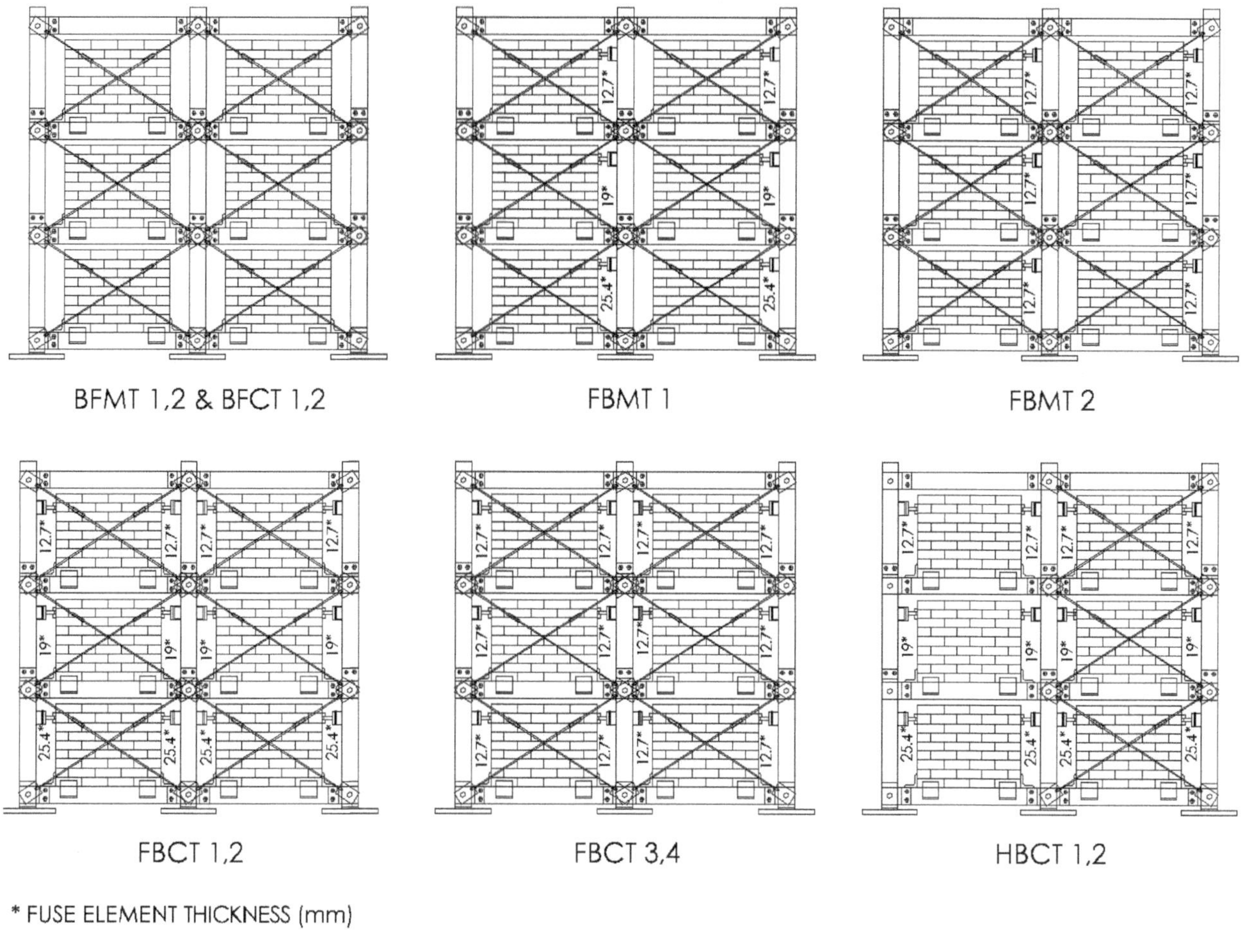

Figure 7. Fuse element locations on test frame.

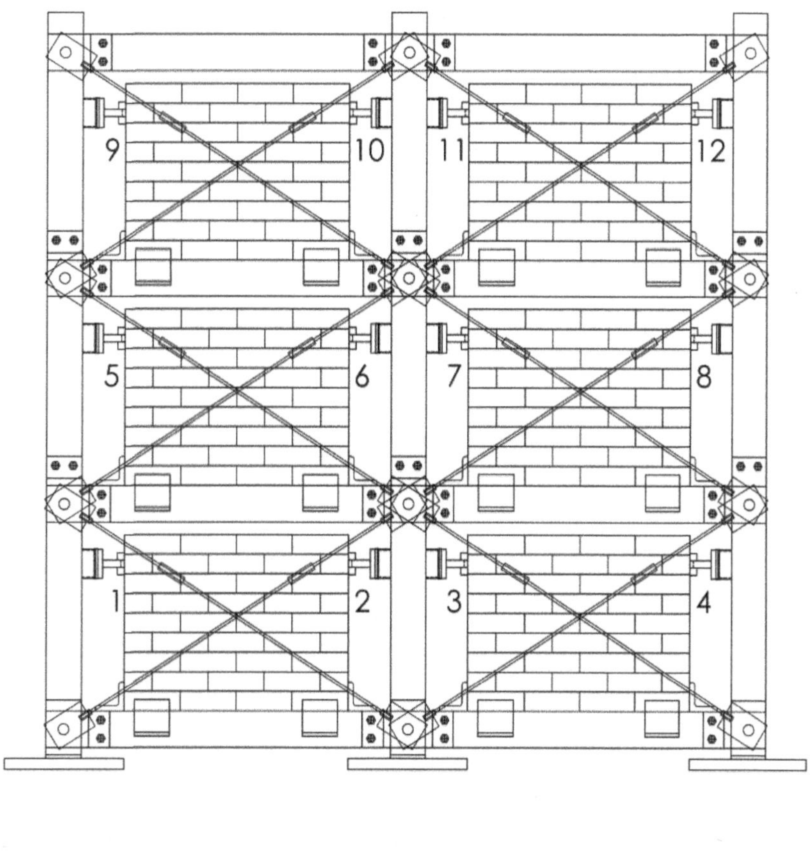

For the parametric tests, monotonic loads were applied at the third story of the test frame using one hydraulic cylinder. As with the cyclic tests, varying fuse element thicknesses and bracing configurations were tested. A test matrix for the parametric test series is given in Table 4. Figure 8 shows typical elevations of the fully braced and half braced specimens. The data collected during these tests included the load applied at the third story level of the test frame by the hydraulic cylinder, and the in-plane deflection of the test frame at each story level.

Table 4. Parametric testing program test matrix.

Test	Masonry Material	Description	Fuse Element Thickness (mm)		
			First Story	Second Story	Third Story
CMUT1	CMU	Fully braced, with fuse elements	22.2	19.1	12.7
CMUT2	CMU	Fully braced, with fuse elements	22.2	19.1	12.7
CMUT3	CMU	Fully braced, with fuse elements	12.7	12.7	12.7
CMUT4	CMU	Fully braced, with fuse elements	12.7	12.7	12.7
CMUT5	CMU	Half braced, with fuse elements	22.2	19.1	12.7
CMUT6	CMU	Half braced, with fuse elements	22.2	19.1	12.7
AACT1	AAC	Fully braced, with fuse elements	19.1	12.7	6.4
AACT2	AAC	Fully braced, with fuse elements	19.1	12.7	6.4
AACT3	AAC	Fully braced, with fuse elements	6.4	6.4	6.4
AACT4	AAC	Fully braced, with fuse elements	6.4	6.4	6.4
AACT5	AAC	Half braced, with fuse elements	19.1	12.7	6.4
AACT6	AAC	Half braced, with fuse elements	19.1	12.7	6.4

Figure 8. Test frame configurations for parametric testing program.

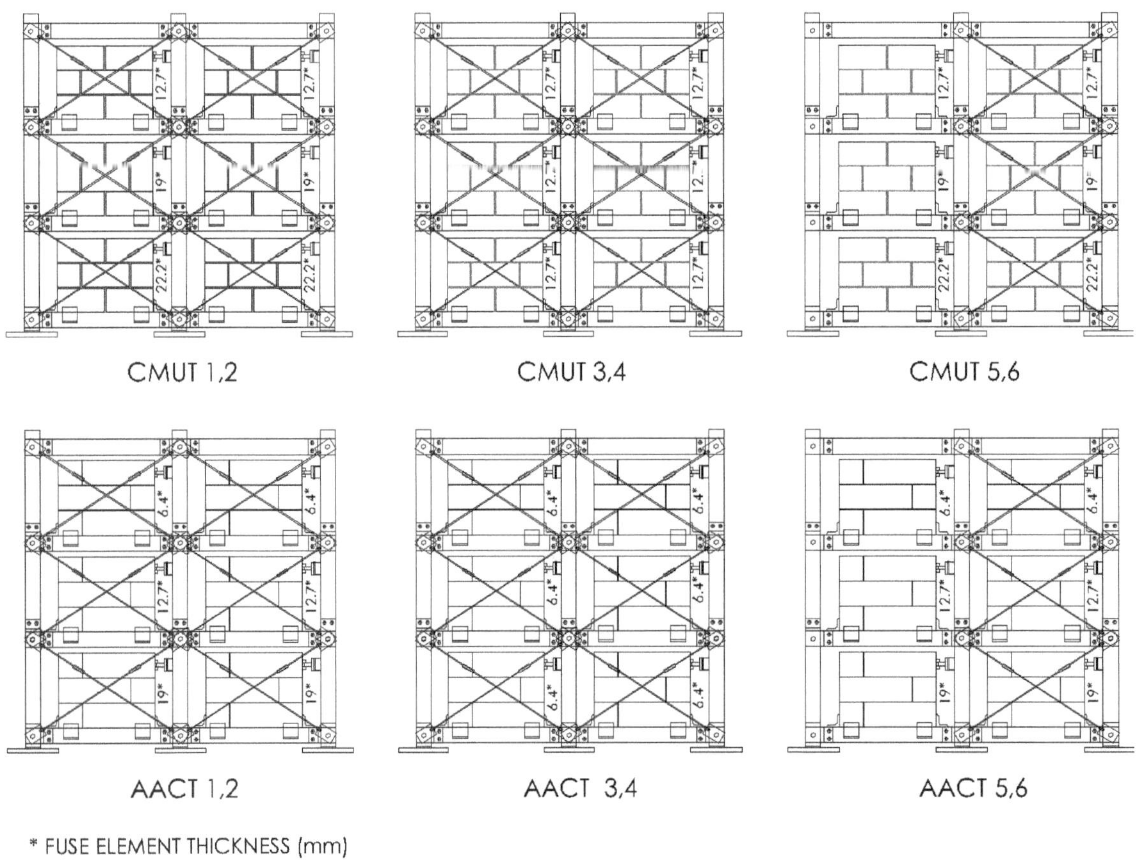

3. Monotonic and Cyclic Tests with Brick Infill Panels

3.1. Bare Frame Monotonic Tests

At the beginning of the cyclic testing program, bare frame monotonic tests were performed without fuse elements, so that the test frame could act independently from the infill walls. The full bracing configuration was used with all the diagonal bracing rods in place. The goal of these tests was to simulate a first mode, inverted triangle test frame displacement shape by applying simultaneous, displacement controlled loads at each story level using three hydraulic cylinders. According to Tomazevic and Klemenc [30], the dominant mode shape for masonry structures is typically the first mode response, which is characterized by an inverted triangle shape. The story displacement histories from BFMT2 (bare frame monotonic test 2), given in Figure 9, show that the displacements applied at each story level during this test resulted in an inverted triangle displacement shape of the test frame. During both bare frame monotonic tests, the force applied by the cylinder at the third story of the test frame was much greater than the load applied at the first and second story levels. For both of these tests, the maximum force applied at the third story level was between 10,680 N and 11,120 N, while the max force applied at the second and first story levels was less than 890 N. Since the geometry and stiffness of the bracing members was the same at each story level, the frame naturally assumed an inverted triangle deflection shape, and very little load needed to be applied at the first and second story levels to maintain this mode shape.

Figure 9. Bare frame cyclic test 2 (BFMT2) displacement history.

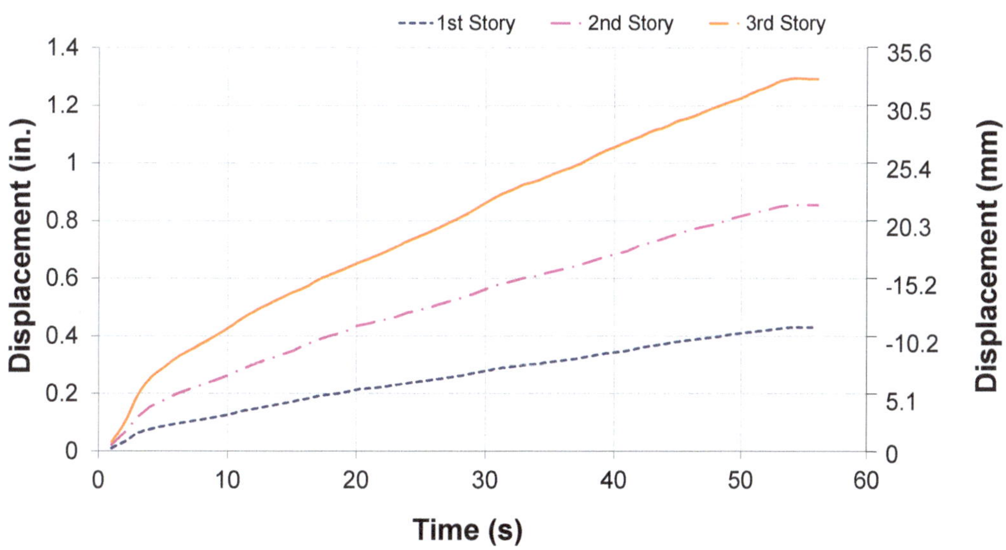

3.2. Full Brace Monotonic Tests

After an appropriate loading procedure was determined for producing a first mode response in the test frame, full brace monotonic tests were performed with fuse elements in place acting on the brick infill panels. For these tests, structural fuse elements were installed on the compression side of each brick infill panel. For FBMT1 (full brace monotonic test 1), the thickness of the fuse elements at the first, second, and third stories were 25.4 mm, 19.1 mm, and 12.7 mm, respectively. For test FBMT2,

12.7 mm thick disks were used at every story level. For both of these tests the full brace configuration was used with all the diagonal bracing rods in place on the test frame.

The purpose of these tests was to define the parameters for the quasi-static load history [31]. These parameters include the deformation level at which the test specimens begin to behave inelastically, the increment in peak deformation between subsequent load steps, and the number of individual cycles performed at each load step. According to the loading protocol [31], after reaching the yield value of the deformation control parameter, δ_y, the increment in peak deformation, Δ, should be constant for subsequent load steps. For the structural fuse system tests, inelastic behavior occurs at the onset of cracking in the lumber disk fuse elements.

During both of the full brace monotonic tests, the desired story level displacement ratios were maintained resulting in an inverted triangle displacement shape of the test frame. As with the bare frame monotonic tests, the force applied by the hydraulic cylinder at the third story level was greater than the force applied by the cylinders at the first and second levels. The maximum punching rod forces measured at each disk location during both tests are given in Table 5. Plots of the force in the steel rod at each fuse element location *vs.* the third story frame deflection for test FBMT1 are given in Figure 10.

Table 5. Full brace monotonic test (FBMT) fuse forces.

Story Level	Test Label	Fuse Thickness (mm)	Fuse Location	Maximum Fuse Force (kN)	Average Fuse Force (kN)
First Story	FBMT1	25.4	Fuse 2	14.55	16.92
			Fuse 4	19.30	
	FBMT2	12.7	Fuse 2	11.30	11.21
			Fuse 4	11.11	
Second Story	FBMT1	19.1	Fuse 6	16.69	17.35
			Fuse 8	18.02	
	FBMT2	12.7	Fuse 6	9.96	10.86
			Fuse 8	11.77	
Thirde Story	FBMT1	12.7	Fuse 10	11.01	10.28
			Fuse 12	9.54	
	FBMT2	12.7	Fuse 10	11.11	10.78
			Fuse12	10.46	

During the full brace monotonic tests, the brick infill panels contributed a significant amount of strength to the overall in-plane load resistance of the test frame. Table 6 summarizes the total in-plane load transferred to the infill panels through the fuse elements at each story level during tests FBMT1 and FBMT2 at the third story displacement intervals of 10.2 mm, 20.3 mm, and 30.5 mm. The greatest percentage of the total in-plane load was transferred through the fuse elements at a third story displacement of 25.4 mm. When the third story displacement reached 30.5 mm, a smaller percentage of the in-plane load was transferred to the infill walls since the fuse elements had experienced more damage by this point in the test. Even after the fuse elements experienced significant damage, some of the in-plane load was still transferred through the fuse elements to the infill walls.

Figure 10. Test FBMT1 punching rod force histories. (**a**) FBMT1 1st story rod histories; (**b**) FBMT1 2nd story rod histories; (**c**) FBMT1 3rd story rod histories.

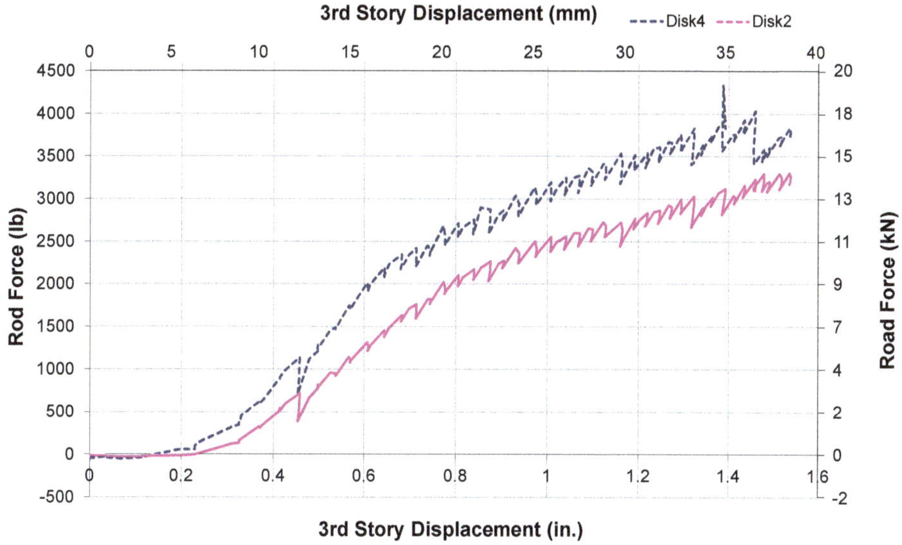

(**a**)

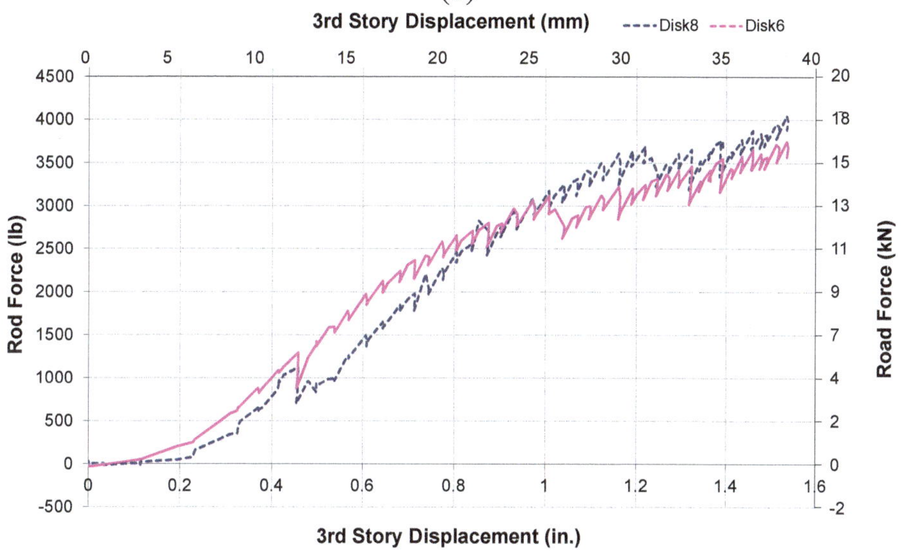

(**b**)

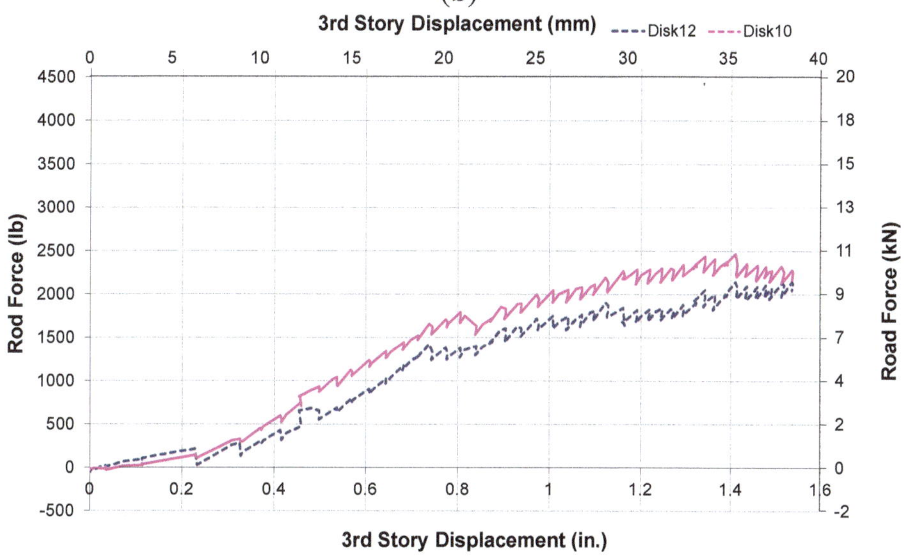

(**c**)

Table 6. Fuse forces transferred to infill panels during full brace monotonic tests.

Test Identification	3rd Story Displacement (mm)	Combined Fuse Force (kN)			Total Fuse Force * (kN)	Total Applied Load (kN)	% Total Load in Fuses (kN)
		1st Story	2nd Story	3rd Story			
FBMT1	10.2	6.24	8.69	4.62	19.55	33.66	58.1
	20.3	20.73	21.87	14.17	56.77	87.91	64.6
	30.5	30.28	29.82	17.71	77.81	144.05	54.0
FBMT2	10.2	5.28	5.16	7.39	17.84	30.99	57.6
	20.3	17.32	17.25	14.42	48.98	79.52	61.6
	30.5	21.33	19.31	20.24	60.88	142.69	42.7

* Combined force at each fuse element location at all three story levels.

During tests FBMT1 and FBMT2, the first fuse element cracking occurred at a third story displacement of between 7.6 mm and 12.7 mm. Based on these results, the value δ_y for the quasi-static loading protocol was set equal to a third story test frame displacement level of 12.7 mm. The upper bound of the third story displacement corresponding to fuse element crack initiation was chosen based on the loading protocol guidelines [32] that at least two load steps be performed before the deformation applied to the test specimen reaches δ_y. In order to prevent damage to the steel bracing rods, the maximum third story displacement level applied to the test frame during the cyclic tests was 35.6 mm. To allow for the maximum number of steps in the cyclic loading history, a third story displacement increment of 7.6 mm was chosen for the increment in peak deformation between load steps. A total of two load cycles were performed during each load step. The quasi-static loading protocol used for the cyclic tests is shown in Figure 11.

Figure 11. Quasi-static cyclic loading history.

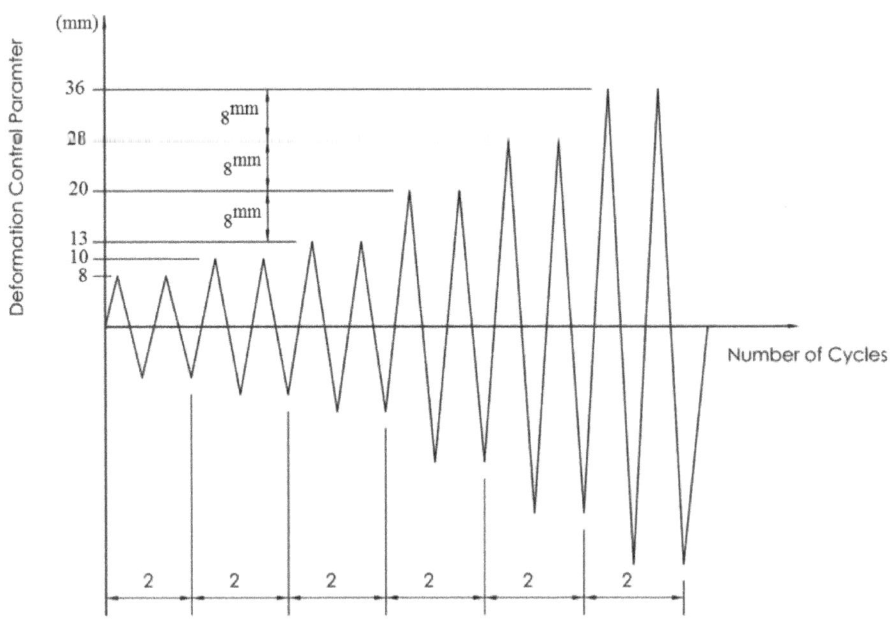

3.3. Bare Frame Cyclic Tests

The bare frame cyclic tests were performed with all of the bracing rods in place on the test frame and no fuse elements. During these tests, displacement controlled loads were applied in both in-plane directions (advance and retract) based on the quasi-static loading protocol shown in Figure 11. The load applied at each story was controlled manually using a separate hydraulic pump attached to each of the load cylinders. The story level displacement histories for test BFCT1 (bare frame cyclic test 1) are shown in Figure 12. The story level load histories for test BFCT1 are shown in Figure 13. The load-displacement, hysteretic behavior at the first, second, and third story levels of the test frame during test BFCT1 is shown in Figure 14. In these figures the load applied at each story of the test frame is plotted *vs.* the displacement measured at that story. The hysteretic plots are fairly symmetrical at story levels two and three for both of the bare frame cyclic tests. This indicates that the work done by the hydraulic cylinders on the test frame at these story levels was similar in both in-plane directions of the load cycle. At the first story level, however, the work done by the cylinders on the test frame was not symmetrical with respect to the direction of the applied load.

Figure 12. Bare frame cyclic test BFCT1 displacement history.

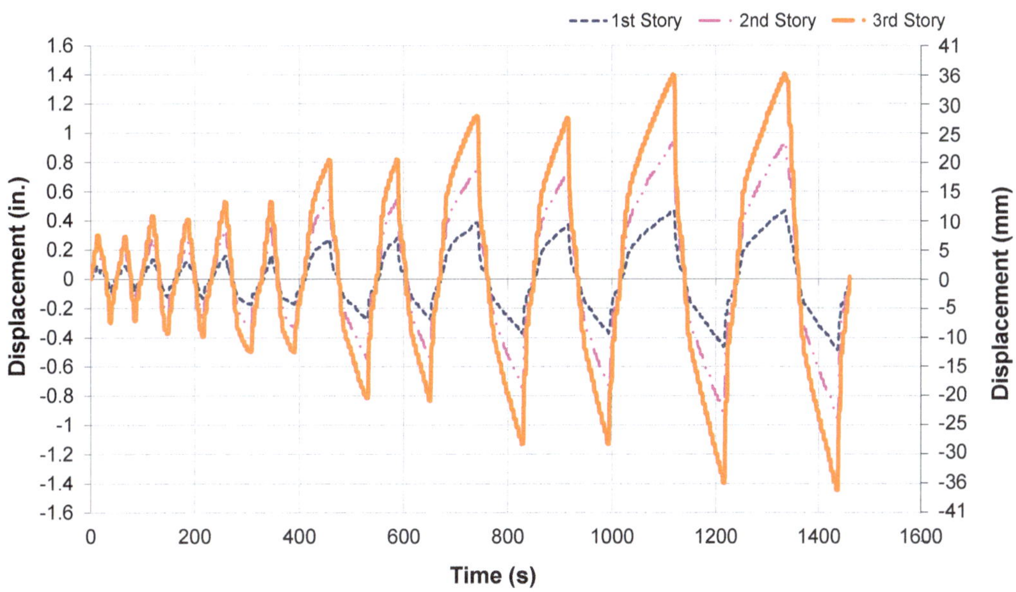

Figure13. Bare frame cyclic test BFCT1 load history.

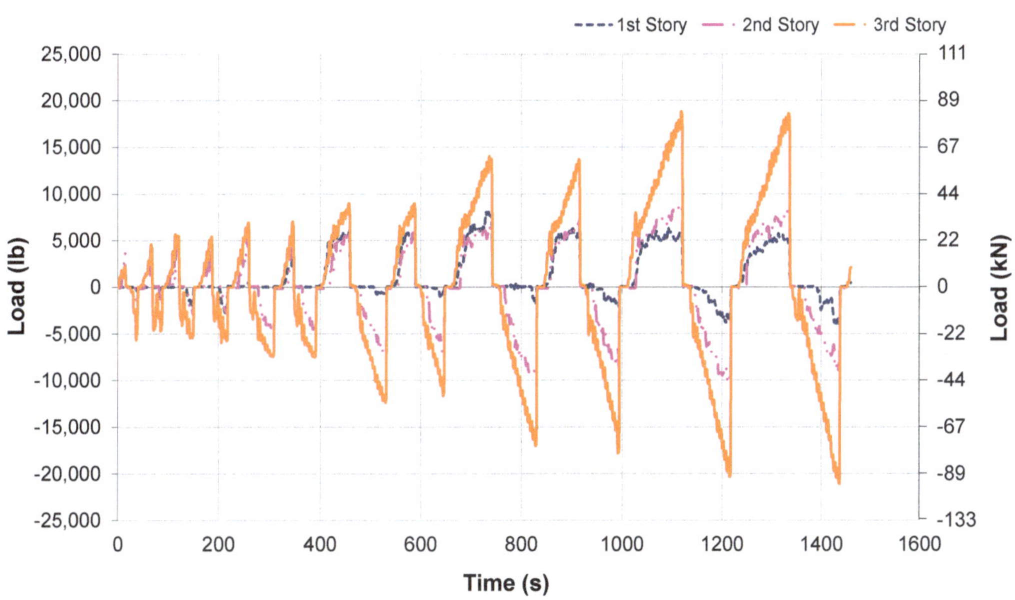

Figure 14. Test BFCT1 load-deflection hysteretic behavior. (**a**) BFCT1 1st story hysteretic behavior; (**b**) BFCT1 2nd story hysteretic behavior; (**c**) BFCT1 3rd story hysteretic behavior.

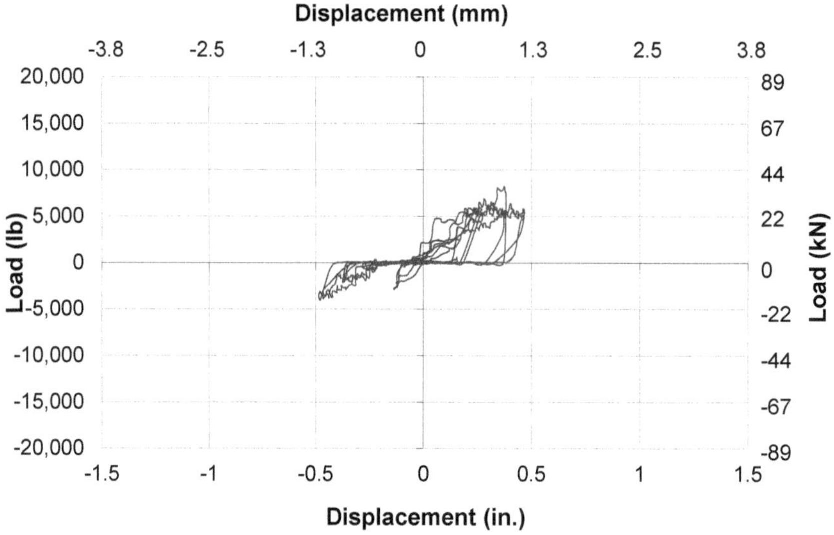

(**a**)

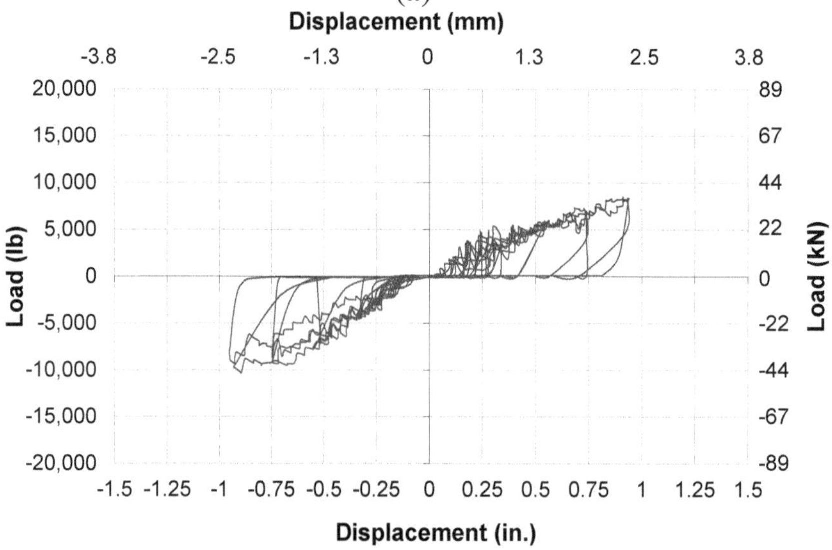

(**b**)

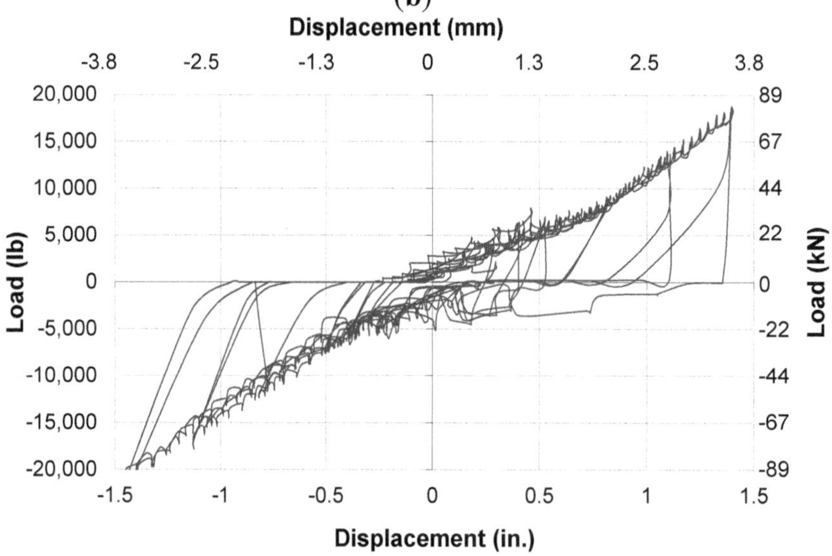

(**c**)

During both bare frame cyclic tests, the load applied to the test frame by the hydraulic cylinders at the third story level was greater than the load applied at the first and second story levels in the advance and retract directions. This behavior was also observed during the bare frame monotonic tests. For the bare frame cyclic tests, however, a greater percentage of the total in-plane load was applied at the first and second story levels than during the monotonic bare frame tests. Although the displacement controlled loads were applied in the same manner during these tests, the quasi-static loading protocol used for the cyclic tests resulted in more load at the first and second stories. These results show that the quasi-static loading history produces a more realistic seismic loading simulation than the monotonic loading protocol.

3.4. Fill Brace Cyclic Tests

The full brace cyclic tests were performed with all diagonal bracing rods in place on the test frame. For these tests, lumber disk fuse elements were positioned on the test frame at the fuse mechanism locations on both sides of the brick infill walls. Tests FBCT1 (full brace cyclic test 1) and FBCT2 were performed with 25.4 mm lumber disk fuse elements at the first story, 19.1 mm fuse elements at the second story, and 12.7 mm fuse elements at the third story. For tests FBCT3 and FBCT4, 12.7 mm fuse elements were used at each story level. The same loading protocol and procedure used for the bare frame cyclic tests were used for the full brace cyclic tests. In-plane loads were applied to the test frame in the advance and retract directions at each story level. The displacement history at the third story level was applied according to the same quasi-static loading protocol used for the bare frame cyclic tests, while the displacements at the first and second stories were based as necessary to achieve an inverted triangle displacement shape.

The response of the test frame to the applied in-plane loads during the full brace cyclic tests was similar to the behavior observed for the full brace monotonic tests. At the start of each test, there was an initial tightening of the system as incrementally increasing displacements were applied to the test frame at each story level in the advance (forward) direction. During this period, the steel bracing rods that were in tension began to tighten, while the steel bracing rods in compression started to buckle. As additional load was applied to the test frame by the hydraulic cylinders, the 22.4 mm dia., half-threaded steel punching rods connecting the fuse elements to the steel test frame began to push on the lumber disk fuses as shear forces were transferred from the steel frame to the infill panels at the upper compression corner of the brick panels. At the same time, the lower compression corners of the infill panels were pressed against the 19.1 mm × 102 mm × 102 mm (3/4 × 4 × 4 in.) steel angles welded to the top of the HSS 127 mm × 127 mm × 9.5 mm (5 × 5 × 3/8 in.) beams.

When the displacement of the third story of the test frame reached the maximum deformation level for that load step, the direction of the load applied by the hydraulic cylinders at each story level was reversed. Load was then applied in the retract (reverse) direction and the test frame was returned to the initial condition of zero displacement at each story level. As additional load was applied in the retract direction, the bracing rods that had previously been in tension began to buckle in compression while the rods that had been in compression during the advance portion of the load cycle began to elongate in tension. The same fuse element behavior observed in the advance direction of the load cycle was also seen in the retract direction. As the displacements at each story level were increased in the retract direction, the fuse elements on the opposite side of the test frame began to transfer shear forces from the steel frame to the infill walls.

For these tests, the target maximum displacement at the third story of the test frame during the first two load steps was 7.6 mm and 10.2 mm, respectively. These values were chosen during the development of the quasi-static loading protocol so that two load steps could be performed before inelastic deformation occurred in the test specimens. During these tests, noticeable cracking did not occur in the lumber disk fuse elements until the displacement at the third story of the test frame reached or exceeded a value of 12.7 mm, during or after the third load step. These results confirmed that the target third story displacement values chosen for the first two load steps were appropriate.

Initial cracking of the lumber disk fuse elements typically occurred during the first cycle of the third load step. The target maximum third story displacement for the third load step was 12.7 mm. Additional cracking occurred in the fuse elements as the magnitude of the third story displacement was increased during subsequent load steps. Two load cycles were completed at each deformation level. After the second cycle, additional fuse cracking did not occur until the displacement of the test frame exceeded the peak third story displacement value of that load step during the next load cycle. This behavior confirmed that two cycles per load step were sufficient to capture the inelastic frame behavior at a given deformation level.

Diagonal bracing rod failures occurred during the final load step of tests FBCT1, FBCT2, and FBCT4. The failures occurred at the ends of the braces near the connection, as shown in Figure 15. Based on the results of rod tension tests performed by Aliaari [19], the bracing rods reach their yield point at a tensile load of around 31.6 kN and have an ultimate capacity of around 37.8 kN. The average total in-plane load applied to the test frame in either direction during both cycles of the final load step was around 141.9 kN. Depending on the proportion of the total in-plane load transferred to the infill walls by the fuse elements during this final load step, the force experienced by the bracing rods at the first story level of the test frame was between 31.1 kN and 35.6 kN. According to this data, the bracing rods were near or at their yield point during the final load step of the cyclic tests.

Figure 15. Typical bracing rod failure.

The displacement histories for full brace cyclic tests for tests FBCT2 and FBCT4 are shown in Figures 16 and 17, respectively. The target quasi-static loading protocol was followed closely throughout the full brace cyclic tests. The load histories for tests FBCT2 and FBCT4 are shown in Figures 18 and 19, respectively. Comparison of Figures 16–19 with Figures 12 and 13 shows the effect of the infills on the bare frame including full brace. The load behavior during these tests was not very similar in the forward and retract portions of the load cycles. This was particularly true at the first and second stories of the test frame. Since these tests were displacement controlled, load was applied at these stories only as needed to maintain the desired displacement shape of the test frame. As with the bare frame cyclic tests, the load applied at the third story of the test frame in either displacement direction was typically higher than the loads at the first and second stories, however, a greater proportion of the total in-plane load was applied at the first and second stories of the test frame during the cyclic tests. The quasi-static loading protocol used for the cyclic tests resulted in greater loads at the first and second stories of the tests frame. This may be due to the cyclic deformation behavior of the lumber disk fuse elements after repeated cycles of damage.

Figure 16. Full brace cyclic test FBCT2 displacement history.

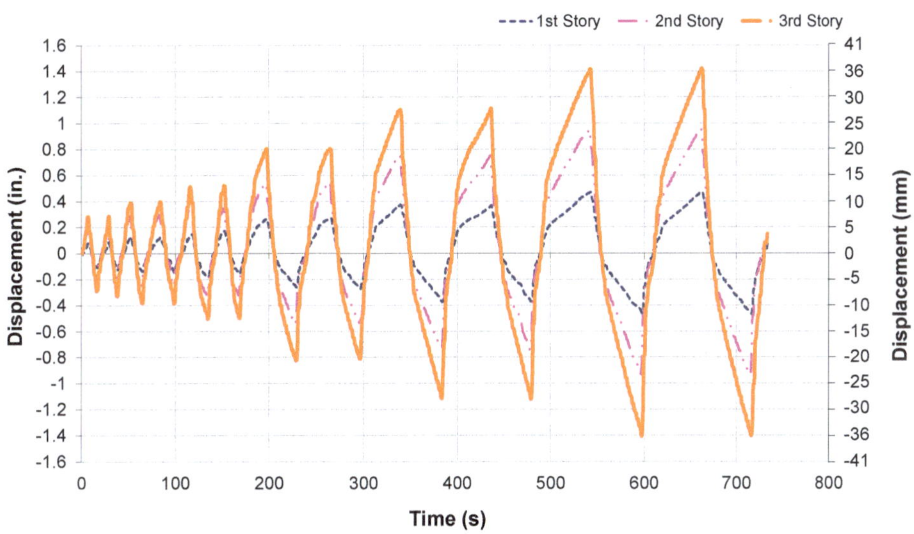

Figure 17. Full brace cyclic test FBCT4 displacement history.

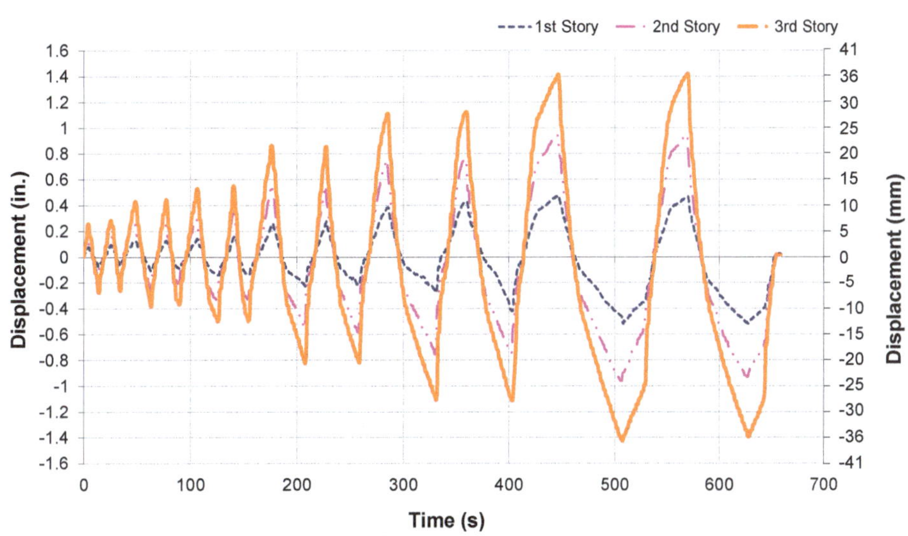

Figure 18. Full brace cyclic test FBCT2 load history.

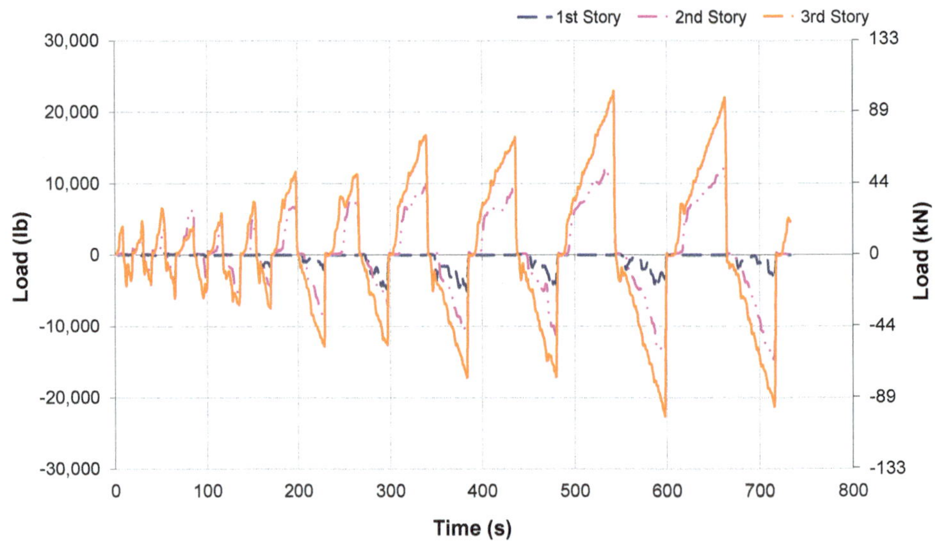

Figure 19. Full brace cyclic test FBCT4 load history.

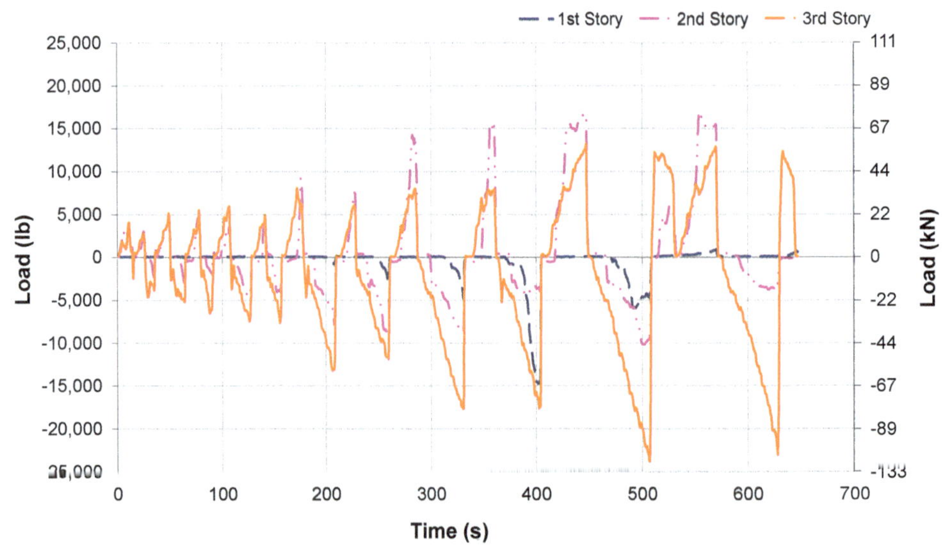

After the completion of each of these tests, the brick infill panels were inspected, and no damage was observed, confirming that the lumber disk structural fuse elements worked well as seismic isolation devices. The results of these tests also show that the response of structural frames to in-plane, cyclic loads can be predetermined by placing fuse elements with different capacities at particular story levels of the frame. By using a structural fuse, the peak loads experienced by the structural frame and infill panels can be limited to acceptable levels.

3.5. Half-Brace Cyclic Tests

For the half-brace cyclic tests, the steel bracing rods on one bay of the test frame at each story level were removed in order to study the behavior of the structural fuse system with more flexible structural frames. For the half-brace cyclic tests, just like the full-brace FBCT1 and FBCT2, 25.4 mm fuse elements were used at the first story, 19.1 mm fuse elements at the second story, and 12.7 mm fuse elements at the third story level. The same quasi-static loading protocol used for the full brace cyclic tests was also used for these tests.

The behavior observed during the half-brace cyclic tests was similar to that of the full-brace cyclic tests. During these tests, cracking did not occur in the lumber disk fuse elements until the displacement at the third story of the test frame reached at least 12.7 mm. As with the full-brace cyclic tests, damage was not observed in the brick infill panels during or after the completion of the tests.

During test HBCT1 (half-brace cyclic test 1), fuse element cracking was not observed until the displacement at the third story of the test frame reached 17.8 mm. Additional cracking occurred in both displacement directions as the load history was continued and the peak displacements were increased at each story level during subsequent load steps. During this test, the greatest fuse element damage occurred in the 12.7 mm lumber disks located at the third story of the test frame. Figure 20 shows the damage to the 12.7 mm lumber disk at fuse location 10. During test HBCT2, the first fuse element cracking occurred at a third story deflection of 12.7 mm. From this point, increased cracking occurred in the fuse elements as the quasi-static load history was continued.

Figure 20. Cracking of 0.5 in. disk at fuse location 10 during test HBCT1.

The load histories for the two half-brace cyclic tests are shown in Figures 21 and 22. The load behavior during these tests was similar to the full-brace cyclic tests. Comparison of these figures with Figure 13 shows the effect of the infill on bare frame with half brace. As with the full-brace tests, the loads applied to the test frame at each story level were not similar in the advance and retract directions. For these tests, very little load was applied to the test frame at the first story level. The magnitude of the loads at the second story, however, was significant. During the advance portion of several of the load cycles the load applied at the second story level was greater than the load at the third story, particularly toward the end of the tests. Similar behavior was seen for the full brace cyclic tests.

In general, the proportion of the total in-plane load applied to the test frame at the second story level was greater for the half-brace tests than the full-brace cyclic tests. This is likely due to the bracing configuration and fuse element distribution. For these tests, the weakest fuse elements were located at the third story level. As increasing damage occurred to the fuse elements during subsequent load cycles, the in-plane stiffness at the third story was reduced.

Figure 21. Half-brace cyclic test HBCT1 load history.

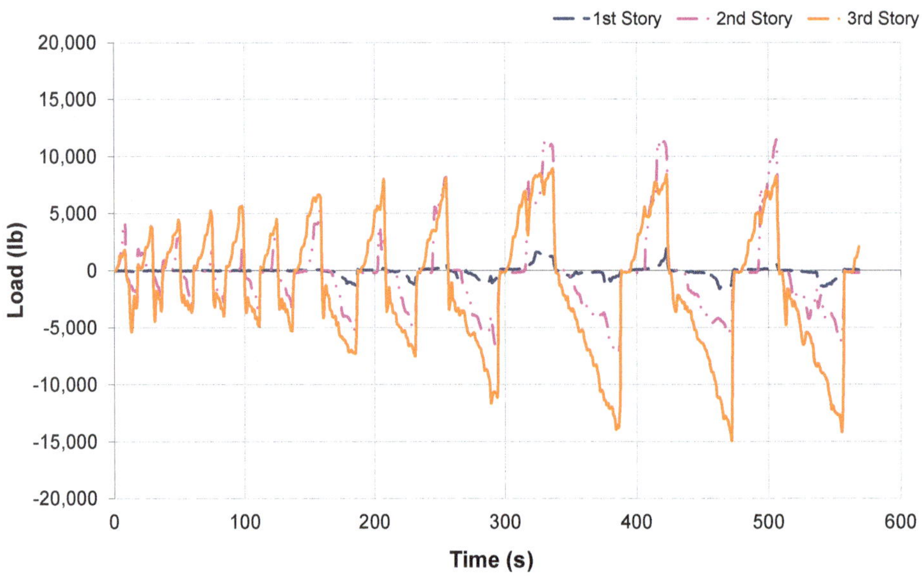

Figure 22. Half-brace cyclic test HBCT2 load history.

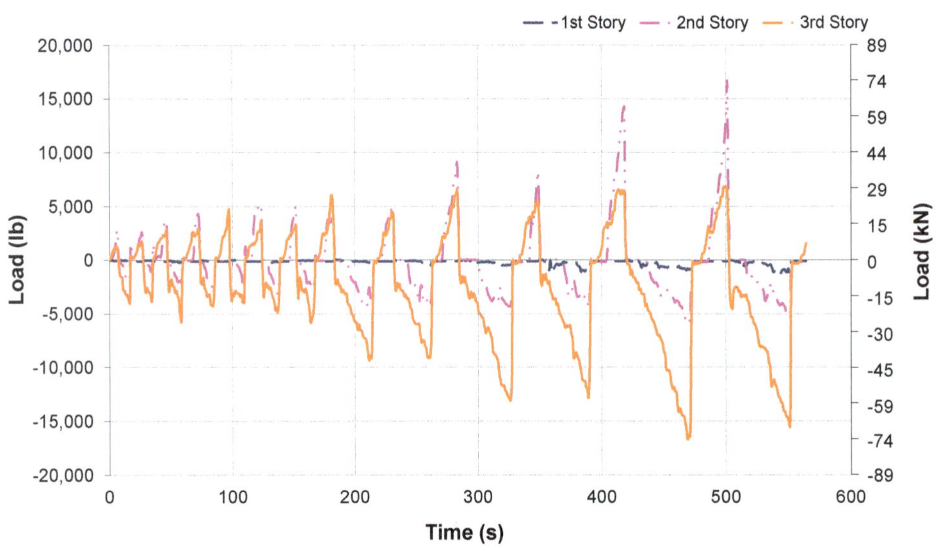

4. Parametric Tests with CMU and AAC Infill Panels

The parametric testing program was performed to evaluate the performance of the structural fuse system for use with various masonry materials. During these tests, in-plane, monotonic loads were applied in the advance direction to the two-bay, three-story steel test frame at the third story level using a hydraulic cylinder and a manually controlled hydraulic pump. These tests were performed with infill panels constructed of concrete masonry units (CMU) and autoclaved aerated concrete (AAC) blocks. The HCT infill panel specimens were not tested since they had very low in-plane strength.

4.1. CMU Monotonic Tests

For tests CMUT1 (concrete masonry unit test 1), CMUT2, CMUT5 and CMUT6, different fuse element thicknesses were used at each story level including 22.2 mm disks at level 1, 19.1 mm disks at level 2, and 12.7 mm disks at level 3. For tests CMUT3 and CMUT4, 12.7 mm disks were used at all of the fuse element locations, as shown in Figure 8. As with the monotonic tests performed in the previous study [28], two different bracing configurations were considered in this testing program. The full-brace configuration was used during tests CMUT1–CMUT4, while the half-brace set-up was used for tests CMUT5 and CMUT6.

The load histories for the CMU parametric tests are shown in Figure 23. During test CMUT4, a problem occurred with the pressure transducer, resulting in a loss of load data for this test. A plot of the load-displacement behavior for each of the CMU parametric tests is given in Figure 24. The load-displacement behavior for tests CMUT1 and CMUT2 was very similar until the displacement at the third story reached around 27.9 mm. At this point of CMUT2, the stiffness of the test frame began to decrease. The stiffness of the test frame during test CMUT3 was slightly less than that of tests CMUT1 and CMUT2. This was expected since 12.7 mm lumber disk fuse elements were used at each story level during this test.

Figure 23. Concrete masonry unit test (CMUT) third story load histories.

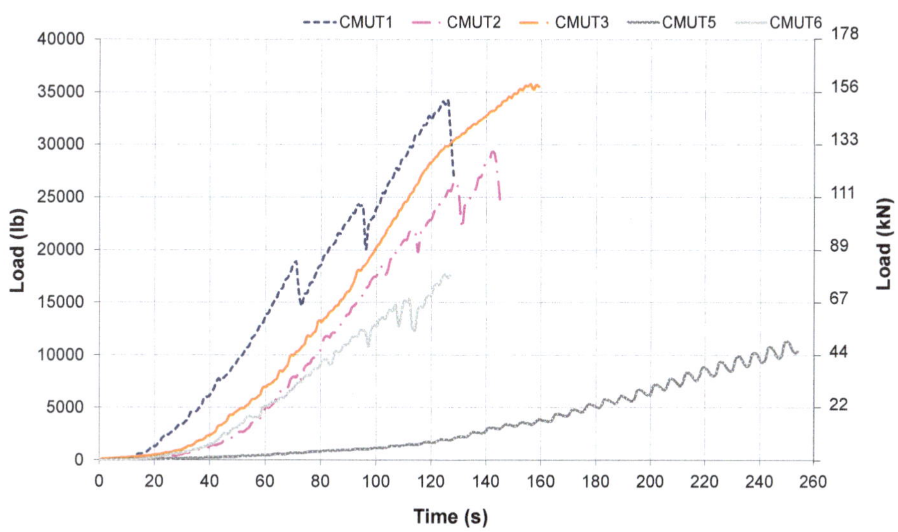

Figure 24. CMUT third story load-deflection behavior.

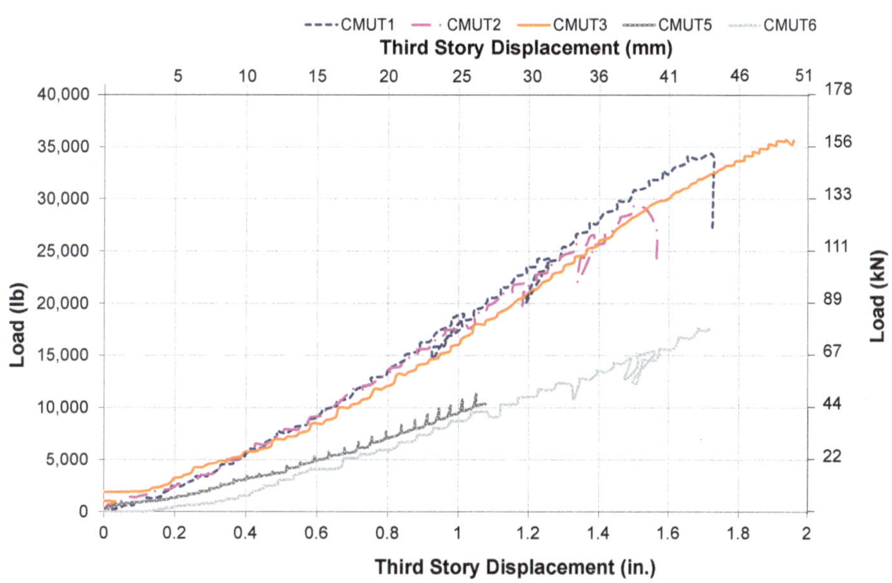

For tests CMUT1 and CMUT2, both of the fuse elements at the third story level reached their maximum capacities resulting in failure of the lumber disks and disengagement of the CMU infill panels. This behavior was expected since for these tests, the weakest fuse elements were located at the third story level. For tests CMUT3 and CMUT4, the same fuse element thickness was used at each story level. During test CMUT3, one of the fuse elements at the first story failed, while for test CMUT4, a lumber disk at the second story failed. This behavior demonstrates the importance of using an appropriate fuse element distribution in multi-story structural frames to control the failure sequence of the fuse elements and avoid a soft story mechanism.

For tests CMUT5 and CMUT6 the half-brace configuration was used, resulting in significantly less in-plane frame stiffness, as shown in Figure 24. Although the slope of the load-displacement curve was slightly smaller for test CMUT6 than for test CMUT5, the overall load-displacement behavior for these two tests was very similar. In general the load-displacement curves for these tests were fairly linear following the first stage of behavior. During this initial stage, the bracing rods were tightening up, and the 22.4 mm dia. punching rods were just beginning to push against the lumber disk fuse elements. Toward the end of the tests, the slope of the load-displacement curves decreased due to the loss of stiffness resulting from failure of the lumber disk fuse elements and disengagement of the CMU infill panels from the in-plane load resistance.

The structural fuse system performed well during the CMU parametric tests. The CMU infill panels participated in the in-plane load resistance of the test frame up to the point where the capacity of the lumber disk fuse elements was reached. At this point, the infill panels were disengaged from the frame preventing damage to the masonry material. Figure 25 shows the damage that occurred to the 12.7 mm thick fuse located at the third story of the test frame during test CMUT2. The CMU infill panels were examined after each test, and no damage was observed in the walls.

Figure 25. Damage to fuse 10 during test CMUT2.

4.2. AAC Monotonic Tests

For these tests, the AAC infill wall panels were positioned on the test frame. For tests AACT1 (Autoclaved aerated concrete test 1), AACT2, AACT5, and AACT6; 19.1 mm, 12.7 mm, and 6.4 mm lumber disk fuse elements were used at the first, second, and third stories, respectively. For tests AACT3 and AACT4, 6.4 mm disks were used at all of the fuse element locations. The full-brace configuration was used for tests AACT1-AACT4, while the half-brace configuration was used for AACT5 and AACT6.

At the end of AACT1, the 25.4 mm × 76.2 mm × 76.2 mm steel bearing plates at fuse locations 6, 10, and 12 were separated from the AAC infill wall panels. This was the result of local crushing failures of the AAC material during the test. The bearing plates were reattached to the wall panels using epoxy. At the end of test AACT3, the 25.4 mm × 76.2 mm × 76.2 mm steel bearing plate at fuse location 10 also had to be re-epoxied onto the AAC infill wall panel due to local crushing of the AAC material, as shown in Figure 26.

Figure 26. Plate separation from AAC wall panel during AACT3.

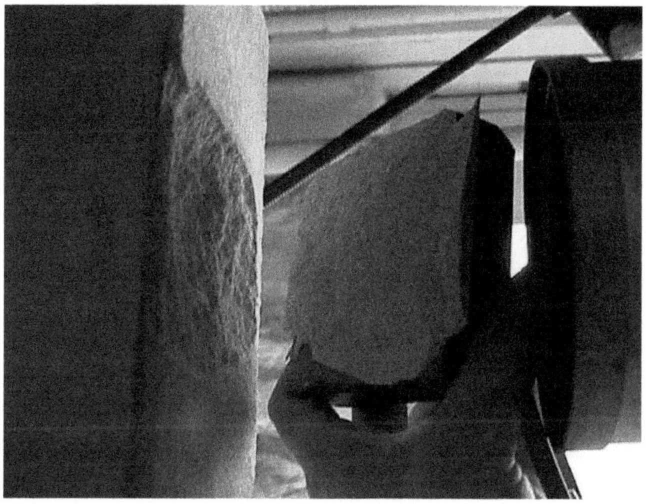

During the AAC parametric tests, in-plane shear failure of the AAC infill panels was prevented by the lumber disk fuse elements. Local crushing failure of the AAC material, however, was observed during these tests at several fuse element locations. This local material failure did not seem to cause a reduction in the in-plane resistance of the AAC panels. Repairs were required to maintain the functionality of the fuse mechanism.

The crushing damage observed during these tests was seen at fuse locations with 6.4 mm, 12.7 mm, and 19.1 mm thick fuse elements. Based on the results of the lumber disk puncture tests, the average maximum capacities of the 6.4 mm, 12.7 mm and 19.1 mm thick lumber disks were 4590 N, 10,657 N, and 19,113 N, respectively. From the in-plane destructive shear tests, the average ultimate shear capacity of the AAC panels was 56.9 kN. Based on this data, the local AAC material failures observed during these tests occurred at load magnitudes that were much less than the ultimate capacity of the AAC wall panels. To avoid this type of premature failure of the AAC material, a different fuse mechanism configuration should be considered for AAC infill panels.

During the AAC fuse element tests, the capacity of the fuse elements used at each fuse location had a significant influence on the in-plane behavior of the test frame. The contribution of the AAC infill panels to the in-plane stiffness of the test frame was successfully controlled by varying the capacity of the lumber disk fuse elements used at each story level. Using appropriate fuse element capacities in the structural fuse system is the key to preventing damage to the infill material and promoting desirable in-plane frame behavior.

5. Summary of Experimental Results

As part of this experimental program, lumber disk puncture tests were performed to study the behavior of the structural fuse elements used during monotonic and cyclic tests [29]. The lumber disk fuse elements demonstrated consistent load-deflection behavior and failure modes with significant post peak load resistance. The relationship between disk thickness and fuse capacity was also fairly linear. Based on the ductile failure modes, consistent load-deflection behavior, low variability in capacity, and linear thickness-capacity relationship observed during these tests, the lumber disks are viable option for use as a structural fuse element.

Masonry wall panel shear tests and prism tests were performed [29] to determine the in-plane shear strengths and compressive strengths of the masonry infill panels tested during the cyclic and parametric structural fuse system tests. Based on the results of similar tests reported in a previous study [19], the average in-plane shear strength of the brick wall panels was 103.4 kN, and the corrected compressive strength, $f_m^{'}$, was 24.8 MPa. The in-plane stiffness of these walls was 7.9 kN/mm. For the CMU wall panel specimens, the average in-plane shear strength and stiffness were 97.9 kN and 15.5 kN/mm, respectively. The corrected compressive strength of the CMU material was 18.5 MPa. The AAC walls had an in-plane stiffness of 7.0 kN/mm and an average capacity of 56.9 kN. The corrected compressive strength of the AAC material was 3.3 MPa. The HCT wall panel specimens were very brittle, and had an average in-plane capacity of only 18.7 kN, with a stiffness of 3.7 kN/mm. Based on the results of these tests, HCT infill panels should be isolated from structural frames and not relied on to carry in-plane lateral loads.

During the full brace and half brace cyclic tests, a greater percentage of the total in-plane load was applied at the first and second levels than during the bare frame tests. As discussed earlier, this was partly due to the cyclic loading protocol used for these tests. As increasing damage occurred to the fuse elements during each load step, the relative in-plane stiffness at each story level changed based on the fuse element distribution and bracing configuration. This resulted in a load distribution to the test frame that was different from what was observed during the monotonic tests. For the cyclic tests performed using the half-brace configuration, an even greater percentage of the total in-plane load was applied at the lower stories.

For building structures subject to in-plane lateral loads, the story forces are distributed to the lateral force resisting frame elements according to their relative in-plane stiffness. It is desirable to have stronger fuse elements and consequently, greater in-plane stiffness at the lower levels of the structural frame to avoid a soft story mechanism. The results of the cyclic testing program showed that having stronger fuse elements at the first and second stories of the test frame resulted in a larger percentage of the total in-plane load being applied at these levels. This behavior was not captured by applying monotonic loads at the top story level of the test frame.

For the cyclic tests where different fuse element thicknesses were used at each story level, the distribution of fuse element damage was more sporadic than for the monotonic tests reported in [19]. In both testing programs the greatest fuse element damage was seen at the third story, where the weakest fuse elements were located. For the cyclic tests, however, significant damage was also observed at the other story levels. For the cyclic tests where the same disk thickness was used at all story levels, the greatest damage was observed in the disks at the first story level. This result was consistent with the test results presented in [19]. Based on this observation, it is important to use higher capacity fuse elements at the lower story levels.

During the cyclic testing program the lumber disk fuse elements successfully isolated the brick infill walls from the test frame as the in-plane story forces increased. No damage was observed to the brick infill walls. The cyclic application of the loads to the fuse elements did not have a significant influence on behavior of the lumber disks. Since a quasi-static cyclic loading history was used, the effect of the loading rate on the behavior of the lumber fuse elements was not evaluated during this testing program.

The results of the parametric tests showed that the structural fuse system can be successfully used to prevent damage to infill walls constructed of various masonry materials. During these tests, the CMU and AAC infill panels participated in the lateral load resistance of the test frame and contributed to the in-plane stiffness. In-plane shear failure of the CMU and AAC wall panels was prevented by the lumber disk fuse elements. During the AAC tests, local material crushing was observed at the location of the fuse mechanism. Although this behavior did not result in brittle failure of the infill panels, it showed that local failure of the masonry material must be considered in the design of the fuse element. In addition to preventing in-plane failure of infill panels, the structural fuse system should be designed and detailed to be durable and require as little maintenance as possible.

6. Summary and Conclusions

A new concept in the performance and design of masonry infill walls is the idea of a structural fuse system. This seismic isolation system allows for composite interaction between infill walls and the structural frame under normal lateral loads. Brittle failure of the infill walls or frame elements is prevented by the introduction of a fuse mechanism which isolates the infill material from the frame under higher loads. Simultaneous, displacement controlled loads were applied at the first, second, and third stories of the test frame to create a first mode response in the structural system; seismic loading conditions were simulated using quasi-static, cyclic loading; a parametric study of the fuse system under static monotonic loads was performed with concrete masonry units (CMU) and autoclaved aerated concrete (AAC blocks); the performance of the lumber disk fuse elements as a viable option for use in a fuse mechanism seismic isolation system was evaluated; and the shear and compressive strength of the masonry materials used in the parametric study were evaluated.

Several conclusions can be made from the results of this study:

- The quasi-static load protocol used is a good approach for use in cyclic load testing of masonry infill wall systems.
- The inverted triangle, first mode deflection shape used to distribute in-plane loads to the test frame is a reasonable approach for applying displacement controlled loads to multi-story frames to simulate seismic loading.
- The quasi-static load protocol used in this testing program resulted in more realistic seismic load behavior than the previously used monotonic protocol.
- The structural fuse system allows the masonry infill walls to contribute to the in-plane stiffness of the test frame at lower levels of lateral load, but prevents damage to the masonry walls by isolating the infill panels from the test frame as the lateral loads increased.

- Infill walls constructed of masonry materials including brick, CMU, and AAC can be safely used to increase the in-plane stiffness of structural frames, resist lateral forces, and limit building deflections, as long as the loads experienced by the infill walls are limited by the use of a seismic isolation device.
- The HCT wall panels tested during this study demonstrated very brittle behavior and low in-plane strength. Therefore, HCT infill walls do not seem to be reliable to carry in-plane lateral loads and should preferably be isolated from structural frames.

Acknowledgments

The invaluable help of BCERL Laboratory Supervisor and Research Associate, Paul Kremer, in all aspects of the experimental work including test setup design, data acquisition system design, and data analysis is gratefully acknowledged.

Author Contributions

The first author Andrew Kauffman was the graduate student working on the research project that led to his M.S. thesis. The second author Ali Memari was the thesis advisor. Both authors worked on the manuscript through the final submitted form.

Conflicts of Interest

The authors declare no conflict of interest.

References

1. Mehrabi, A.B.; Shing, P.B.; Schuller, M.P.; Noland, J.L. Experimental evaluation of masonry-infilled RC frames. *J. Struct. Eng. ASCE* **1996**, *122*, 228–237.
2. Davidson, E.B.; Wang, L.R.L. A Study of The Cyclic Lateral Resistance of Low Rise Masonry Wall Panels. In Proceedings of the 3rd North American Masonry Conference, Arlington, TX, USA, 3–5 June 1985; pp. 48.1–48.15.
3. Thomas, F.G. The strength of brickwork. *Struct. Eng.* **1953**, *33*, 35–46.
4. Benjamin, J.R.; Williams, H.A. The behavior of one-story brick shear walls. *J. Struct. Div. Proc. Am. Soc. Civ. Eng.* **1958**, *84*, 1723–1729.
5. Holmes, M. Steel frames with brickwork and concrete infilling. *Proc. Inst. Civ. Eng.* **1961**, *19*, 473–478.
6. Drysdale, R.G.; Hamid, A.A.; Baker, L.R. Infill Walls and Partitions. In *Masonry Structures: Behavior and Design*; Prentice Hall College Division: Boulder, CO, USA, 1993; pp. 509–538.
7. Flanagan, R.D.; Bennett, R.M.; Adham, S.A.; Fischer, W.L. Masonry Infill Performance during the Northridge Earthquake. In Proceedings of the 7th North American Masonry Conference, South Bend, IN, USA, 2–5 June 1996; pp. 906–915.
8. Negro, P.; Colombo, A. Irregularities induced by nonstructural masonry panels in framed buildings. *Eng. Struct.* **1997**, *19*, 576–585.

9. Chaker, A.A.; Cherifati, A. Influence of masonry infill panels on the vibration and stiffness characteristics of R/C frame buildings. *Earthq. Eng. Struct. Dyn.* **1999**, *28*, 1061–1065.
10. Meharbi, A.B.; Shing, P.B. Seismic Resistance of Masonry-Infilled RC Frames. In Proceedings of the 8th North American Masonry Conference, Austin, TX, USA, 6–9 June 1999; pp. 465–476.
11. Shing, P.B.; Mehrabi, A.B. Behaviour and analysis of masonry-infilled frames. *Prog. Struct. Eng. Mater.* **2002**, *43*, 320–331.
12. Moghaddam, H.A. Lateral load behavior of masonry infilled steel frames with repair and retrofit. *J. Struct. Eng. ASCE* **2004**, *130*, 56–63.
13. Kakaletsis, D.J.; Karayannis, C.G. Influence of masonry strength and openings on infilled R/C frames under cyclic loading. *J. Earthq. Eng.* **2008**, *12*, 197–221.
14. Uva, G.; Porco, F.; Fiore, A. Appraisal of masonry infill walls effect in the seismic response of RC framed buildings: A case study. *Eng. Struct.* **2012**, *34*, 514–526.
15. Porco, F.; Fiore, A.; Uva, G.; Raffaele, D. The influence of infilled panels in retrofitting interventions of existing reinforced concrete buildings: A case study. *Struct. Infrastruct. Eng.* **2014**, *2014*, doi:10.1080/15732479.2013.8627.
16. Pujol, S.; Fick, D. The test of a full-scale three story RC structure with masonry infill walls. *Eng. Struct.* **2010**, *32*, 3112–3121.
17. Stavris, A.; Koutromanos, I.; Shing, P.B. Shake-table tests of a three-story reinforced concrete frame with masonry infill walls. *Earthq. Eng. Struct. Dyn.* **2012**, *41*, 1089–1108.
18. Sattar, S.; Liel, A. Seismic Performance of Reinforced Concrete Frame Structures with and without Masonry Infill Walls. In Proceedings of the 9th US National and the 10th Canadian Conference on Earthquake Engineering—Reaching Beyond Borders, Toronto, Canada, 25–29 July 2010; pp. 1–10.
19. Aliaari, M. Development of Seismic Infill Wall Isolator Subframe (SIWIS) System. Ph.D. Thesis, Penn State University, University Park, PA, USA, 2005; p. 388.
20. Manos, G.C.; Mpoufidis, D.; Demosthenous, M.; Triamataki, M. Influence of Masonry Infill Panels on the Response of R.C. Structures Subjected to Lateral Loads. In Proceedings of the 5th North American Masonry Conference, University of Illinois at Urbana-Champaign, Urbana Champaign, IL, USA, 3–6 June 1990; pp. 93–107.
21. Polyakov, S.V. *Masonry in Framed Buildings: An Investigation into the Strength and Stiffness of Masonry Infilling*; National Lending Library for Science and Technology: Yorkshire, UK, 1963.
22. Colombo, A.; Negro, P.; Verzeletti, G. Infilled Frames: Certainties and Uncertainties. In Proceedings of the 11th European Conference on Earthquake Engineering, Paris, France, 6–11 September 1998; Balkerna: Rotterdam, The Netherlands, 1998; p. 10.
23. Paulay, T.; Priestley, M.J.N. Masonry-Infilled Frames. In *Seismic Design of Reinforced Concrete and Masonry Buildings*; Wiley: New York, NY, USA, 1992; pp. 587–592.
24. Bertero, V.; Brokken, S. Infills in seismic resistant building. *J. Struct. Eng. ASCE* **1983**, *109*, 1337–1361.
25. Aliaari, M.; Memari, A.M. Analysis of masonry infilled steel frames with seismic isolator subframes. *Eng. Struct.* **2005**, *27*, 487–500.
26. Aliaari, M.; Memari, A.M. Development of a seismic design approach for infill walls equipped with structural fuse. *Open Civ. Eng. J.* **2012**, *6*, 249–263.

27. Memari, A.M.; Aliaari, M. Seismic Infill Isolator Subframe (SIWIS) System for Use in Buildings. In Proceedings of the ATC-17–2 Seminar on Response Modification Technologies for Performance Based Seismic Design, Los Angeles, CA, USA, 30–31 May 2002; pp. 189–200.
28. Aliaari, M.; Memari, A.M. Experimental evaluation of a sacrificial seismic fuse device for masonry infill walls. *ASCE J. Archit. Eng.* **2007**, *13*, 111–125.
29. Kauffman, A. Cyclic In-Plane Lateral Load Testing of Masonry Infill Walls with Structural Fuse Elements. Master's Thesis, Penn State University, University Park, PA, USA, 2009; p. 338.
30. Tomaževic, M.; Klemenc, I. Verification of seismic resistance of confined masonry buildings. *Earthq. Eng. Struct. Dyn.* **1997**, *26*, 1073–1088.
31. Krawinkler, H. *Guidelines for Cyclic Seismic Testing of Components of Steel Structures*; Applied Technology Council: Redwood, CA, USA, 1992.
32. Krawinkler, H. Cyclic loading histories for seismic experimentation on structural components. *Earthq. Spectra* **1996**, *12*, 1–12.

2

Climate Based Façade Design for Business Buildings with Examples from Central London

John Napier

Department of Architecture, University of Lincoln, Brayford Pool, Lincoln LN6 7TS, UK;
E-Mail: jnapier@lincoln.ac.uk

Academic Editor: David Dernie

Abstract: There is a disconnection between commercial architecture and environmental thinking, where green features can be included as part of a strategy for gaining approvals and marketing projects, but those features are not reviewed after completion and occupation of the building and knowledge is not shared. High levels of air conditioning are still considered unavoidable. Elaborate double skin façades and complex motorized shading systems are adopted; often masking an underlying lack of basic environmental thinking. This article returns (in principle) to the physics of comfort in buildings and the passive strategies which can help achieve this with a low energy and carbon footprint. Passive and active façade design strategies are outlined as the basis of a critical tool and a design methodology for new projects. A new architectural sensibility can arise based on modeling the inputs of sunlight, daylight and air temperature in time and space at the early stages of design. Early but sound strategies can be tested and refined using advanced environmental modeling techniques. Architecture and environmental thinking can proceed hand in hand through the design process.

Keywords: climatic design; day-lighting; façade design; shading; solar heat gain; cross disciplinary; design workflow; nearly zero carbon; business buildings

---

1. Introduction

This article brings together passive design thinking and the urban business building, which can be the building block of a compact city served by public transport. The business building of the paper's title is a multi-story building, most likely for office or institutional use. Both the design and its dense occupation reflect the high value of the land on which it sits.

The starting point is that while all buildings should be targeting a net zero energy profile, the majority of city buildings are dressed in cladding (whether traditional or highly glazed) which ignores orientation and passive solar design thinking. A fabric first approach to façade design can help reduce air conditioning loads.

A series of façade design strategies are reviewed, the results of which can form both a critical tool and a design methodology for new projects.

Examples shown in case studies of recent buildings and projects primarily from central London; links are made to other climatic conditions to encourage a cross fertilization of ideas to and from this focus.

A notional design sequence is set out to indicate how climate based façade thinking can be incorporated into the accelerated design processes common today.

2. Context and Literature Review

2.1. Environmental Design in Architecture

Reyner Banham pioneered in foregrounding the role of environmental systems in architecture [1]. The systems he foregrounded, however, were principally the energy intensive mechanical systems of air supply and air conditioning. The air conditioned sealed container was already the predominant mode of North American commercial architecture. Strong structural frames allowed large expanses of glass which together were seen as signs of the technological age.

From Louis Kahn came the concept of servant and served space and separately the playful influence of Archigram led to Piano and Rodgers' design of the Pompidou Centre in 1977 and the continuing work of Richard Rodgers and his partners.

In Britain, the need for air conditioning was not so apparent with its milder climate. In "*The Selective Environment: an Approach to Environmentally Responsive Architecture*" [2], the role of the r as a modifier of climatic conditions was made pre-eminent. The façade would filter sunlight, daylight and fresh air from the exterior as appropriate according to the variation of the seasons either by the actions of the occupant's or by mechanical means. The effect was intended to significantly reduce the energy budget for the buildings designed. In the earlier "*The Architecture of Energy*" [3], the environmental opportunities of covered atria as thermal buffer zones mediating outside temperatures and facilitating natural ventilation were presented. For low energy performance to be achieved in practice it is helpful if the narrow definition of comfort criteria is relaxed somewhat within working spaces as well as buffer spaces, allowing temperatures to fluctuate somewhat, especially in summer [4].

The commercial sector adopted the fashion for atria as dramatic social spaces but often without the buffer zone thinking, imposing the more narrow comfort criteria which are standard in the industry.

The European Buildings Energy Performance Directive states that new buildings in the European Community should be by 2020 "*Nearly Zero Energy Buildings*" [5]:

> "*... The nearly zero or very low amount of energy required should be covered to a very significant extent by energy from renewable sources, including energy from renewable sources produced on-site or nearby.*"

Definitions of "energy balance" for a lettable building under this concept requires energy used for heating, cooling, and lighting (but excluding unregulated energy use for computers and equipment) to be matched by on or near site renewable energy [6]. This "unregulated energy use" will, however, still need to be sourced and paid for. In addition, this usage may impact the thermal energy balance of the building and in summer lead to additional cooling requirements. In addition, the energy balance definition should allow for some weekly as well as seasonal climatic fluctuation; an array of energy storage batteries should be provided in a Net Zero Energy Building to smooth out the variations in heating/cooling required and the intermittent availability of renewable energy. Larger fluctuations can be achieved by importing or exporting to the electricity grid.

In parallel with this initiative, "*Net Zero Energy Buildings*" [7] catalogues a series of buildings which generate energy at or near the building in question through Photovoltaic (PV) panels. The case studies are well supported by technical explanation and performance data. A typical characteristic of the buildings identified, however, is that they are in the low density edge of town locations and supported by generous car parking arrangements.

2.2. Façade Design Research Trends

A strong international research network gathers annually under the umbrella of the Energy Forum with a conference called "Advanced Building Skins" [8]. To stand out in this environment, research initiatives often propose highly intricate dynamic façade concepts which would require building managers to invest considerable knowledge, skill and time to maintain the optimum tuning of systems to the fluctuations of the weather as well as the specifics of the local microclimate. Maintenance of dynamic systems is an additional responsibility for which many facilities managers may not be trained. The same dynamic systems are often proposed as an addition to an over glazed façade.

The Advanced Building Skins conference strongly promotes the idea of Building Integrated Photovoltaic (BIPV) panels. The advantage of on-site generation is that little energy will be lost from the point of supply to the point of use. Not only installed at roof level, PV panels are becoming part of an increasing number of façade systems. Because of the improved financial efficiency of the systems there is less need for them to attempt to be optimally oriented towards the sun, which is in any case a moving target. Indeed, a vertical surface may gather more solar radiation in winter and this may be an advantage.

PV panels can be incorporated seamlessly into façades as a direct substitute for opaque cladding materials in floor zone, spandrel and core areas. The additional cost of the installation is thus offset by the cladding materials which would otherwise have been used (often of high quality). It is shown elsewhere in this paper that the amount of clear glazing needed in façades is less than is often provided.

A substantial trend in both research and practice is the use of double skins in façade assemblies [9,10]. A double skin provides a buffer zone for management of thermal extremes and an acoustic and wind protected zone which can support opening window panels. If movable blinds are located in this zone, solar heat gain can be excluded from the building interior. In winter, the same solar heat gain can be retained within the cavity as an insulating layer between the colder air outside and the building interior.

Since this is an additional investment in the cost of the façade, it merits additional scrutiny as to where and when it can be of most benefit.

2.3. Legislation and Marketing Initiatives

Environmental assessment methods such as the Building Research Establishment Environmental Assessment Method (BREEAM) [11] and Leadership in Energy and Environmental Design (LEED) [12] have been an important force in bringing attention to many environmental issues within the development industry. The industry, however, has become wise to these initiatives, prioritising "easy wins" in the point scoring systems over the longer term benefits which may carry short term costs. The value of the ratings of these systems is most pronounced at the stage of marketing the project; post occupancy management is only recently being included as of importance within these systems. There is a sense that the systems promote green features ahead of green thinking.

The UK Building Regulations propose target (carbon) emissions rates for new buildings other than dwellings in Part L2A [13]. This target is regularly reviewed and the target made more onerous. In addition, recommendations are made for maximum glazing percentages.

2.4. Climatic and Passive Design

A scientific approach to passive design was set out by Olgay in 1963 [14]. The work contains many charts, graphs and data; the architecture referred to, however, is in the main, low in density and unremarkable in design. Passive design thinking, which evolved from this, tailors building form, orientation, day-lighting and ventilation strategies to the individual site and its microclimate.

Architecture, whether formal or vernacular, when studied globally, reveals many climatically specific responses [15,16] to shading and daylight, ventilation and breeze and thermal capacity. Cross fertilisation of such thinking is informing passive design today. The extent of similarity and difference between climatic regions can inform practice which is increasingly globalised. The definitions and boundaries for these climatic zones, for the purpose of an architectural article coalesce around the characteristics of the indigenous architecture as well as approximate latitude bands and temperature/humidity profiles. With London as a focus, a building type with a need for cooling and a climate expected to change over the lifetime of buildings designed today, reference is made to warmer climates, in the tropics and the desert, where markedly different architectures have evolved.

2.4.1. Temperate Climate

London is at a latitude of 51 degrees North, and Britain is classified as a Marine Mild Winter Climate (Cfb) meaning temperate, without a dry season, with a warm summer, under the Koppen climate classification system [17]. Although there is a significant seasonal change due to the high latitude, winters are warmer than comparable latitudes due to the warming effect of the Gulf Stream ocean currents. This may become destabilized with substantial volumes of cold fresh water from Greenland's melting ice cap being discharged into the path of this current; the benign thermal flywheel effect could then potentially be significantly reduced. Summers are expected to be hotter and drier; and winters generally warmer and wetter [18]. In addition, like other major cities, London is also subject to an Urban Heat Island effect where temperatures in the city can be significantly higher than in the adjoining countryside [19,20].

Despite the seasonal temperature variations, but because of high occupational densities and deep plans, many of London's business buildings are in net cooling mode for much of the year. The potential of a warming climate indicates that more attention should be made to manage cooling loads, the most obvious being due to solar heat gain through extensive glazed areas.

A possible source of ideas can be found by looking at warmer climates and the architectural responses indigenous to those areas.

2.4.2. Tropical Humid Climate

Tropical humid climates, straddling the equator and extending towards the Tropics of Cancer 23 degrees North and of Capricorn 23 degrees South are much less affected by seasonal change caused by changing solar geometry. Seasons are mostly distinguished by being wet or dry, although the timing of these seasons is less predictable than in the past.

In addition, the temperature does not vary substantially over the daily cycle, limiting the opportunity for night time cooling, unlike in desert climates.

Passive design in the tropics relies on the shading of spaces and façades from the heat of the sun and harnessing cooling breezes through appropriate parts of the building. Circulation systems in buildings such as hotels and educational establishments can be thermally open, shaded from sunlight and where possible cooled by local breezes. The extent of fully enclosed (and frequently air conditioned) space is thus kept to a minimum.

Building orientation if predominantly facing north and south can allow for façade shading to exclude the heat from direct sunlight during the middle of the day whilst allowing generous glazed areas to admit reflected light into interiors. East and west façades should be provided with much less glazing, as low angle sun in the early morning and late afternoon can transmit substantial heat gain into building interiors. The relatively consistent seasonal solar geometry allows for simple rules of thumb to inform the location and shading of glazed elements in building façades.

2.4.3. Desert Climate

Many hot dry regions can be found some distance from the equator (in the Middle East, North Africa and in Australia, for example) where a significant seasonal climate can be experienced, due in part to the change in solar geometry. Clear skies bring substantial solar radiation (and heat gain to building interiors if not managed) in daytime and a steep drop in night time temperatures. Daytime temperatures can often exceed 40 °C.

Traditional development in hot, dry regions has clustered buildings close together so that shaded spaces can be created between them for streets and courtyards; streets have been oriented to suit prevailing winds and "windcatchers"; passive down draught cooling systems to bring breezes into the interior especially at night time together with thermal mass to smooth out temperature swings over the daily cycle.

Much of this has been lost in more contemporary development with typically wider streets and standalone buildings. However, passive solar design is being re-evaluated and traditional techniques can be enhanced with new technical knowledge and systems, as can be seen at the Masdar development in Abu Dhabi (see Figure 1).

Figure 1. Shading by adjoining building, façade modelling and limited glazed openings, Masdar, Abu Dhabi, Foster and Partners.

Improved façade design, which will exclude most sunlight and introduce reflected light, will lead to savings in mechanical equipment and the associated space requirement as well as reducing electricity bills over the years of operating the building.

Diffuse and reflected light can be harnessed by building design to reduce the reliance on artificial lighting; traditional Arabic architecture has evolved many techniques (in particular, the mushrabaya screen) which can be adapted to newer design solutions. Solar geometry tends to be more complex (and subject to more seasonal variation) than in the Tropical regions as the proximity to the equator is usually not so great; hence, the need for smaller glazed areas and the interest in dynamic shading systems. Care should be taken in desert regions to avoid the effect of sand storms rendering intricate systems less movable over time.

Several desert regions are rich in fossil fuel energy reserves; restraint in the use of these can help reduce global warming. These same regions are blessed also with abundant solar energy; the opportunity for generating substantial renewable energy exists.

2.4.4. Changing Climates—Mitigation and Adaptation

Design for mitigation of climate change often aligns with low energy building design which has a more immediate connection with clients and users' interests.

There is a need to consider adaptation measures, which may be needed more incrementally over time as the extent of climate change in specific areas becomes clearer. Pioneering research in this area adopts a spectrum of climate scenarios with which to test the future proofing of live projects [18]. Façade design can be based on a robust chassis on which additional shading devices can be fixed in the future as the economic case for them becomes more pressing.

Climate data for design prediction needs to be at least up to date and methodologies for forecasting future climate are reviewed in a recent paper [21].

2.5. Low Energy Building Design Strategies

Low energy design (helping mitigation of climate change) should aim first of all to reduce building energy consumption for heating, cooling and electric lighting. In larger commercial buildings (especially deep plan buildings) there is a greater need for cooling than heating over the seasonal cycle. This is in part due to high occupancy rates and in part due to artificial lighting loads.

Day-lighting and passive ventilation design set limits on the depth of buildings at around 13.5 m. Deeper buildings have to be lit and ventilated by atria at this spacing or rely full time on artificial systems. For deep buildings with air conditioning the influence of the façade on the performance of these buildings is less because of the reduced surface area per unit of floor space accommodated. A typology of building shapes and related environmental systems is indicated graphically in Figure 2 below.

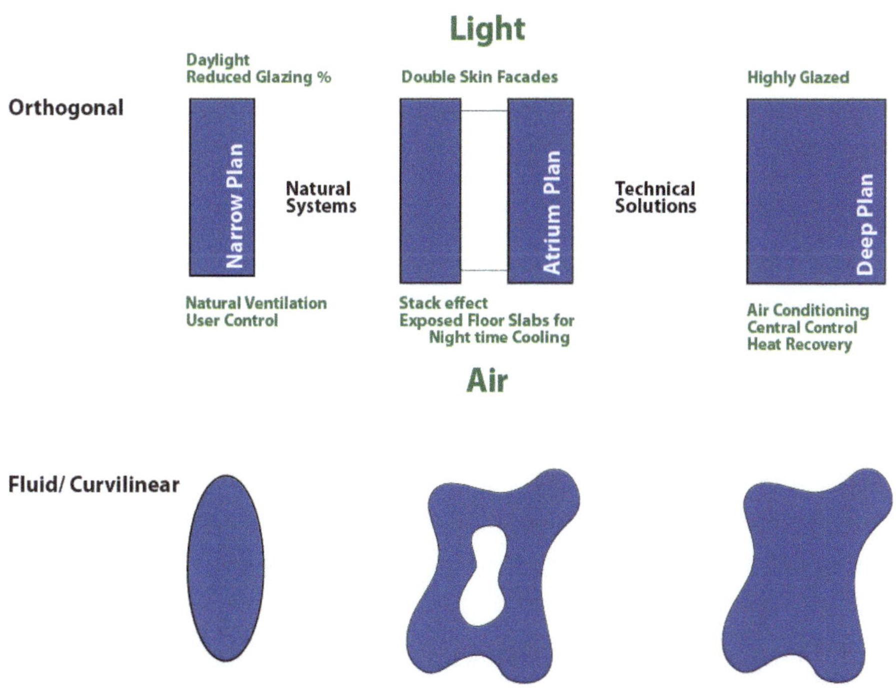

Figure 2. Typology of environmental systems and spaces (workspace).

Building form which has been designed substantially for natural lighting will generally be supportive of natural ventilation strategies. Narrow depth plans with courtyards or atria can support combinations of single sided and cross ventilation. Deeper plan areas or more densely occupied areas will benefit from mechanical ventilation and/or air conditioning. Chilled beams and ceilings provide cooling (through the use of water as a transfer medium) without humidity control. Chilled beams can save space in floor to floor height and can work in conjunction with the exposure of structural thermal mass in moderating and delaying temperature swings which can be cooled with night time air ready for the next day.

Natural ventilation requires façade openings in the form of windows, panels or grilles. In taller buildings or buildings in urban environments, a buffer zone is sometimes formed to protect these openings from driving wind, rain, temperature extremes and external noise. This can be done in the form of a glazed screen or double skin façade which also serves to screen the visual complexity of windows and grilles.

This type of thinking is still at an early stage in the commercial market which has up to now favoured sealed buildings and air conditioning or mechanical ventilation.

2.6. The Role of Daylight in Buildings

Everyone agrees that light is a fundamental ingredient of good design, but agreement on the amount and quality of light, both natural and artificial, that makes good design is less easy to find. Daylight is also an important contributor to the reduction in the need for artificial lighting with its associated energy costs and contribution to heating loads. Façade design has a big role to play in both of these issues.

Climate Based Daylight Modelling (CBDM) challenges the industry's reliance (through BREEAM and LEED among others) on the Daylight Factor and its "Standard Overcast Sky" as a design standard [22].

> *"Practitioners encounter guidelines and recommendations for target DF values that they know are likely to result in over-glazed buildings with excessive solar gain and/or heat loss."*

CBDM measures absolute illuminance from both sun and sky, direct and indirect light as a dynamic simulation in time from regional climatic datasets. Although a more sophisticated an analytical system [22],

> *"... Development of more accessible, user-friendly CBDM applications will be needed, with an inevitable compromise between precision/flexibility and ease-of-use."*

Urban quarters which work well for public transport and the pedestrian are likely to, at the same time, restrict access to sunlight and daylight to their buildings because of their proximity to one another, this effect varies throughout the day and throughout the seasons. Light available to building interiors is likely to have a higher proportion of reflected light than in less compact areas. Access to sunlight and daylight on building façades can be assessed using environmental software such as IES and Ecotect [23,24]. This reduced access to light contradicts to some extent the otherwise accepted view that glazing percentages in commercial buildings are too high.

For buildings with specific needs, light can be introduced from above (north light only if required), from the side (windows) and from reflected sources such as window reveals, light shelves, light scoops, *etc.* As Rafael Moneo states [25]:

> *"... What stands out is the importance attributed to light, the true protagonist of the compact architecture."*

As these strands of thinking come together it becomes clear that the façade of a building within the scope of this study should introduce more (day) light into the building but less (solar) heat.

3. Façades: Climatic Design Principles and Strategies

3.1. Glazing Percentage and Orientation

A better combination of glass and solid, shading and reflectance, view and solar control is required from mainstream building design if a significant reduction in carbon emissions and an improvement in occupant comfort and productivity is to be achieved. Internal blinds are no defense against the heat of the sun as that heat will have already entered the buildings' thermal zone; the heat reflected back towards the glass will be retained within the building (this powerful effect is popularly known as the greenhouse effect and is not always welcome in workplace interiors).

Solid façade elements give scope for good insulation levels which will help reduce extremes of heat and cold influencing the heating and cooling loads.

If there is to be less clear glass then how should it be distributed within a building enclosure? The desire for full height glazing in all areas should be subject to the strongest scrutiny. The view of the underside of office desks from outside benefits no one. Circulation areas and meeting rooms, if they could be defined are areas where more generous glazing can be appreciated. This is a difficulty created by the speculative nature of property markets where no options are ruled out but space is lacking in any definition.

East and West façades are the hardest for which to manage solar heat gain as the low angle sun in morning or evening is not controlled by horizontal shading. South facing façades (in the Northern hemisphere) can admit good quantities of light even if there are projecting external shading systems; much of this daylight will have been reflected and the direct heat of the sun reduced. North facing light is weaker and more even in its distribution.

In conclusion, buildings are more climatically designed (and hence, more likely to be energy and carbon efficient) when their façades are specific to their orientation. Buildings may, in fact, have primary façades which welcome daylight and sunlight and secondary façades which restrict this, adding directionality to the way the building addresses its surroundings. Although well known to many, it is surprisingly rare in recent practice. Façades with larger amounts of glass should have measures in place to manage the exposure to the extra light and heat (and associated fluctuations of these) which will result. This is analogous to wearing sun cream if you intend to spend long periods in the sun.

3.2. Double Skin Façades

A double skin façade provides a minimal depth glazed thermal buffer zone for the moderating of fluctuating thermal conditions between inside and outside. Many variations of this idea have been developed in terms of the size of the compartments and the environmental operation of the system, which can be free flowing (gaps between glass panels), motorized automatically or with local override. The internal façade line can be fully or partly glazed and may be sealed or incorporate opening windows. The cavity is an ideal thermal position to incorporate blinds and shading systems [26].

European examples will tend to be associated with narrow plans and the opportunity to open windows behind this screen, which will have screened air flow and traffic noise. UK and U.S. examples, however, will more likely wrap sealed buildings with a double skin façade and their mechanical systems may be engaged with the import and export of air to and from the buffer zone.

A new type of cross disciplinary design team is emerging where building physics specialists work alongside façade specialists (who may be consultants or may be façade contractors); architectures may specialise in this field to join teams such as this. An account of the elite design groups who have the skills and/or potential to maximise the benefits of double skin façades is given by Trubiano in an earlier issue of this journal [27].

What is missing is much sharing of knowledge of performance post occupancy in this fast growing field. Projects are marketed with an array of green features and excellent BREEAM or LEED ratings (and this will help gain planning and financial approvals). Visualisations with blue and red arrows will show the route that air is expected to take but seasonal or daily control strategies are rarely published and comparisons of similar façades with different orientations would be of value. Are the double-skin façades wrapped all the way round primarily for visual consistency or are they effective in each orientation but under different control regimes?

The additional investment in two layers of material of high embodied energy (aluminium and glass) may suit some clients with high profile projects but it is not clear that wider dissemination of this system is a good use of resources.

The customisation of the intricate web of digitally driven environmental systems in double skin façades and mixed mode ventilation systems is an art in itself; a new breed of facilities manager may evolve to bring together the opportunities presented by innovative design and the community of building occupants who may have differing views of comfort in buildings [28].

Without wider sharing of good practice (currently inhibited by commercial confidentiality), the potential benefits of these exciting developments may not be matched in practice over time.

3.3. Glazing Coatings and Specifications

In the time when fully glazed façades were the norm, specialist glazing was the only line of defense against solar heat gain. Sophisticated glazing assemblies (double and triple glazed) together with specialist coatings have been developed to control heat gain while maximizing the transparency of the window system. Low-e glass will help retain heat in the interior of the building if this is required.

These glass assemblies still have an important role to play but are not enough in the context of very low energy buildings where physical solar shading should be provided before the summer sun's rays reach the façade and a reduced glazing percentage is the simpler method of maintaining indoor conditions.

3.4. Shading and Reflecting—Static and Dynamic Systems

Static shading devices range from overhangs and deep reveals within more solid walling systems to maintenance walkways and lighter filigrees of projecting louvered aluminium. Solar shading can also be of folded metal providing an opaque projection still within the work "package" of a curtain wall specialist. Such projecting devices can often hang over the building boundary and not sacrifice the opportunity to maximize the commercial value of a site.

More lacey constructions of perforated or twisted metal can contribute to a decorative effect informed by the new conditions of digital fabrication. Such "veils" continue the analogy of the curtain (wall) which has been part of the language of façade design since structural frames freed this part of a building from load bearing responsibilities. In some cases, they link back to the two directional shading devices of the Middle Eastern mashrabiya tradition.

Horizontal shading systems can double up as "light shelves". The shelf effect enables reflection and diffusion of daylight deeper into the interior of the building. The light reflected from the top of the sky is brighter than that of a lower angle.

Dynamic shading at its simplest includes movable shutters, external blinds and awnings; the traditional hand operated moving systems which have been used in warmer countries for generations. Robust designs for these products have evolved to deal with wind, rain and the heat of the sun.

Roller Blind systems are now available with motorized controls which can be programmed by the BMS to be deployed in multiple zones to suit the varying patterns of solar exposure. These blinds are programmed to be retracted on the onslaught of extreme conditions of wind and rain. Tilting louvers can also be operated by motorized systems.

More substantial shutter systems based on folding, sliding, or folding and sliding elements are options based usually on the size of window module. These can operate in the horizontal plane or sometimes in the vertical plane. The shutter material may take advantage of the ease with which metal sheets can be perforated to allow for some transparency and air flow. Blinds which are fixed in position but can be rotated by pre-programmed motorized systems are well established in the market.

Complex moving elements are found in the Al Bahar towers recently completed in Abu Dhabi. Here, an origami like arrangement of folding panels based on a hexagonal geometry makes reference to the region's Islamic patterns. As an (otherwise highly glazed) elliptical tower, a wave of open panels can track incrementally across the curved surface providing changing scenery at different times of the day [29].

This is a field of innovation and experimentation where different methods of solar shading are being invented and patented and different aesthetic solutions are being tried and tested.

3.5. Building Integrated Solar Energy Systems

If parts of the façade are not needed for day-lighting the interior, are there other uses to which the surfaces in question can be put?

The cost and efficiency of PV solar generation of energy has decreased significantly in recent years. Not only installed at roof level, PV panels are becoming part of an increasing number of façade systems with issues of efficiency and aesthetics coming to the fore. The advantage of on-site generation is that little energy will be lost from the point of supply to the point of use.

A study has shown that energy generated on east and west façades (in conjunction with south and on the roof) can contribute to a smoother energy generation profile over the day with a closer match to the demand than if panels were all optimally orientated. This study was based on unobstructed façades, different to those of the compact urban quarters of this study. An exploration of solar access of the different façades at design stage may lead to an adoption of BIPV on an exposed façade and at upper levels on more shaded façades. An aesthetic logic for this could be sought [30].

The same study indicates an approximate ratio of the maximum energy which can be generated in Northern Europe compared with Southern Europe of 50% for an equivalent installation; notwithstanding this, the direct substitution of quality surface materials in a building façade for PV panels is an attractive financial possibility. PV panels may also be incorporated within double skin façades as shading devices within a glass assembly.

4. Case Studies

The value of the preceding sections of the paper lies both in helping support design strategies and as a critical toolbox in reviewing completed buildings from both an environmental and an architectural viewpoint.

Projects have been chosen for short reviews as case studies based on the context of the urban settings, the general nature of the business spaces provided, their potential for influencing other projects, and differing approaches to embedding environmental thinking within a variety of design positions and their design quality as perceived by the writer. The projects are reviewed within the context of the previous discussion.

Two of these projects are in the Kings Cross railway land development. This large scheme fits the criteria for a well-designed dense and connected urban quarter with a strong pedestrian priority.

4.1. Ropemaker Place, London

Ropemaker Place is located on the boundary of the City of London's financial district and the London Borough of Islington (where a general planning height restriction of 30 m applies). The building steps up in height form the North (LB Islington) to 20 stories at the South East corner where it faces the 30-storey tower known as City Point and its open plaza. The surrounding streets are narrow and the building is well connected to public transport, London buses, London Underground and the upcoming Crossrail link, all at Moorgate.

As a deep plan building for the London financial market, the design of Ropemaker Place by Arup Associates has taken a strict approach to limiting solar heat gain.

The percentage of clear glazing is very low. The building is clad principally in opaque insulated glass panels with coloured inserts of varying pattern across and up the façades. Projecting "windows" are arranged above desk level in long rows, which are then bookended by opaque glazing.

The projecting windows are angled to restrict the entry of sunlight: on the south elevation, the window is tilted downwards with a hood like canopy at the top, whereas on the east and west façade's the window glass is angled laterally towards the North with solid ends facing the South (Figures 3 and 4). These latter windows appear in serrated rows of strip windows within the flat expanses of opaque glazing. This angled window treatment is stated to reduce solar heat gain by 27% from an equivalent flat façade.

The rigorous exclusion of sunlight supports the low energy strategy for this deep plan building; the occupants may miss the amenity of the sun's rays reminding them of the daily cycle.

The substantial opaque areas of the façade could have been an opportunity for energy generation; however, an artist designed array of subtly changing coloured panels is the preferred solution. The use of glass as a design feature for the opaque cladding is highlighted in The Architects Journal [31] for its relatively high embodied energy; it seems that the new architectural thinking is camouflaged to fit with the expectations of the market.

The author was a member of the design team for this project as part of his work with Arup Associates.

Figure 3. South facing inclined windows, Ropemaker Place, London, Arup Associates.

Figure 4. Serrated windows on East façade (West façade similar), Ropemaker Place, London, Arup Associates.

4.2. One Pancras Square, Kings Cross, David Chipperfield Architects

This commercial building makes a powerful historic reference to the cast iron age while acknowledging the benefits of a reduced amount of glazing to the building's interior. Cast iron column casings form an apparent structure at an interval more frequent than the regular steel frame underneath (see Figure 5). The columns are in the round as a portico on the prominent southern elevation and on the east and west half round as part of a beam and slab arrangement which shades the glazing whose percentage is reduced in the process (Figure 6) [32]. This percentage is approximately 40% consistently around the main floors of the building. Two Pancras Square has been newly completed and no performance data is expected at this stage.

Figure 5. South facing portico with cast iron columns. One Pancras Square, London, David Chipperfield Architects.

Figure 6. Typical façade detail showing glazing percentage. One Pancras Square, London, David Chipperfield Architects.

4.3. Two Pancras Square, Kings Cross, Allies and Morrison

This is a building by a practice which has adopted a contextual approach from its inception. A modern tripartite composition of solid and void achieves the reduced glazing percentage indicated in the paper in a polite and restrained manner. The exposed east façade will later be part of a tall urban street as the adjoining site development gets underway (see Figure 7). Allies and Morrison were the architects of the master plan for this development [33,34] which combines density with a traditional network of streets working with the existing conditions and features from the railway and canal era.

Figure 7. Two Pancras Square, London, Allies and Morrison (**right**) with One Pancras Square (**left**).

The almost imperceptible incremental increase in window size rising up the building can be seen as an aesthetic transition from the 6 m wide column spacing at ground level to the thinner pergola-like framing at roof level 3 m apart. A daylight driven response would plan for more daylight at the lower parts of the building where shading by adjoining buildings will be more evident. An example of this can be seen in the north elevation of the CH2 building in Melbourne, an urban ecological building of which a successful performance review has been carried out and published [35].

The glazing percentage varies from approximately 50% at the lower level (see Figure 8) to 60% at the top of the building. Two Pancras Square has been newly completed and no performance data is expected at this stage.

Figure 8. Typical façade detail showing glazing percentage. Two Pancras Square, London, Allies and Morrison.

4.4. Kings Place, London, Dixon Jones

This landmark building houses music venues, gallery and conference space, a canal side bar and a restaurant underneath commercial offices currently occupied by the Guardian newspaper among others [36]. A five-storey height, wave-like projecting glass screen shields the building from traffic noise while presenting a consistent front to the adjoining major new development (see Figure 9). The screen, which is thermally open at the bottom and at the top, allowing warm air to disperse due to stack effects, faces west whilst also wrapping round the south west corner into part of the south façade (see Figure 10). This screen allows a more highly glazed inner façade than would otherwise be justified in low energy terms. Other façades are composed of punched window openings in a more traditional manner (see Figure 11). The building shows how environmental thinking can be incorporated into a powerful urban and architectural composition [37].

Figure 9. Kings Place, London, Dixon Jones (West Elevation).

Figure 10. Close-up of double skin façade, Kings Place, London, Dixon Jones.

Figure 11. North West corner, showing differing façade system facing North, Kings Place, London, Dixon Jones.

On visiting the site on a sunny winter morning, almost all of the internal blinds on the south façade were in the down position; even the fixed cavity shading did not limit the penetration of the low angled winter sun (see Figure 12).

Figure 12. Blinds in drawn position, South façade, Kings Place, London, Dixon Jones.

4.5. Education Executive Agency and Tax Offices, Groningen, UN Studio

This building is not in a dense urban context and has significant provision of underground car parking. It is included here because it demonstrates an incremental adjustment to solar orientation and shading within and around a curvilinear building outline. This would be difficult to conceive, let alone analyze and calculate without recent advances in three-dimensional modeling, parametric software and computer power [38]. A composition based on two ellipses; the pointed spandrel projection increase and decreases incrementally as the façade faces from east to south and to west in particular at the same time as the window height is raised and reduced again. The top surface of the projecting overhang is able to reflect light (minus the solar heat gain) into the interior, acting as a light shelf.

The project is a sophisticated example of parametric design thinking incorporated into a curvilinear building. UN Studio is a research led practice which makes strong connections between environmental led design and parametric form finding [39].

4.6. 101 Bishopsgate, London, KPF

This building (see Figure 13) is highly glazed on three façades facing the North, East and West, whilst the south face of the building is occupied by lifts, toilets and stairs which shield the office space from the southern solar heat gain. These core elements are in part clad in a building height array of PV panels, showing that façade areas that are not required to act as windows can generate renewable energy at the point of source for building users. The cost of these panels can be offset by the cost of the cladding which would have been required if the PV panels were not there. In this case, development of the adjoining site to a similar height may in the near future obstruct the solar access of these panels during the middle of the day.

Figure 13. PV panel cladding at lift cores, 101 Bishopsgate, London, KPF Architects.

4.7. White Collar Factory Old Street, London, AHMM

In Derwent London's and AHMM Architects' "White Collar Factory" concept [40], a looser and more robust approach to housing innovative enterprises is emerging. The cost of building and accommodating business is moving from the (formerly suspended) ceiling to the façade, where opening windows and shading/screening devices are included as part of a reduced percentage of overall glazing. Exposed structure at ceiling level can contribute through its thermal capacity to the flywheel effect of smoothing out temperature changes over the daily cycle in conjunction with night time cooling. Changes in computing patterns are leading to a reduced emphasis on under floor cabling as Wi-Fi enables a more fluid and mobile way of working.

The construction of this project was at the foundation stage at the time of writing.

5. Implementing Climatic Based Façade Design

The façade is not just the thin zone outside where the architect is encouraged to package a predetermined interior floor space and present it in its external surroundings. The façade should modify the interior environment, significantly lightening the air conditioning (or other mechanical system) loads, whilst admitting daylight, limiting solar heat gain, controlling glare and managing privacy. In order to do this, glazing can be arranged in appropriate ways to respond to the cardinal points of the compass and the surrounding buildings. Wall surfaces not actively in use as windows can be clad with PV panels and help generate electrical energy.

Effective climatic design requires a set of environmental objectives which are agreed with the client and the design team from the earliest stages of a project. An up to date template for task management in the complex procurement and digital collaboration world incorporating key sustainability checkpoints now exists in the form of the RIBA Plan of Work 2013 [41]. Within this framework, the right issues can be identified at the right time and a proactive contribution from the key specialists can be expected. Allowance is made for sharing key data at the right time and supporting the commissioning and handover stage more effectively.

These check points include identifying during concept stage and confirming during design stage a seasonal control strategy and considering resilience in response to future changes in climate. For British practice, the Part L analysis should commence early so that environmental thinking is part of the decision making process and not a response to an already confirmed design.

An example of how climatic based design thinking (focusing on the façade design) can be incorporated into the design sequence for an urban business building can be seen in Table 1.

Robust design based on Table 1 can be embedded in the project before a full team is appointed and before the more frontloaded and accelerated Building Information Modelling (BIM) activity is too far advanced. Environmental software such as IES or Ecotect can be used to support these early stages. Façade design can be then tested in more detail and refined using such software in parallel (and off the critical path) during a design development phase. A variety of approaches to incorporating environmental design within the new shared digital environment are reviewed in detail in BIM Design [38].

Table 1. Design sequence for embedding environmental thinking within early façade design ideas (Environmental and climatic activities highlighted in green).

1	Building form	Review the plot boundaries, surrounding buildings and context and planning restrictions
		Review orientation, access to sunlight and daylight, shading from adjoining buildings
		Will additional light and ventilation routes be introduced via an atrium or courtyard?
		Develop initial building form based on the above
2	Energy balance	Formulate an initial view of the relative importance of heating and cooling, sunlight and daylight over the seasons, based on building form and expected occupancy
3	Orientation, glazing %	Study visibility, principle direction of arrival, entrance positions, clients profile
		Review and analyse which surfaces of this building volume will be illuminated by the sun, at what times/seasons, at what frequency and for how long?
		Review and analyse how much daylight is accessible to these building surfaces, both directly and indirectly
		Based on the above, is there a reason for increasing the glazing % above recommended amount (40%) for any (or all) of the elevations?
		Should the proposal have directionality with primary and secondary façades?
4	Glazing concepts	Develop glazing concepts for the building and its individual façades based on the above
		Incorporate solar shading and/or double skin façade concepts as appropriate responding to orientation and solar access.
5	Vertical differentiation	If sunlight and daylight are less accessible at lower levels of the façades should the design be adjusted to suit this?
		Is there an opportunity to harness solar energy on the various façades, and is this more pronounced at upper levels of the building?
6	Detail of glazing concepts	Consider extent of solar shading and /or operation of any double skin façade system proposed. Review solar angles at winter/summer/mid-season. Review energy implications of solar heat gain to energy balance in these three seasonal conditions.

6. Summary and Conclusions

The set of issues identified and examined in this paper can be used both as a tool for critical evaluation of buildings and as part of a design methodology for new projects.

Post occupancy testing, however, of double skin façades, in particular, is urgent and results should be published in order to help improve practice. It can then become clearer whether these systems should be adopted for specific orientations or more generally around buildings and if so, how they can be customized for the different orientations.

Architectural and sustainable strategy must go hand in hand in the design of the façades of buildings, and both passive and active design thinking should inform this process. In addition to facilitating views and the appearance of the building, each façade (and each part of that façade) can contribute to daylight, solar control, energy generation, insulation and/or thermal buffering according to its exposure to sunlight and daylight over the daily and seasonal cycles on a three-dimensional grid. In addition to the façade systems which automatically open and close vents and blinds according to the weather, each part of the façade plays its part in several aspects of building physics according to where it is.

Using advanced modeling techniques, the relevant design information can be made available and early ideas tested; the challenge is to invent new aesthetic systems to effectively use that information.

Acknowledgments

Thank you to Hugh Byrd, Professor of Architecture at the University of Auckland (and of the University of Lincoln from 2013 to 2014) for supporting and adding direction to the early ideas for this article. The 8th Energy Forum conference "Advanced Building Skins" in Bressanone, Italy brought me in touch with a wide range of academics and practitioners of various disciplines, helping me learn to what extent my thinking was original and where it formed part of ongoing discussions around the world. Three design practices have included me in project teams for urban business buildings, principally in London and Manchester; much of my knowledge of and interest in this subject comes from my work in these practices—Arup Associates, Denton Corker Marshall and Sheppard Robson. With Denton Corker Marshall I had the privilege of working with Gartners, now part of the worldwide Permasteelisa cladding group.

Conflicts of Interest

The author declares no conflicts of interest.

References

1. Banham, R. *The Architecture of the Well Tempered Environment*; University of Chicago Press: Chicago, IL, USA, 1984.
2. Hawkes, D.; McDonald, J.; Steemers, K. *The Selective Environment: An Approach to Environmentally Responsive Architecture*; Taylor and Francis: London, UK, 2001.
3. Hawkes, D.; Owers, J. *The Architecture of Energy*; Longman: London, UK, 1982.

4. Chappells, H.; Shove, E. Debating the future of comfort, environmental sustainability, energy consumption and the indoor environment. *Build. Res. Inf.* **2005**, *33*, 32–40.
5. Concerted Action—Energy Performance of Buildings. Available online: http://www.epbd-ca.eu/themes/nearly-zero-energy (accessed on 15 February 2014).
6. Voss, K.; Musall, E.; Lichtmess, M. From low energy to zero energy: Status and perspectives. *J. Green Build.* **2011**, *6*, 46–57.
7. Voss, K.; Musall, E. *Net Zero Energy Buildings; International Projects of Carbon Neutrality in Buildings*; DETAIL Green Books: Munich, Germany, 2011.
8. Various Authors. Advanced Building Skins. In Proceedings of the 8th Energy Forum, Bressanone, Italy, 5–6 November 2013.
9. Schittich, C. *Building Skins*; Birkhauser Edition Detail: Munich, Germany, 2006.
10. Oesterlie, E.; Lieb, R.-E. *Double—Skin Facades; Integrated Planning*; Prestel Verlag: Munich, Germany, 2001.
11. BREEAM. Available online: http://www.breeam.org/ (accessed on 4 December 2014).
12. LEED. Available online: http://www.usgbc.org/leed (accessed on 4 December 2014).
13. The Building Regulations. Part L2A. Available online: http://www.planningportal.gov.uk/uploads/br/BR_PDF_AD_L2A_2013.pdf (accessed on 7 July 2014).
14. Olgyay, V. *Design with Climate: Bioclimatic Approach to Architectural Regionalism*; Princeton University Press: Princeton, NJ, USA, 1963.
15. Dahl, T. *Climate and Architecture*; Routledge-Taylor & Francis Group: London, UK, 2010.
16. Oliver, P. *Dwellings, the Vernacular House World Wide*; Phaidon: London, UK, 2003; pp. 128–147.
17. Peel, M.C.; Finlayson, B.L.; McMahon, T.A. Updated world map of Koppen-Geiger climate classification. *Hydrol. Earth Syst. Sci.* **2007**, *11*, 1633–1644.
18. Gething, B.; Puckett, P. *Design for Climate Change*; RIBA Publishing: London, UK, 2013.
19. GLA. *London's Urban Heat Island; A Summary for Decision Makers*; Greater London Authority: London, UK, 2006.
20. Kolokotroni, M.; Ren, X.; Davies, M.; Mavrogianni, A. London's Heat Island; Impact on current and future energy consumption in office buildings. *Energy Build.* **2012**, *47*, 302–311.
21. Robert, A.; Kummert, M. Designing net-zero energy buildings for future climate, not the past. *Build. Environ.* **2012**, *55*, 150–158.
22. Mardaljevic, J.; Heschong, L.; Lee, E. Daylight Metrics and Energy Savings. *Light. Res. Technol.* **2009**, *41*, 261–283.
23. IES Software. Available online: http://www.iesve.com/ (accessed on 17 December 2014).
24. Ecotect Software. Available online: http://usa.autodesk.com/ecotect-analysis/ (accessed on 17 December 2014).
25. *Rafael Moneo International Portfolio 1985–2012*; Moneo, R., Ed.; Edition Axel Menges: Stuttgart-Fellbach, Germany, 2013; p. 197.
26. Poirazis, H.H. *Double Skin Facades for Office Buildings, Literature Review*; University of Lund: Lund, Sweden, 2004.
27. Trubiano, F. Performance based envelopes: A theory of spatialized skins and the emergence of the integrated design professional. *Buildings* **2013**, *3*, 689–712.

28. Napier, J. Manchester civil justice centre: Procuring and managing an institutional building with a mixed mode ventilation system—A case for post-occupancy evaluation. *Buildings* **2013**, *3*, 300–323.
29. Cilento, K. Al Bahar Towers Responsive Facade/Aedas. Available online: http://www.archdaily.com/270592 (accessed on 14 December 2014).
30. Esteban, S.; Izard, J. Performance of photovoltaics in non optimal orientations: An experimental study. *Energy Build.* **2015**, *87*, 211–219.
31. Harman, H.; Brundle, M. Ropemaker Place. *Arch. J.* **2010**, *12*, 22–31.
32. David Chipperfield Architects. Available online: http://www.davidchipperfield.co.uk/ (accessed on 1 November 2014).
33. Allies and Morrison. Available online: http://www.alliesandmorrison.com/ (accessed on 1 November 2014).
34. Makower, T. *Touching the City; Thoughts on Urban Scale*; John Wiley and Sons: Hoboken, NJ, USA, 2014; pp. 118–119.
35. Council House 2, Our Green Building. Available online: http://www.melbourne.vic.gov.au/sustainability/ch2/Pages/CH2Ourgreenbuilding.aspx (accessed on 17 December 2014).
36. Kings Place. Available online: http://www.kingsplace.co.uk/home (accessed on 1 November 2014).
37. Dixon Jones. Available online: http://www.dixonjones.co.uk (accessed on 1 November 2014).
38. Garber, R. *BIM Design, Realising the Creative Potential of Building Information Modelling*; John Wiley and Sons: Hoboken, NJ, USA, 2014; pp. 224–241.
39. UN Studio. Available online: http://www.unstudio.com (accessed 1 November 2014).
40. White Collar Factory. Available online: http://www.whitecollarfactory.com (accessed on 15 February 2014).
41. RIBA Plan of Work 2013. Available online: http://www.ribaplanofwork.com/ (accessed on 15 February 2014).

3

Measuring the Weathertight Performance of Flashings

Mark Bassett [*,†] **and Greg Overton** [†]

BRANZ Ltd., P.O. Box 50908, Porirua City 5240, New Zealand; E-Mail: greg.overton@branz.co.nz

[†] These authors contributed equally to this work.

* Author to whom correspondence should be addressed; E-Mail: mark.bassett@branz.co.nz

Academic Editor: George Baird

This paper is an extended version of a paper "Weathertight performance of flashings for taller buildings" presented at the "Building a Better New Zealand" Conference 2014, Auckland.

Abstract: Residential buildings are now better engineered to manage rainwater following the leaking building problem in New Zealand. The next challenge is to improve the weathertightness of medium-rise buildings which often use joint details widely applied on low-rise buildings but are subject to higher wind pressures and surface runoff rates. This study begins to address this challenge by measuring the water leakage performance limits of the following common flashings with static and dynamic rain and wind loads to see how their performance might be improved: (a) Horizontal H and Z jointers between direct fixed sheet claddings; (b) The window head flashing in a cavity wall; (c) A horizontal apron flashing at the junction between a roof and wall. At this stage, water penetration resistances have been measured but the data has not yet been discussed in the context of wind pressures and rain loads on mid-rise buildings. All of the joints were found to resist water leakage to pressures equivalent to the hydrostatic head of the upstand, so long as there were no air leakage paths through the joint. When vents were added, or openings were present that might arise due to construction tolerances, then the onset pressure for leakage was found to fall by as much as 50%. Vents, of course, are essential for ventilation drying in rainscreen walls and even with vents present, the onset of leakage was at generally at least twice the 50 Pa wet wall test pressure applied in New Zealand. Opportunities were found to improve the way vented joints deal with runoff by enlarging the gap between the cladding and flashing. This prevented the outer joint volume from filling with water and occluding the vents. The apron flashing was found to cope better than a window head joint with runoff, because of the larger 35 mm vertical gap between the cladding and apron.

Keywords: weathertight joints; flashings; upstands

1. Introduction

Most water leaks in residential buildings occur at joints between claddings and components such as windows, *etc.* This is illustrated in Figure 1 using data from a survey of leaking buildings in New Zealand by Bassett *et al.* [1]. Over 60% of leakage sites were at junctions between claddings and other components and less than 40% were associated with roof and wall claddings. Contributing causes were found to include the omission of traditional metal flashings at junctions between components along with limited protection from eaves in newer building designs. A similar fraction "26%" of water entry points around window and door junctions with claddings were identified by Morrison Hershfield Ltd. [2] in a survey of leaking buildings in British Columbia, Canada. In that survey, 90% of water leakage sites were found to be at junctions between materials and components or at penetrations through the cladding.

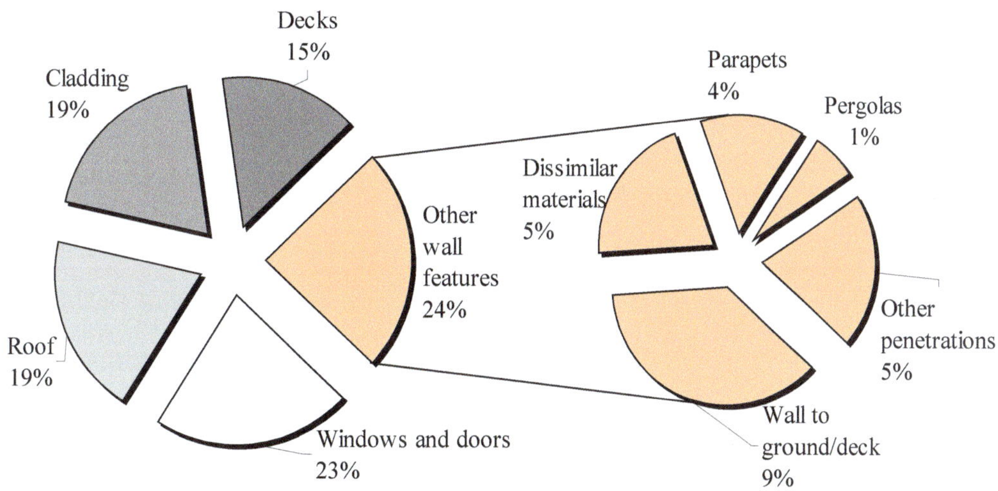

Figure 1. Water entry sites in New Zealand leaking buildings.

The Department of Building and Housing (now part of the Ministry of Business, Innovation and Employment) responded to the leaking building problem by modifying the compliance document, E2/AS1 External Moisture, of the New Zealand Building Code (NZBC) to improve the standard of water management in wall and roof designs (Department of Building and Housing [3]). Most of the changes were based on the 4Ds (deflection, drainage, drying and durability) approach developed in Canada by Hazleden and Morris [4] and incorporated a water-managed cavity behind the cladding. Additional important changes include flashings around the perimeter of windows and doors.

A new test procedure E2/VM1 "Verification Method" (Department of Building and Housing [5]) was also developed to apply the following tests and criteria to claddings and flashings on cavity walls:

- Water should not systematically reach the dry side of the cavity during any of the tests (including above the upstand on horizontal flashings above cavity closers);
- The first test in E2/VM1 is a static and cyclic pressure test of the complete wall system;
- The second test introduces defects in the cladding around flashings to confirm that water is successfully controlled by drainage paths including the back of the cladding and flashings at junctions between components or different claddings;
- Finally, the "wetwall test" applies 50 Pa pressure difference across the cladding and checks that water leaks through the cladding are confined to the drainage paths and do not bridge the cavity.

These tests apply the following criteria to flashings in cavity walls:

- The flashing forms part of the drainage path for water on the back of the cladding;
- The flashing must rainscreen the joint and prevent water from reaching the underlay with a rain intensity of at least 3 $L/m^2 \cdot min$ and a 50 Pa air pressure difference between the wetwall and the cavity.

Requiring the wall underlay to remain dry during testing is quite stringent and specific to New Zealand. It is the primary reason for this investigation of flashing upstand heights in claddings. There are many questions about the performance of flashings that are not answered by the E2/VM1 procedure and which this research is starting to address:

- How far should the flashing upstand extend upwards behind the cladding to deal with wind driven rain?
- Do the upstand heights of flashings need to be adjusted to cope with rain bouncing of a roof—particularly important for apron flashings between roof and wall?
- Do cavity closers have a role in managing air and water entry through the joint?
- How should cladding to flashing clearances be sized to control runoff?
- Do hems folded into metal flashings improve the weathertight performance of flashings?

Perhaps the most pressing need for research on the leakage performance of joints is in support of timber framed walls above the height limit for E2/AS1 (three stories maximum but nominally 10 m above ground). Many of these buildings adapt E2/AS1 claddings and flashings to timber framed infill panels that are exposed to higher wind pressures and surface runoff rates than low rise residential buildings.

There are few scientific investigations of the water leakage performance of joints and how their design might be improved to cope with extreme exposure. One of the earliest investigations (Ishikawa [6]) measured the leakage characteristics of joints in a metal curtain wall and concluded that the key elements were a large external opening to prevent a water film from bridging the gap, and an airtight internal joint to support wind pressures. Further studies by Herbert [7] and [8] measured water entry rates through labyrinth type joints exposed to driving rain in field trials, showing that the important attributes of successful joints were an airtight inner wall and sloped drainage to deal with water entering the joint. A more recent study (Lacasse *et al.* [9]) measured leakage rates through specific defects in walls such as missing lengths of sealant, and used the leakage function of wind pressure and rain load to estimate the moisture loads that have to be managed by vapour diffusion and ventilation drying within the wall. Water leakage rates were measured as a function of runoff rates and static wind pressure and fitted to an empirical relationship that was then used to estimate moisture entry loads in a range of North American climates. A similar approach to developing water leakage functions for joints is followed in this study but the next step of comparing leakage rates with the capacity of joints to control leakage with drainage and ventilation drying is outside the scope of this paper. A future study will follow the procedures developed by Sahal and Lacasse [10] and determine the rain loads, wind pressures and runoff rate for claddings for taller buildings in New Zealand.

2. Measured Rain Leakage Characteristics

Equipment illustrated in Figure 2 was used to measure the water leakage characteristics of joints between building components. It consists of a pressure chamber connected to a fan and a fluctuating piston that together apply a steady pressure and a superimposed fluctuating pressure across the specimen. The pressure amplitude can be changed by adjusting the piston stroke although below 0.2 Hz there was simply not enough travel to reach large pressure amplitudes. In all of the fluctuating pressure tests described in this paper, the pressure amplitude was 2:1 as required in E2/AS1. The wall specimen measured 0.7 by 0.7 m (area 0.49 m^2) and the length of joint under investigation was 0.53 m.

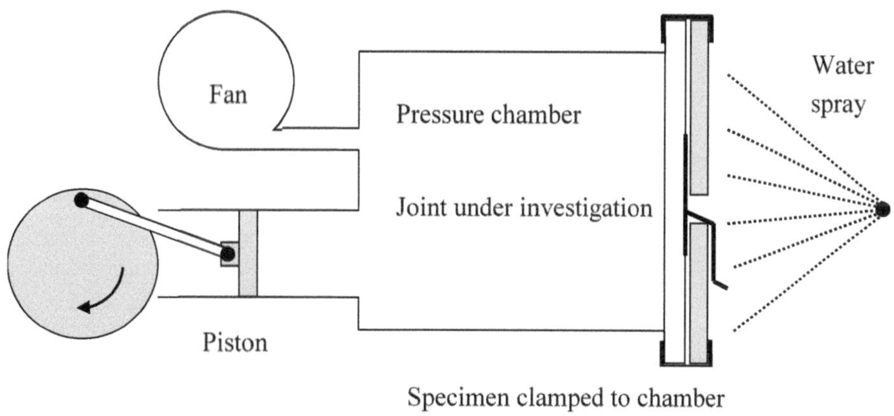

Figure 2. Equipment for measuring the weathertightness characteristics of joints.

A range of water spray nozzles were used to wet the sample area with between 3.4 $L/m^2{\cdot}min$ (corresponding to the minimum rain load called for in E2/VM1) and 0.08 $L/m^2{\cdot}min$. This is similar to the rain loads applied to large wall assemblies (0.02–3 $L/m^2{\cdot}min$) by Bassett *et al.* [11] and provides a wide enough range of rain intensity to develop rain leakage functions that include most rain events. In these measurements the runoff rate was found to be a more useful measure of water delivered to the joint.

The leakage rates measured through junctions using this equipment establish performance limits for joints rather than actual leakage rates that might apply in mid-rise buildings in New Zealand. In a later project will use these leakage characteristics to propose flashing upstands dimensions for these buildings.

3. Typical Flashed Joints Seen in Residential Buildings

There are a large number of joints described in E2/AS1 but the three junctions illustrated in Figure 3 capture most of the essential features of flashings and their application in walls with cavities or direct fixed claddings.

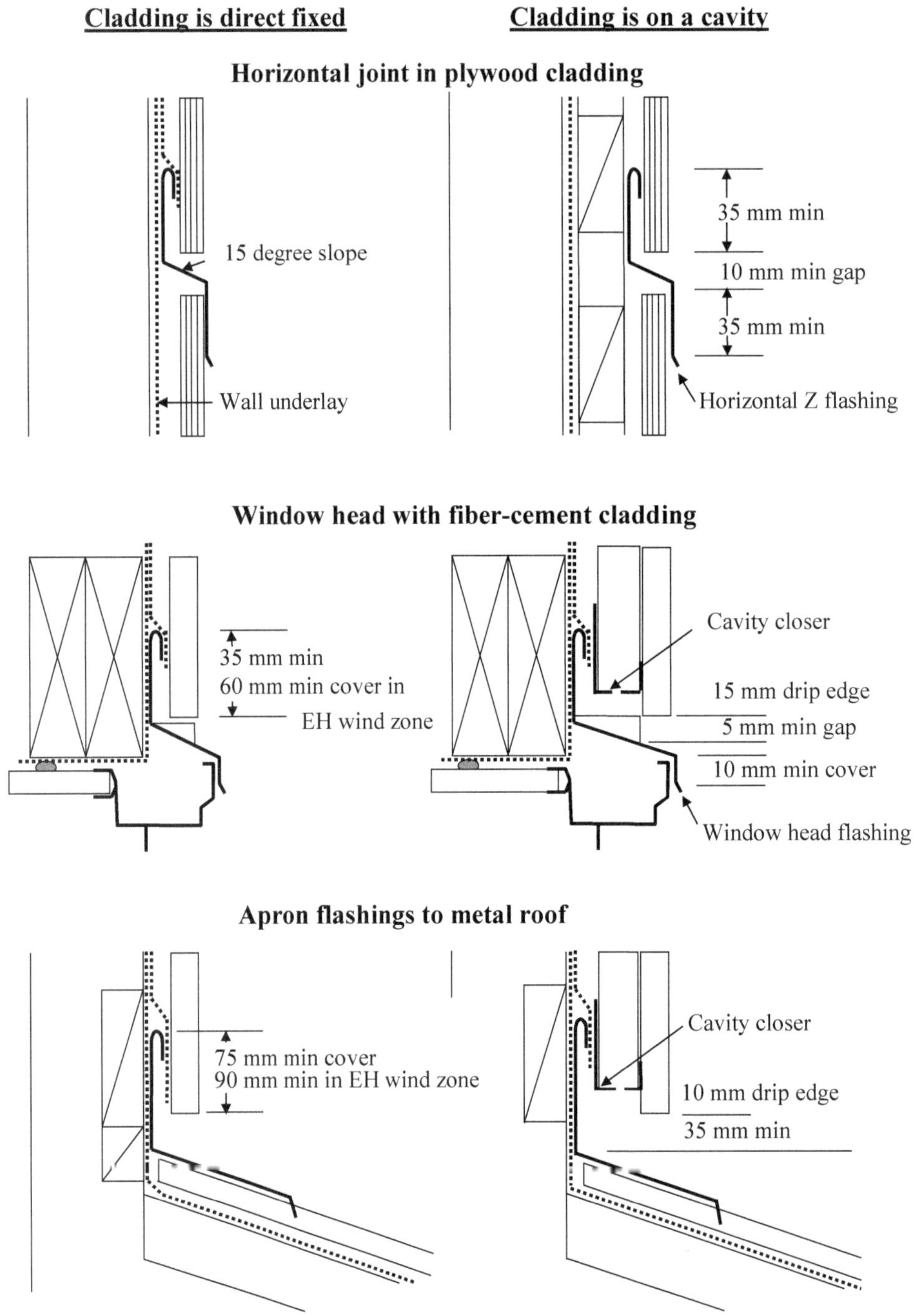

Figure 3. A selection of flashings used in NZ buildings and taken from E2/AS1.

The three joints chosen for what is a preliminary study are:

- A Z jointer between direct fixed plywoodwall panels;
- A window head flashing in a cavity wall;
- An apron flashing between a roof and a higher section of cavity wall.

Upstand heights in these joints have evolved from field experience although recent changes to extend the applicability of E2/AS1 to an "extra high" wind zone were responsible for precautionary increases to upstand heights in walls with cavities and direct fixed claddings −35 to 60 mm for window head flashings and 75 to 90 mm for roof to wall apron flashings. Another recent change required mandatory hems to the top of flashing upstands used in extra high wind zones. In lesser wind zones, the hem can be traded for an additional 25 mm of upstand height. These were largely precautionary changes ahead of applicable field experience or laboratory results of the type that this study aims to provide. Another significant change that came with more widely adopting cavity construction was the provision for vents in a cavity closer. Vents are potential air leakage paths and an entry point for air carried spray. This project set out to link cladding overhang and clearances in joints with the effectiveness with which the joint rainscreens against water entry.

4. Leakage Characteristics of a Horizontal Cladding Jointer

The leakage characteristics of a plastic H jointer were measured between two cladding panels fixed to cavity battens as illustrated in Figure 4.

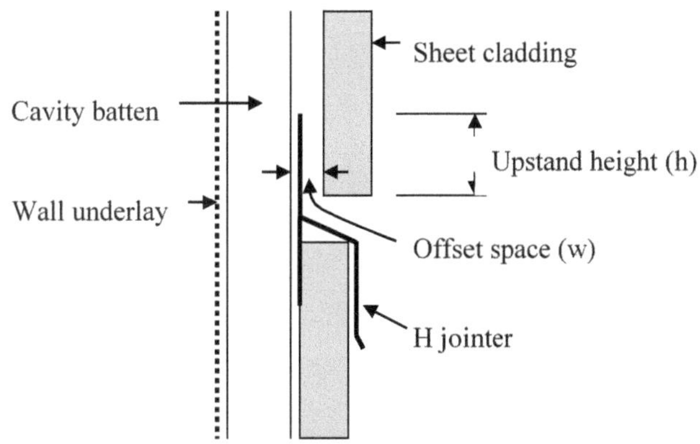

Figure 4. Sectional view through an H jointer between two sheets of wall cladding.

This is similar to the Z flashing illustrated in Figure 3 between plywood panels on a cavity wall except that it does not include a hem now required in E2/AS1 in the extra high wind zone. There are two main leakage paths in the H jointer as follows:

- Leakage between the up-turned leg of the jointer and the cladding. In particular the significance of the offset space width (*w*), the height of the upstand and the presence or absence of a hem folded into the upstanding leg;
- Leakage over the top of the lower cladding past the lower legs of the H jointer—not investigated here.

4.1. Significance of Offset (Air Gaps) between the Jointer and the Cladding

The air pressure difference corresponding to the onset of leakage through the H jointer was measured for a range of upstand heights by cutting down the upper leg of the jointer to 15, 35 and 60 mm. A 530 mm wide section of jointer (spanning between cavity battens fixed to framing) was sprayed with water at 3.4 L/m^2 and the air pressure difference adjusted over the range 0–500 Pa until leakage was detected above the jointer. The pressure at which water reached the top of the upstand are shown in Figure 5 as 140, 294 and 500 Pa for the three upstand heights 15, 35 and 60 mm.

The pressures needed to spill water over the upstand were a little higher than this (147, 300 and 588 Pa) because of the additional pressure needed to overcome surface tension. When the offset gap *w* was increased beyond 2–3 mm, air carried water leakage was observed bubbling above the upstand.

Figure 5 shows air carried water leaks occurring at lower pressures as the offset gap is increased for all three upstand heights.

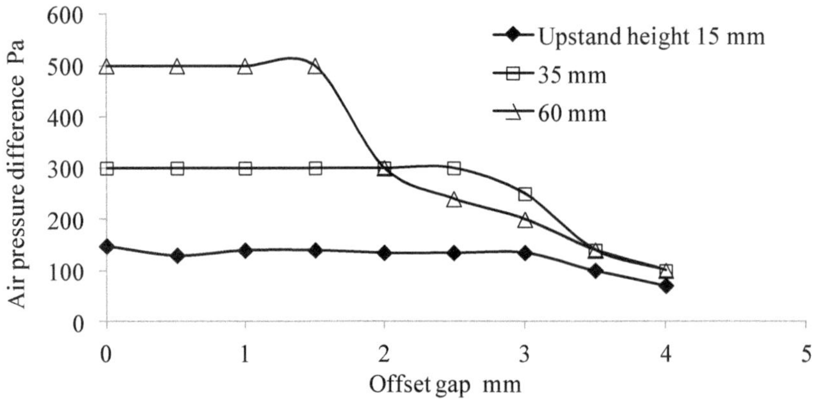

Figure 5. Air pressures at which water leakage occurrs through the H jointer as a function of upstand height and the offset gap width.

4.2. Significance of a Hem on the Upstand of the Jointer

The PVC H jointer was replaced with folded aluminium flashings having upstand heights of 35 and 60 mm and with and without a hem formed as shown in Figure 3. In this case water leakage rates were measured gravimetrically over a period of one minute by weighing an absorbent strip placed just above the flashing. In all cases the cladding was held against the flashing upstand ($w = 0$) although the flashings with a hem will have formed a 3 mm gap between and most of the flashing upstand and the back of the cladding. With a 1.3 g/m·s runoff rate over the joint, static pressure leakage characteristics for the 35 and 60 mm high flashings were as shown in Figure 6 to be similar to those measured earlier for the PVC H jointer with the onset of leakage at 300 and 500 Pa. The onset of leakage was found to be independent of the hem, but once again, it was possible to reach onset leakage pressures as high as 370 and 580 Pa by carefully spring loading the hem of the 35 and 60 mm high flashings against the cladding. Although these measurements were conducted using standard construction materials, it is likely that the tolerances achieved in the laboratory were tighter than would be seen in buildings, and that onset leakage pressures in the 100–300 Pa range indicated in Figure 5 for offset gap widths above 2 mm are more likely where no special care is taken to hold the flashing against the cladding. Higher upstand dimensions do bring weathertight performance advantages but only where air carried water leakage can be eliminated with a tighter fit between flashing and cladding. Onset leakage pressures of 100–300 Pa are higher than the 50 Pa "wet wall" test pressure applied in E2/VM1 to the field of the cladding and on this basis there is little argument for increasing upstand heights.

Results of a more detailed study of the 35 mm high flashings applying static and fluctuating pressures are presented in Figure 7. Here, the water leakage rate is plotted against the peak air pressure difference and appears to indicate leakage rates at lower pressures as the frequency of the applied pressure increases. In fact, there was insufficient adjustment in the stroke of the fluctuating piston to keep the amplitude of the applied pressure independent of frequency and so the following additional analysis has been required.

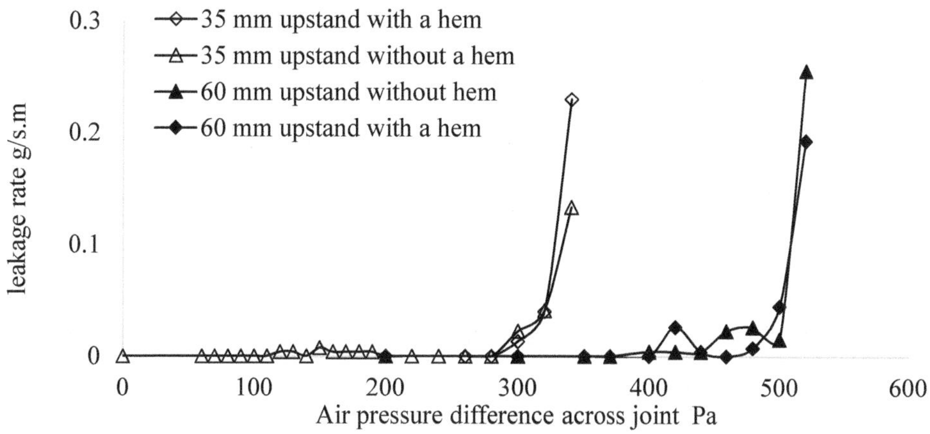

Figure 6. Steady pressure leakage characteristics of flashings with 35 and 60 mm upstands with and without hems.

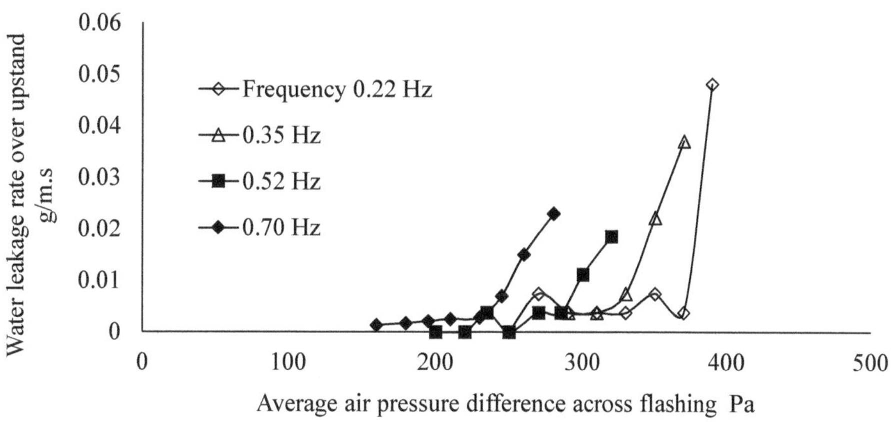

Figure 7. Dynamic leakage characteristics of flashings with a 35 mm upstand and a hem.

Observations showed that water accumulating inside and relaxing out of the joint was in phase with with pressure fluctuations up to 0.7 Hz, suggesting that the leakage rate at any time in the cycle might simply be calculated from the steady pressure leakage rate function and the applied sinusiodal pressure as follows:

The applied pressure difference $\Delta p = A + B \sin \omega t$

Where Δp = the applied pressure difference (Pa); A = the average pressure (Pa); B = the amplitude of pressure fluctuation (Pa); ω = the angular velocity of the fluctuating pressure (rad/s); L = the instantaneous joint leakage rate (g/m·s) and average leakage rate (L_{av} g/m·s), where the static leakage function is a function of the pressure $L = f(\Delta p)$ and the average leakage rate $L_{\text{av}} = \frac{\omega}{2\pi}\int_0^{2\pi/\omega} f(A + B \sin \omega t)dt$.

Calculated and measured leakage rates are plotted in Figure 8 for the 35 mm flashing with a hem. For these measurements, the hem was spring loaded against the back of the cladding to improve measurement repeatability, and average leakage rates were calculated numerically using the appropriate steady pressure leakage function.

The slope of the fitted line in Figure 8 is 1.06 with $r^2 = 0.75$, indicating that leakage rates calculated on the basis of there being no significant inertia in the physical system agree reasonably well with measured data up to a frequency of 0.7 Hz. Higher frequencies than this tend to fall outside the power spectrum density of atmospheric turbulence (Kareem and Cermak [12]) and outside the range of frequencies considered relevant in pressure equalisation studies (Burnett and Straube [13]). Consequently the frequency of dynamic pressures applied in weathertightness tests is generally below 0.7 Hz, e.g., 0.2 to 0.3 Hz in E2/AS1. A model that factors in the supply rate of water to the joint, the inertia of water filling the joint and the relaxation time for water draining from the joint has not been pursued.

A similar dynamic pressure response was seen with the same height flashing (35 mm) without a hem. In this case, leakage rates were higher because it was less easy to clamp the flashing against the cladding over its entire length. Once again, there was no evidence for frequency-dependent leakage characteristics below 0.7 Hz.

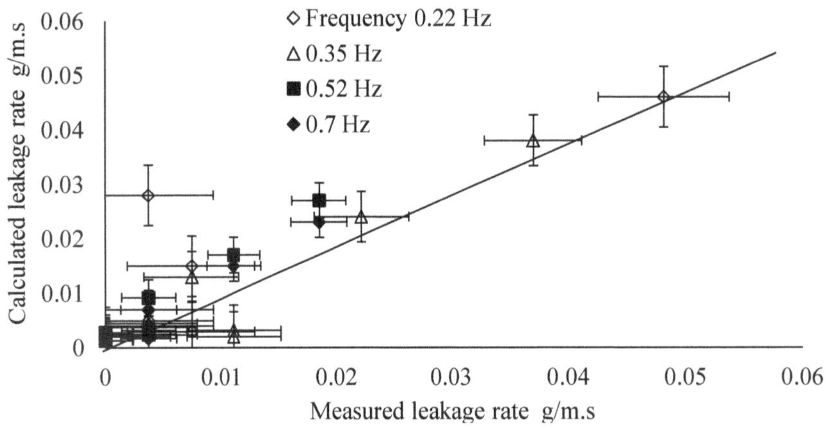

Figure 8. A comparison of measured and calculated leakage rates with fluctuating air pressures and showing experimental uncertainties.

5. Leakage Characteristics of Cavity Closer in a Window Head Joint

A window head joint was assembled as in Figure 9 but with the capacity to adjust the position of the upper cladding in relation to the window head flashing. This allowed for some variations in joint dimemsions, in particular, the gap between cladding and flashing (*g*).

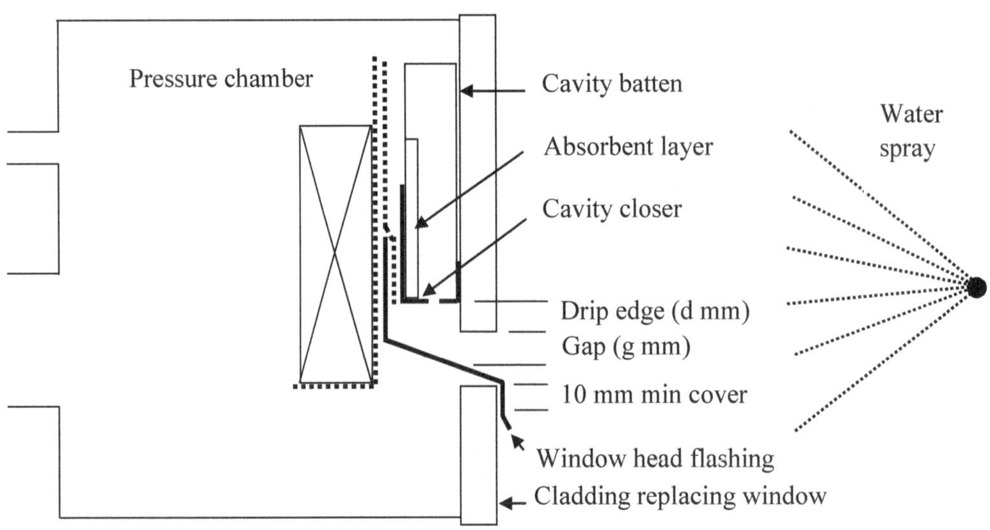

Figure 9. Experimental window head joint with cavity closer and variable joint dimensions.

There are four potential water leakage paths above the window flashing in Figure 9 as follows:

- Between the head flashing upstand and the cavity closer which is similar to the leakage path examined earlier;
- Up through vent openings in the cavity closer as examined here;
- Past the stop ends on the head flashing—not dealt with here;
- Under the flashing over the lower cladding (or window head—not dealt with here).

The upstand height of the head flashing above the base of the cavity closer is shown in Figure 3 to be 20 mm (45 mm in an extra high wind zone) but in practice, the upstand on commercially available cavity closers is around 75 mm, significantly increasing the effective upstand against water leaking through vent openings in the cavity closer. The 1500 mm^2/m vent area at the base of the cavity closer was made up of a series of slots 3 mm wide by 13 mm long and exceeded the minimum vent area of 1000 mm^2/m required by E2/AS1. Water leakage rates through the cavity closer were measured as a function of the steady air pressure across the joint and plotted against the distance travelled up the upstand. This was achieved by segmenting the absorbent layer into 9 parallel strips that could be individually weighed and assigned to a height above the cavity closer. The water spray rate was 3 $L/min{\cdot}m^2$ as required by E2/VM1 and Figure 10 plots deposition rates against height above the cavity closer.

It is clear that pressures above the 50 Pa wet wall test pressure were needed to drive water through the cavity closer and onto the dry side of the cavity although the leakage onset pressures are considerably lower than for the H jointers discussed earlier. In fact the onset leakage pressures for leakage above 35mm and 60 mm are in the range 100–125 Pa compared with 300 and 500 Pa for the H jointer, indicating that the airtightness of the joint and other aspects of joint configuration are important.

In practice, a significant proportion of wind pressure on a wall will lie across other components such as the internal lining and underlay, especially above a window which effectively partitions the wall cavity for pressure moderation across wet joints in the cladding. Secondly, it is important to acknowledge the value of ventiltion drying in both cavity walls and with direct fixed weatherboard claddings. While the vents associated with window head flashings might reduce onset pressures for water leakage, the pressures are still well above the "wet wall" test pressure of 50 Pa in E2/VM1. The potential for ventilation drying in cavity walls and behind direct fixed weatherboard walls is shown in WALLDRY-NZ (Bassett [11]) to provide the secondry water management needed to cope with even quite leaky claddings and to offset some loss in onset leakage pressures across window head flashings. For joints such as the window head flashing studied here, the water leakage measured and plotted in Figure 10 have yet to be compared with the capacity for ventilation drying and drainage from the joint.

It was observed that water leakage past the cavity closer depended on the runoff rate past the window head. High runoff rates tended to fill the joint, restricting the air flow into the cavity and allowing for wind pressures to carry water through vents in the cavity closer. This suggested that the water leakage rate may depend on dimensional factors that might be optimised to improve the water leakage characteristics of the head joint. The following factors that have been investigated in sequence:

- Dependency on runoff rate over the joint achieved with a range of water spray nozzles;
- The gap between cladding and flashing (*g*);
- Whether the pressure was applied statically or dynamically.

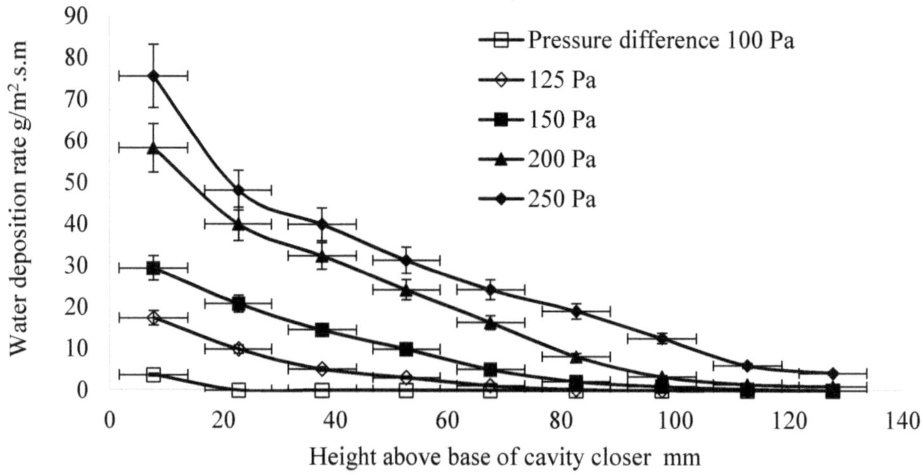

Figure 10. Deposition rate of water as a function of height above the cavity closer.

5.1. Dependency of Onset Leakage Pressures on Runoff Rate

The net water leakage rate into the cavity closer was measured at runoff rates 0.6–12 g/m·s over the window head joint (equivalent to surface flow rates of 0.08 to 1.7 L/m^2·min on the limited wall area above the head joint). Figure 11 plots the steady pressure difference at which water first penetrated the cavity closer against the runoff rate over the window head joint. During these measurements, the joint dimensions were fixed at those shown in Figure 9 with a 5 mm gap between cladding and the head flashing. Two water leakage regimes were observed. Above a runoff rate of 2 g/m·s, water tended to bridge across the joint, partially filling the space below the cavity closer and allowing water to be carried past the cavity closer at relatively low pressure differences. At runoff rates below 2 g/m·s the joint drained out, leaving an unobstructed air path to the cavity closer. With the vent unobstructed there was some water leakage above the cavity closer at higher pressures due to air carried spray, but the leakage rates were much smaller than those that were measured when the runoff rate exceeded 2 g/m·s.

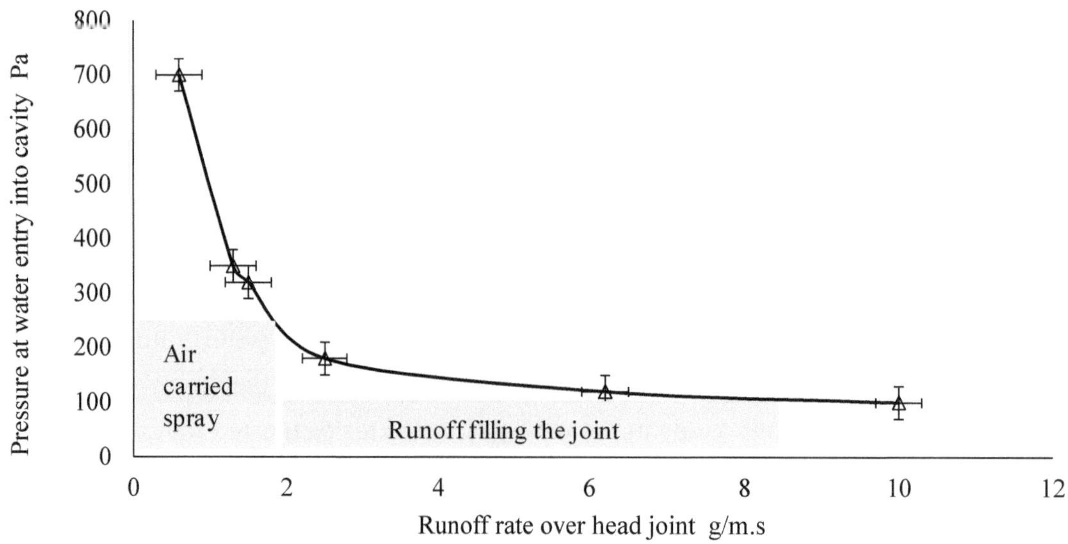

Figure 11. Pressure difference at the onset of leakage through a window head joint as a function of runoff rate over the joint.

5.2. Dependency of Onset Leakage Pressures on Gap between Cladding and Head Flashing

The dependency on runoff rate in Figure 11 suggests that the cladding to head flashing dimension might be an important factor in the weathertight performance of this joint and a separate sequence of leakage measurements were carried out to investigate this possibility with the gap width *g* ranging from 1 to 9.6 mm. Figure 12 shows how the onset leakage pressure increased with larger gap (*g*) dimensions leading to the joint draining out more effectively. This could be worth exploring more in the context of taller buildings exposed to higher wind pressures and surface runoff rates. Below a gap width of 5 mm the leakage performance of the joint appears to improve marginally, but it has to be remembered, this is at the expense of free drainage from the joint.

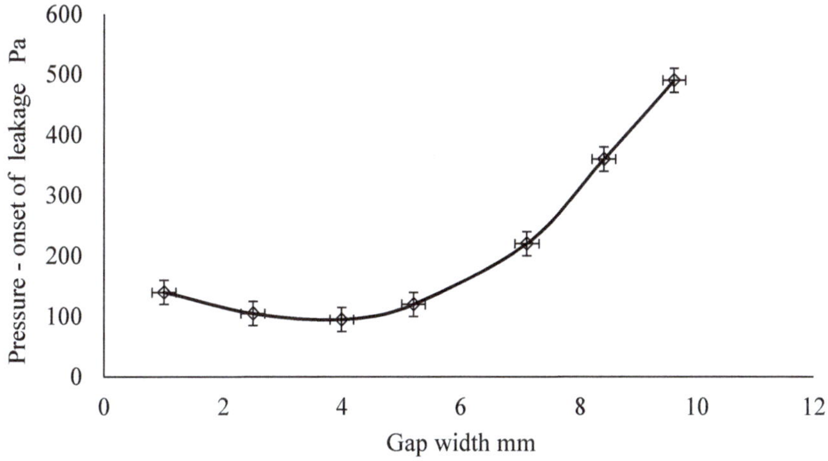

Figure 12. Dependence of onset leakage pressure on gap width for the window head joint and the runoff rate set at 2 g/s·m.

Table 1 gives a pictorial view of leakage through a window head joint with varying surface run-off rates and gap dimensions between upper cladding and head flashing. It shows that increasing the cladding to head flashing gap (*g*) improved the capacity of the joint to deal with high run-off rates, but this might also be achieved in other ways, for example, by adding a small kick-out to the lower edge of the cladding to deflect water from the opening.

Table 1. A pictorial view of leakage through a window head joint with varying surface run-off rates and gap dimension (*g*) between upper cladding and head flashing.

Case	Water Leakage Description
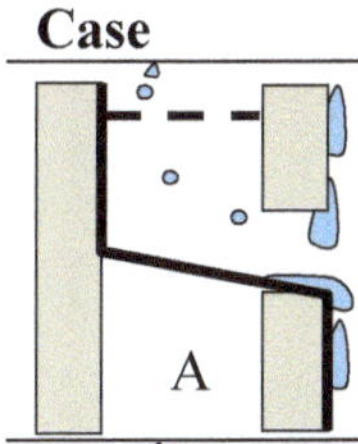 A	**Case A:** Gap width (*g*) = 5 mm and the run-off rate is less than 2 g/m·s. The gap is not significantly occluded by run-off, and the flashing drains out effectively. Above a wind pressure of 200–300 Pa, air flow through the joint entrains small droplets, which pass through into the cavity.
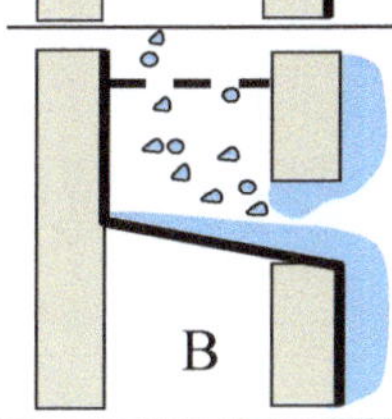 B	**Case B:** Gap width (*g*) = 5 mm and the run-off rate is larger than 2 g/m·s. The gap is now occluded by run-off, and sufficient wind pressure is supported at the outer joint to drag water into the joint. Higher air velocities through the occluded opening drag water past the cavity closer at pressure differences around 100 Pa.

Table 1. *Cont.*

Case	Water Leakage Description
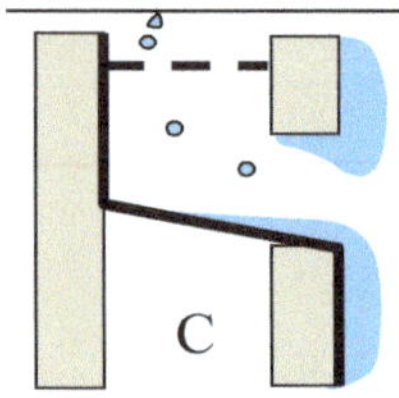	**Case C:** Gap width increased to (g) = 8 mm and the run-off rate is the same as Case A (below 2 g/m·s). The gap is no longer occluded by run-off, and the flashing drains out effectively. Beyond a wind pressure of 350 Pa, air flow through the joint entrains small droplets with leakage rates (0.015 g/m·s at 360 Pa).
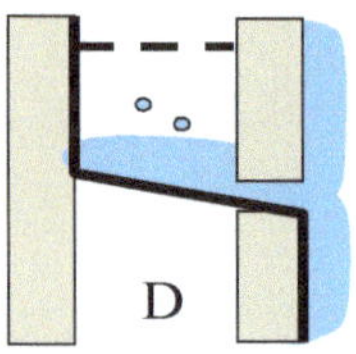	**Case D:** Gap width decreased to (g) = 3 mm and the run-off rate is unchanged at 2 g/m·s. Now the gap is almost entirely occluded by run-off, and the flashing does not drain until both the wind pressure and run-off rate relax. Water leakage past the cavity closer is initially low but increases rapidly when the pressure difference exceeds the hydrostatic head distance (d) to the cavity closer.

6. Leakage with Dynamic Applied Wind Pressures

Water leakage of the window head joint was measured with dynamically applied air pressures in the 0–0.6 Hz range. During these measurements, the geometry of the joint was as shown in Figure 9 and the run-off rate over the joint was constant at 2.5 g/m·s. The pressure amplitude was adjusted to fall to a zero minimum where possible, but at low frequencies, there was not enough travel in the oscillating piston to achieve the full pressure amplitude. The water leakage rates into the cavity closer have been plotted in Figure 13 against peak air pressure for five frequencies in the range 0–0.6 Hz.

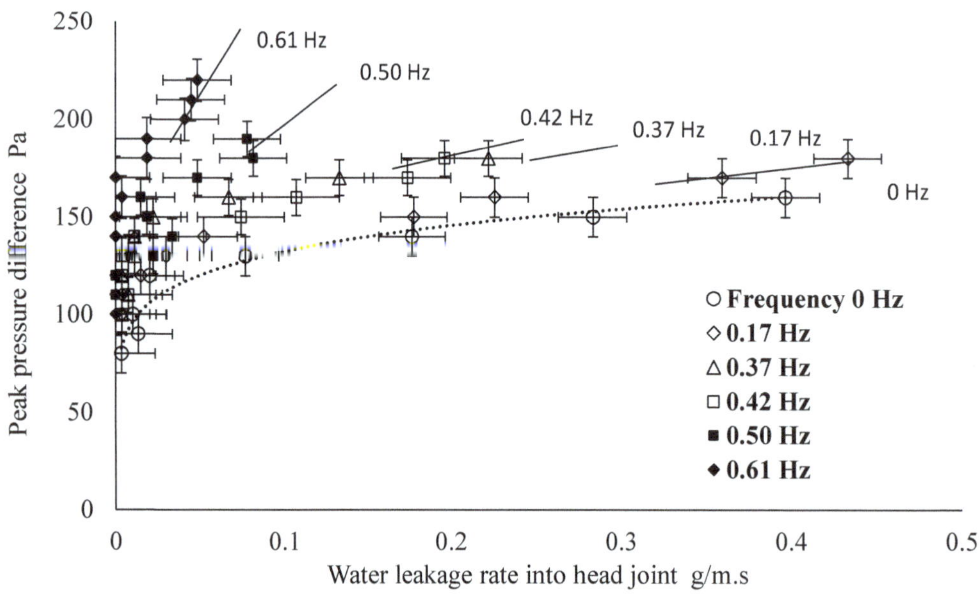

Figure 13. Frequency dependent water leakage into a window head joint.

There is a clear trend for the leakage characteristic to converge on the static pressure result at low frequencies, but it diverges towards much lower leakage rates at higher frequencies. This is a different result to that for H jointers between direct-fixed wall panels. Here, the leakage rate appeared to be in phase with varying wind pressures and was insensitive to frequency sensitivity in the range 0–0.7 Hz.

A two-phase flow analysis of the water leakage process in construction joints is outside the scope of this study, but the question of whether the leakage rates shown in Figure 13 can be derived from the instantaneous leakage rates as for H and Z jointers has been addressed in Figure 14. Here, the instantaneous leakage rates over a full cycle have been calculated from the steady pressure results as described earlier and the calculated values compared with measured leakage rates. It is clear that leakage rates for frequencies below 0.4 Hz can be calculated from the steady pressure leakage data, but at higher frequencies, inertia and the time constant of the joint cavity filling and drainage processes will have to be accounted for. There is room for further development of the leakage rate dependency on fluctuating pressures.

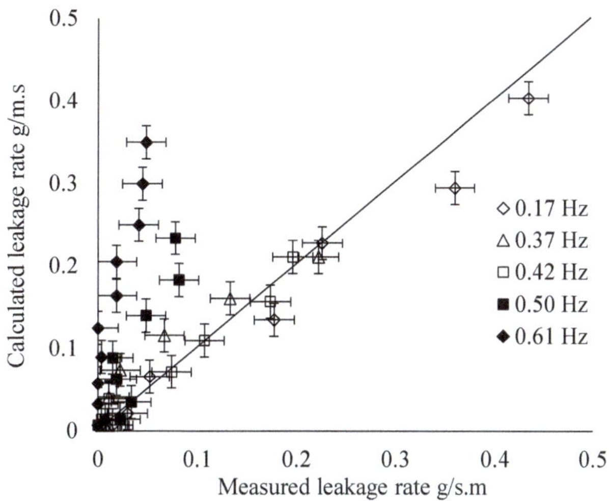

Figure 14. A comparison of measured and calculated leakage rates using a simplified model.

7. Leakage Characteristics of an Apron Flashing to Roof Joint

An apron flashing was assembled as shown in Figure 15 with the cavity wall details given in Figure 3. The apron flashing was simplified to a single piece in folded aluminium with an upstand height (distance from base of cladding to top of hemed upstand) of 75 mm.

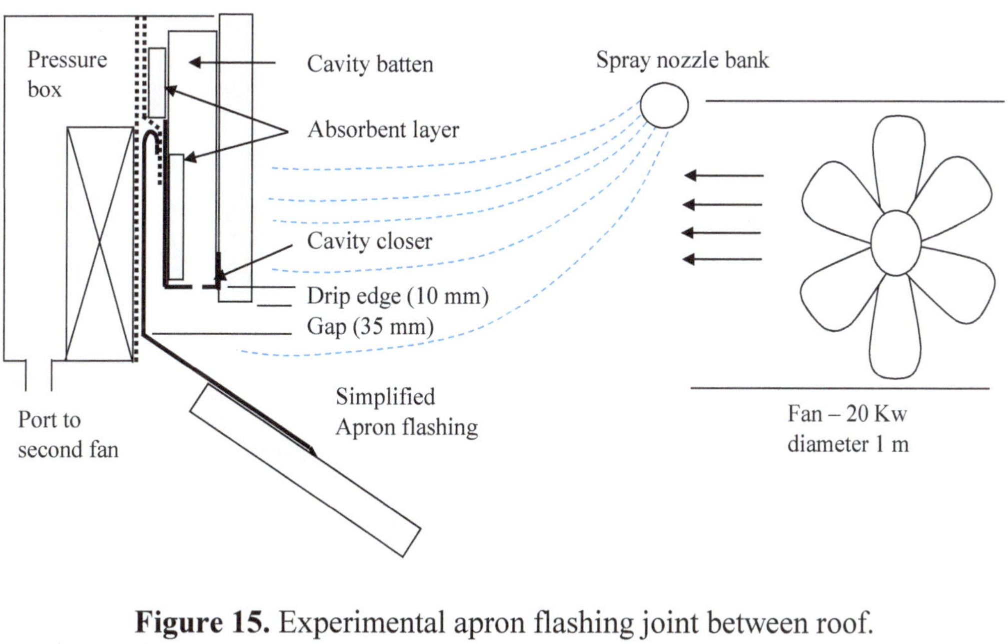

Figure 15. Experimental apron flashing joint between roof.

The complete wall specimen measured 2.4 m by 2.4 m and a portion of the wall cladding was made removable to retreve and weigh the absorbent layer. The leakage characteristics of the joint were measured statically as a function of runoff rate following the methods used for the window head flashing. Then the static spray bar was replaced with a 1 m diameter fan capable of driving water spray at the joint with air velocities up to 17 m/s. This allowed the leakage due to wind carried rain to be compared from that due to runoff.

Four potentially significant water leakage paths past the apron flashing in Figure 16 are as follows:

- Between the flashing upstand and the cavity closer;
- Up through vent openings in the cavity closer;
- Past the ends of the apron flashing—not dealt with here;
- Under the flashing over the roof deck—not dealt with here.

Leakage characteristics of a 1.2 m section of joint were measured with a 75 mm flashing upstand, a cavity closer upstand of 75 mm against the flashing and a smaller 18 mm upstand against the back of the cladding. The vent area in the cavity closer was 1500 mm^2/m made up of a series of slots 3 mm wide by 13 mm long. It was the same cavity closer present in the window head joint studied earlier. Runoff rates over the joint were in the range 3–60 g/m·s.

Water leakage rates were measured inside the cavity closer against the taller leg at the back of the cavity, and directly above the flashing. No water entry was detected above the 75 mm flashing upstand, due to the hem fitting tightly against the cavity closer and effectively closing off air leakage paths in this area. As with the window head flashing, water entered through ventilation holes in the cavity closer and leakage rates were measured at four runoff rates and air pressure differences in the range 0–1300 Pa and plotted in Figure 16. The joint behaved like a window head joint with a large gap between cladding and flashing that prevented the joint from filling up with water at even the highest runoff. The measured leakage rates were small compared to those measured into the window cavity closer and were entirely due to small droplets of spray carried by air flows through ventilation holes in the cavity closer.

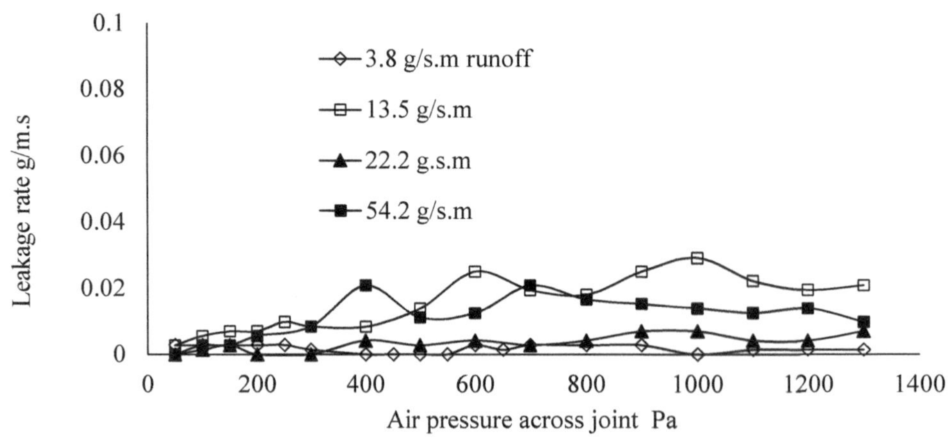

Figure 16. Water leakage rates through the cavity closer above a saddle flashing.

Wind Driven Water Leakage past the Apron Flashing

The specimen was positioned in front of the large fan and the static air pressure measured at the joint. At maximum fan speed this was 168 Pa (equivalent to an air speed of 16.7 m/s and well short of the pressures applied in E2/VM1 for buildings located in very high and extra high wind zones). The E2/VM1 test sequence does not simulate wind carried driving rain and so the measurements described here are at best exploratory. The sprays positioned above the fan outlet delivered 16 L/min or an average of 2.8 $L/m^2 \cdot min$ over the specimen.

The rate of water entry into the cavity closer was measured as a function of the air pressure difference across the joint and the height of water penetration on the plane of the wall underlay. As in the earlier measurements, there was no leakage past the top of the 75 mm apron flashing but significant quantities of water entered through vents in the cavity closer. The air pressure difference across the cavity closer was adjusted from the baseline 168 Pa in steps of 200 Pa to 1168 Pa using the second fan shown in Figure 15. Figure 17 shows the water entry rate plotted against height above the base of the cavity closer. This time, the water entry rate is plotted on a log scale to illustrate the large difference between water entering the joint and carried through the cavity closer to higher levels by air flows.

There were two processes at work carrying water into this joint. The first, involved rain drops bouncing off the apron flashing and carried by momentum into the base of the cavity closer. From here they travelled no further unless an air pressure difference greater than 100 Pa was present to drive water further into the joint. The second process involved smaller droplets entrained in air flows which travelled higher into the joint. The first process delivered large quantities of water and the second delivered much smaller water flows that were similar to those measured earlier with static sprays. It is clear that wind driven rain is a significant issue with apron flashings that will require further investigation in the context of tall buildings.

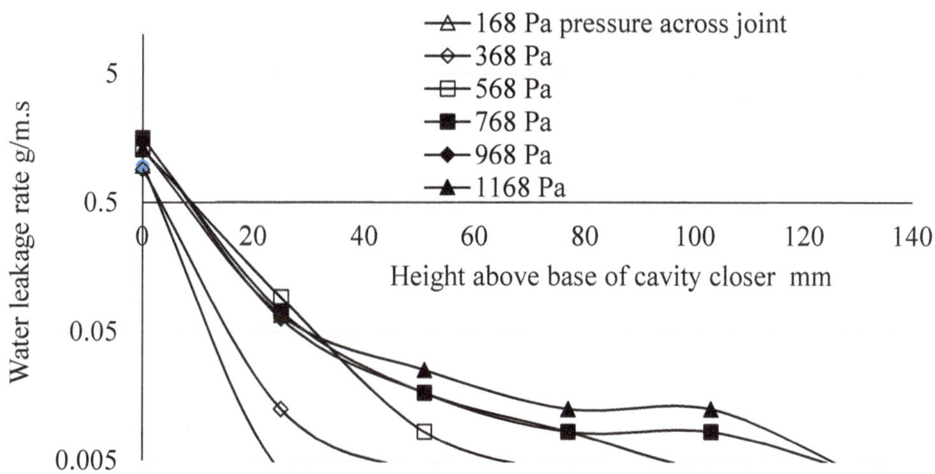

Figure 17. Mass flow of water reaching various heights above the base of the cavity closer of an apron flashing.

8. Conclusions

The water leakage characteristics of three flashed joints from the New Zealand Building Code approved document E2/AS1 have been measured. At this stage, the leakage characteristics are in a raw form and have yet to be interpreted in terms of weather data for mid-rise buildings in New Zealand. Water leakage rates were measured as a function of rain load, air pressure difference (statically and dynamically applied) and, in one case, with the rain load driven by high wind speeds. In all of the joints studied, the water leakage performance limit was found to depend on the presence of air leakage paths, either in the form of vents in cavity closers or due to some misfit between flashings and claddings. For this reason, it is likely that field performance will be well short of the leakage onset pressures measured here for ideal joints. Bearing in mind this limitation, the following conclusions were drawn:

- **H jointers between sheets of cladding**. With the flashing close fitting, the joint water penetration up to pressures close to the head of water equivalent of the up-stand height. In more realistic building applications where the gap between flashing and cladding might exceed 2–3 mm, air carried water leaks were seen at lower leakage onset pressures (100–300 Pa) with little dependency on up-stand height and the presence of a hem. Water leakage rates through tight-fitting joints were measured with dynamic pressure in the range of 0–0.7 Hz. There was no detectable effect of inertia, with measured water leakage rates agreeing with calculations using static pressure leakage characteristics.
- **Window head flashing in a cavity wall.** The onset of water leakage through this joint occurred at 100 Pa as a result of air carried leakage through vents in the cavity closer. Leakage past the flashing upstand was much less significant when tightly fitted against the cavity closer. Both onset leakage pressures were much higher than the 50 Pa "Wet Wall" test pressure adopted in E2/VM1. The leakage onset pressure was also found to depend on the runoff rate over the joint and on the gap between the cladding and head flashing. In fact this gap might need to be increased to cope with runoff rates on tall facades. The dynamic leakage characteristics were quite different to those at static pressures, unlike the case for H jointers. Leakage rates at frequencies above 0.4 Hz were over predicted by the simple instantaneous pressure model and will therefore require a model involving the inertia of water in the joint and the filling and drainage time constants.
- **Apron flashing between wall and roof**. The 35 mm gap between the base of the wall cladding and the apron flashing prevented high runoff rates from accumulating water in the space below the cavity closer. As a consequence, only very small air carried water leaks through the cavity closer were detected up to very high air static pressure differences. Leakage rates a hundred times greater were measured with rain driven at the joint by 17 m/s wind speeds. The new water entry process involved large droplets bouncing off the apron and entering the base of the cavity closer where water could be carried deeper into the joint by air flows. No leakage was detected between the apron flashing upstand (75 mm above the base of the cladding) while it fitted tightly against the cavity closer upstand.

This study identified opportunities to improve the leakage characteristics of joints to handle runoff and higher wind pressures for buildings outside the scope of E2/AS1. This is, however, only a start for more comprehensive studies that will be needed to include fluctuating wind pressures and wind driven rain into the water leakage functions for joints. These studies will also compare measured leakage rates with the capacity for drainage and ventilation drying.

Acknowledgments

This work was funded by the Building Research Levy. The assistance of Mark Hearfield, and Roger Stanford is gratefully acknowledged.

Author Contributions

Both authors contributed to the experimental work and the first author wrote the paper.

Conflicts of Interest

The authors declare no conflict of interest.

References

1. Bassett, M.; Clark, S.; Camilleri, M. Building Weathertightness Failures—Associated Risk Factors. In Proceedings of the BETEC Symposium on Building Science Applications at Syracuse University, New York, NY, USA, 5 September 2006.
2. Hershfield, M. *Survey of Building Envelope Failures in the Coastal Climate of British Columbia*; Report for Canada Mortgage and Housing Corporation; Morrison Hershfield Ltd.: Ottawa, Canada, 1996.
3. *Approved Document E2/AS1 External Moisture*; Department of Building and Housing: Wellington, New Zealand, 2005.
4. Hazleden, D.G.; Morris, P.I. Designing for Durable Wood Construction: The 4D's. In *Durability of Building Components 8*; NRC Research Press: Ottawa, Canada, 1999.
5. *Approved Document E2/VM1 External Moisture*; Department of Building and Housing: Wellington, New Zealand, 2005.
6. Ishikawa, H. An Experiment on the Mechanism of Rain Penetration through Horizontal Joints in Walls. In Proceedings of the 2nd International CIB/Rilem Symposium on Moisture Problems in Buildings, Paper 2.3.1., Rotterdam, The Netherlands, 10–12 September 1974.
7. Herbert, M.R.M. *Open-Jointed Rain Screen Claddings*; Building Research Establishment Current Paper 89/74; Building Research Establishment (BRE): Watford, UK, 1974.
8. Herbert, M.R.M.; Harrison, H.W. *New Ways with Weatherproof Joints*; Building Research Establishment Current Paper 90/74; Building Research Establishment (BRE): Watford, UK, 1974.
9. Lacasse, M.A.; O'Connor, T.J.; Nunes, S.; Beaulieu, P. *Report from Task 6 of MEWS Project Experimental Assessment of Water Penetration and Entry into Wood-Framed Wall Specimens*; Research Report IRC-RR-133; Institute for Research in Construction, National Research Council of Canada: Ottawa, Canada, 2003.
10. Sahal, N.; Lacasse, M.A. Proposed method for calculating water penetration test parameters of wall assemblies as applied to Istanbul, Turkey. *Build. Environ.* **2008**, *43*, 1250–1260.
11. Bassett, M.; McNeil, S.; Overton, G. WALLDRY-NZ: An Educational Tool for Rain Loads and Ventilation Drying in Cavity Walls. In Proceedings of the 5th International Building Physics Conference, Kyoto, Japan, 28–31 May 2012.
12. Kareem, A.; Cermak, J.E. Wind Tunnel Simulation of Wind-Structure Interactions. In *International Instrumentation Symposium, 24th, Albuquerque, N. Mex., May 1–5, 1978, Proceedings. Part 2*; Instrument Society of America: Pittsburgh, PA, USA, 1978.

13. Burnett, E.; Straube, J. *Vents, Ventilation Drying and Pressure Moderation*; Canada Mortgage and Housing Corporation Research Report; Canada Mortgage and Housing Corporation: Ottawa, Canada, 1995.

4

Detroit Works Long-Term Planning Project: Engagement Strategies for Blending Community and Technical Expertise

Toni L. Griffin [1,*], Dan Cramer [2] and Megan Powers [2]

[1] J. Max Bond Center on Design for the Just City, Spitzer School of Architecture, The City College of New York, 141 Convent Avenue, New York, NY 10031, USA

[2] Grassroots Solutions, 2828 University Avenue SE, Suite #150, Minneapolis, MN 55414, USA; E-Mails: dan@grassrootssolutions.com (D.C.); megan@grassrootssolutions.com (M.P.)

* Author to whom correspondence should be addressed; E-Mail: tgriffin@ccny.cuny.edu

External Editor: Sharon E. Sutton

Abstract: In January 2013, civic leaders, community stakeholders, and residents came together to release *Detroit Future City: 2012 Detroit Strategic Framework Plan*, a guiding blueprint for transforming Detroit from its current state of population loss and excessive vacancy into a model for the reinvention of post-industrial American cities. Three years prior, the U.S. Census had reported that the city had lost 24% of its population over the last decade and had experienced a 20% increase in vacant and abandoned property, bringing total vacancy to roughly the size of Manhattan. In addition to physical and economic challenges, Detroiters had also acknowledged significant barriers to effective civic engagement. Foremost among these barriers were a profound sense of immobilization, planning fatigue, and a general perception of cynicism about planning and engagement efforts. These challenges were compounded by historic racial dynamics and tension. This case study elaborates on the comprehensive and innovative civic engagement executed in a citywide planning process called the Detroit Works Project, which took place from late 2010 through late 2012. For the citywide planning process to be successful and sustainable, civic leaders and project funders committed to a planning initiative that would be different from previous efforts, in large part because the "owners" of the process would be diverse and inclusive across all community sectors. The case study, written by three of the key consultants from the project, describes four key civic engagement strategies deployed in the creation of the strategic framework: (1) addressing profound challenges of culture, race, and politics by deliberately building trust; (2) elevating community expertise by fostering a sense of ownership of the process; (3) blending technical and community expertise; and (4) viewing civic engagement as an ongoing two-way conversation rather than a series of large-scale episodic events. This article elaborates on important lessons that other communities might learn from Detroit's planning initiative in relation to these strategies. It concludes with a brief summary of the results and implications of the civic engagement process.

Keywords: Detroit; Detroit future city; civic engagement; community expertise; community; participation; stakeholder; civic infrastructure

---

1. Introduction

Increasingly, authentic and inclusive civic engagement has grown to be a core element of many planning processes [1] Though definitions of civic engagement (sometimes called "community engagement") vary, many people understand it as inclusive of both political and non-political processes [2]. Participating in civic life can include activities like voting, attending public meetings, advocacy work, and placemaking efforts, among others. Within the urban planning space, those who spearhead initiatives and planning processes have seen firsthand the tremendous benefits of engaging a wide variety of stakeholders around planning concerns that affect them and their communities. Not only does meaningful civic engagement inform the process and help ensure that planning recommendations and decisions reflect community priorities and experiences, but it can also improve the traction and sustainability of an initiative by building credibility and trust, fostering transparency, and, ultimately, helping to raise visibility and spark demand and support for the plan. Scholars now refer to this type of engagement as "robust civic engagement", where the overarching goals are to build a civic community, where civic infrastructure is supported, and long-term community capacity and leadership is enhanced [3]. In short, civic engagement, whether at the neighborhood or citywide scale, strengthens both the planning process itself as well as its lasting outcomes.

Certainly, designing and executing a successful civic engagement process in ways that are meaningful to both organizers and participants can be challenging. The number and diversity of voices participating, the quality of data and input, the credibility of organizers and presenters, and the presence of political trust and transparency are just some of the factors that can influence whether or not an engagement process is effective and substantive [4].

This article explores civic engagement in the successful multi-year comprehensive planning process of the citywide Detroit Works Project: Long-Term Planning (DWPLTP) initiative. The purpose of the project was to develop a citywide strategic framework that represented a shared vision for Detroit's economic recovery, creating more sustainable land use patterns, revitalizing neighborhoods, reconfiguring infrastructure and city services, utilizing public land assets, and increasing civic capacity. This descriptive case study of the project, written by three of the project's key consultants, offers the perspective of practitioners who were anything by objective observers of DWPLTP but rather were deeply invested in the process. We hope that our critical analysis of the four key civic engagement principles that underpinned our work and the tangible evidence of its outcomes substantiates the fairness of our interpretation.

The case study characterizes the four civic engagement principles as they evolved and shaped the resulting strategic framework, which we believe could be instructive for other communities. The four principles are:

1. Recognize the profound challenges that culture, race, and politics can present to a civic engagement process. Address these challenges through a deliberate strategy of building trust and enthusiasm with and within the community;
2. Elevate community expertise by fostering a sense of ownership of the process among local civic leaders, as well as adopting an expansive view of community that includes the perspectives not only of residents, but also of multiple sectors and segments of the community;
3. Create an intentional approach for blending technical and community expertise on a citywide scale. Acknowledge experiential and anecdotal inputs as valid, and integrate them with technical research and analysis to create a more accurate and inclusive case for change and implementation;
4. View civic engagement as an ongoing two-way conversation rather than a series of large-scale episodic events. Provide a wide range of opportunities and methods for people to interact with, learn about, provide feedback to, and in other ways shape the resulting plan.

2. Project Overview

2.1. Origins of the Strategic Planning Process and Key Players

The Detroit Works Project was initiated by newly elected Detroit Mayor Dave Bing in mid-2010. It stemmed from a desire by the mayor, civic leaders, and the Kresge Foundation (the principal funder of the project) to better address the city's growing challenges of vacancy, abandoned buildings, and population loss. The Foundation enlisted urban planner Toni L. Griffin, who had extensive experience working for other large cities on urban redevelopment and urban design, to lead the technical work for the initiative. Griffin assembled a world-class technical and civic engagement team that included the Detroit firm Hamilton Anderson Associates, along with a number of other consultants and technical experts (*see the Appendix for the complete list of Planning Team members*).

The duration of the entire Detroit Works Project was from mid-2010 to the completion of the strategic framework in January 2013. After the first phase of the project in late 2010, and upon completion of initial existing conditions audits and early rounds of civic engagement, project leaders recalibrated the process, bringing on additional technical expertise to address specific challenges like infrastructure and local expertise to lead a more robust civic engagement component. In mid-2011, a dedicated civic engagement team was assembled to complement the technical team, led by the Detroit Collaborative Design Center at the University of Detroit-Mercy. The local nonprofit organization Michigan Community Resources and the engagement consulting firm Grassroots Solutions joined the civic engagement team as well. In addition, a volunteer group of more than a dozen local community and organizational leaders, called the Process Leaders, was enlisted to help guide and advise civic engagement for the project as activities began to ramp up. Finally, a team of communications consultants led by Lovio George and Canning Communications helped to craft and disseminate regular project communications and a media strategy. The entire DWPLTP was managed by a multi-sector Steering Committee with representatives of city government, philanthropy, nonprofits, business, and other local institutions.

2.2. Engagement Tactics for the Strategic Planning Process

The civic engagement team, with support from the Process Leaders, worked with the technical team to design a set of engagement methods not only to gather community input to inform the strategic framework, but also to build trust and enthusiasm in the process, as well as provide opportunities for community stakeholders to learn about and discuss urban planning issues that were relevant to their lived experiences. To this end, an innovative mix of in-person and virtual civic engagement tactics were layered and deployed throughout the process, with an average of three to four engagement opportunities per week. Some of these tactics were designed to accomplish one objective—such as gathering community input—while other tactics helped advance all the objectives (gathering input, trust-building, and community learning/dialogue). Following is a list of tactics and their primary purpose.

2.2.1. In-Person Engagement Tactics

1. Large-scale town hall meetings (community learning);
2. Topic-based summits (gathering input, community learning/dialogue, and idea generation);
3. Community conversations (gathering input, building trust, and community learning/dialogue);
4. The "Roaming Table" (building trust);
5. Open houses and drop-in visits to the project's Home Base office (gathering input, building trust, and community learning/dialogue);
6. Attending or presenting at existing community meetings (gathering input and community learning/dialogue);
7. Street team door-knocking and leafleting (building trust);
8. Technical team working sessions (synthesizing input).

2.2.2. Virtual Engagement Tactics (Including both Online and Phone)

1. Telephone town halls (building trust and community learning);
2. "Detroit 24/7" online planning game (gathering input, idea generation, and community learning/dialogue);
3. E-newsletters (building trust and community learning);
4. Home Base "hotline" calls (community learning);
5. Website updates and social media (building trust and community learning);
6. "Detroit Stories" video history project (building trust and community learning/dialogue);
7. Earned media, such as print, radio, and television media and communications features (building trust and community learning).

Evaluation staff members of the civic engagement team carefully documented and created a quantitative analysis of the engagement inputs collected, capturing common themes within the key challenges, opportunities, and ideas expressed. This analysis was shared with the technical team during the development of the strategic framework. More information that describes the engagement methodology, participation and demographics, as well as content and themes of community input can be found in an extensive 300-page DWPLTP Civic Engagement Appendix [5]. In addition, several of the key engagement tactics that helped to shape civic engagement as an ongoing two-way conversation are described at greater length in Section 3.2.4 (Strategy #4) of this article.

2.3. References for This Case Study Analysis

This descriptive case study offers four key strategies used in Detroit that may prove instructive to other cities in making urban planning processes inclusive and viable, especially in a complex economic and sociopolitical environment. The four key strategies were shaped by a comprehensive audit [6] of civic engagement conducted by Grassroots Solutions in the city of Detroit previous to the Detroit Works Project. Completed in 2010, the audit surfaced themes and insights about Detroit's context, past planning efforts, engagement assets, challenges, and opportunities. Several of these insights are shared in Section 3.2.1 (Strategy #1) of this article.

3. Results and Discussion

3.1. Setting the Stage

In March 2010, after months of an interim mayor and a special election, newly elected Detroit Mayor Dave Bing delivered his first State of the City address to a hopeful citizenry eager to stand behind local leaders who represented a new era of integrity and access. Mayor Bing was well aware of the profound fiscal, operational, economic, and social justice challenges he inherited, and he had already begun conversations with civic and philanthropic leaders about organizing a process to address these concerns comprehensively. In his first public address to the city he noted that "*we have an opportunity to reinvent Detroit like never before.... Any plan will involve direct participation from our community because that's where our real strength is—in the commitment and connection Detroiters share with our city and each other. Every Detroiter has a voice and a role in this process* [7]."

This statement lifted the community and raised expectations that the new administration would be transparent, inclusive, and willing to work with communities towards positive change. But Detroiters were not new to civic participation, and they had certainly been let down by hollow public commitments of inclusion numerous times before. A healthy amount of skepticism was present.

The city's leadership had been plagued by corruption for many decades, and city government had not been viewed as a reliable partner in recent years. The Detroit civic community of business, nonprofits, philanthropy, and community development sectors had retreated into their own silos to effect change on the ground [8]. As a result, there were many "messengers of change" in Detroit, each with different levels of credibility and currency within the larger community. In addition, the local economy was in grave trouble. The city had experienced a 24% loss in population over the previous decade, as well as a 20% increase in vacant and abandoned property—equivalent to 20 square miles (roughly the size of Manhattan) [9]. Finally, the city's rate of physical deterioration and struggle to provide resources and deliver basic services had intensified in recent years. It is within this context that DWPLTP began.

3.2. Process and Strategies

Echoing the mayor's pronouncement that every Detroiter would have a voice and role in the citywide planning process, the project funders—the Kresge Foundation and the Ford Foundation—also expressed a strong interest in designing inclusive and meaningful civic engagement. After months of collaboration to develop a scope of work, create a project schedule, and assemble technical and civic engagement teams, city officials, community leaders, consultants, and foundation leaders launched the project in September 2010. It kicked off with five town hall-style meetings to introduce the planning process and articulate how it would fulfill Mayor Bing's State of the City address commitment of including the voices of Detroiters. The purpose of the collaborative project was to develop a strategic framework—essentially a blueprint for decision-making that could be used by multiple sectors—reflecting the best of Detroit's imagination and innovation to create a more prosperous, connected, sustainable, and socially-just city.

Residents' response was tremendous in this early phase of the project; Detroiters were eager for the opportunity to engage with the new city leadership. Nearly 5000 people participated in the first five town hall meetings alone, which proved to be logistically challenging and inhibited the kind of interactive dialogue and community learning that could take place. It was immediately apparent that the community was not yet ready to turn its attention towards visioning Detroit's future in the face of urgent day-to-day challenges and inefficiencies. Many attendees wanted to talk more or first about pressing problems in their neighborhoods such as vacant property, crime, and unreliable trash collection. In addition to frayed relationships with city officials, there were concerns and fears about "outsiders" leading the meetings and planning process—individuals with whom trust had not yet been established. Meeting participants wanted instead to interact and have dialogue with local Detroiters and people they knew.

Given the high levels of frustration, especially among residents, about how immediate concerns were being addressed in the city, project leaders recognized the need to make an important shift. The mayor and project funders split the project into two tracks—one to concentrate on immediate needs and interventions called "Detroit Works Short-Term Actions", to be led by the mayor's senior staff, and another to focus on creating a shared and comprehensive vision for long-term transformation called "Detroit Works Project: Long-Term Planning" (DWPLTP) [10].

From the outset of the DWPLTP process, project leaders sought to recalibrate and increase engagement opportunities to optimize the degree of community knowledge, learning, and interactive dialogue. In other words, the project aimed to build stakeholders' awareness and learning about planning issues and solicit their personal views of such issues [11]. In essence, project leaders sought to achieve greater authenticity, scale, credibility, and value through civic engagement efforts. The following elaborates on the civic engagement strategies that evolved as we applied, and adapted, the four key civic engagement principles in the field.

3.2.1. Strategy #1: Overcoming the Challenges of Culture, Race, and Politics

The engagement audit from early 2010 had highlighted and confirmed many of the engagement assets that existed in Detroit, such as the resiliency of residents, a vibrant faith community, a legacy of organizing, and robust neighbor-to-neighbor efforts and networks. In particular, audit interviewees considered the people of Detroit to be the city's greatest asset, describing them as resilient, tenacious, loyal, and deeply attached and invested in the future of the place they call home [12]. In addition to the audit, around this time there was an increase in philanthropic investments in Detroit, and energetic and innovative entrepreneurial approaches were beginning to have positive impacts in several neighborhoods. Perhaps most notably, a common sense of readiness for transformational change was emerging among people working and living in the city.

However, the audit also surfaced many historic dynamics and dysfunctions around effective civic participation that remained a stark reality in Detroit, including three important challenges of note. First, and not least among these barriers to effective engagement, was a generally weakened civic infrastructure. The local nonprofit sector was reportedly under-resourced for engagement efforts and projects, which affected the ability of organizations to achieve both scale of reach and depth of outcomes. Much of the existing local organizing or engagement capacity seemed to be constrained by narrow program parameters, shoestring budgets, and limited staff. As one person noted, "*In order to get individual voices you have to go door-to-door, but this takes increased resources, and who has the capacity* [13]?" In addition, established power centers within various local sectors sometimes impeded new approaches, ideas, and voices. Organizations and sectors in Detroit were often described as "siloed"—with tense or frayed relationships common among sectors—as well as highly turf conscious [14]. Due to many of these capacity and infrastructure challenges, taking engagement efforts to a multi-neighborhood or citywide scale had been historically an intimidating endeavor.

A second challenge was the common feeling in Detroit of being "planned to death." The severity of the economic crisis and the physical devastation within the city was palpable. As one participant said, "*I tend to be little more than skeptical until I see some (any) evidence of change, first. Planning sessions do not make me hopeful*". There was a sense of collective cynicism about systemic change, even as people expressed urgency and impatience for tangible progress. However, long-term solutions are both complex and daunting, and, therefore, it was not always easy to gain traction on implementation, leaving residents to feel like nothing was happening [15]. These kinds of factors combined to fuel a culture of profound immobilization among Detroiters [16]. As one leader noted early in the process, "*It's easy to get bogged down in big problems*".

Lastly, the challenges of racial disparity and racial tension were a constant undercurrent in the city. Despite a majority population of 82% African Americans in Detroit in 2010 [17], there was a strong sense among residents that there had not been an adequate presence of African-American voices in positions of leadership and power within city institutions, corporate businesses, and the philanthropic sector. Similarly, many people articulated the need for more proportional access and representation by African Americans and other communities of color as active participants in decision-making about Detroit's future. Broadly speaking, achieving racial equity had not been a leading driver for planning decisions in the city. As one person noted, "*There is deep and historical racial distrust. The divisions that we have here are significant. The economic disparities around race are so huge it makes it hard to get past it* [18]". Compounding this tension was a common fear of "outsiders [19]", stereotypically defined as nonresident, influential, wealthy, professional, and white, running the process as well as outsiders "taking over the city". When considered in the context of place attachment, the bond between people and place, race, class and gender identity have a large influence on how the "insiders"—typically the marginalized people of a community—feel a lack of empowerment or position in affecting future decisions about their community [20]. The insider-outsider dynamic was racially based (black *versus* white populations, which sometimes translated to class differences); geographically-based (city *versus* suburbs *versus* out-of-town); and tenure-based (recent *versus* long-standing residents).

To be sure, issues of race, culture, and politics often complicate and strain attempts to develop and implement citywide planning processes. However, what was unique about the case of Detroit's planning process was that the civic engagement team acknowledged these challenges, considered public sentiment after the tense first set of large-scale meetings, and was determined to craft a deliberate strategy to overcome such infrastructural, racial, and political barriers by intentionally and intensively focusing on building and rebuilding trust. This helped the project achieve the objectives of scale and credibility. Civic engagement team leaders realized that building trust explicitly within and across all segments of the Detroit community could improve significantly the community stakeholders' impressions of, participation in, and ultimate buy-in of the effort. Building trust meant focusing on inviting people into the planning conversation both earlier and more inclusively than had been done in the past, and it required investing time and resources in outreach and cultivating relationships. This trust building felt at times like a case of one step forward and two steps back. One participant in the project in mid-2012 put it this way: "*I came into this with the utmost respect for DWP, and I still have that, but there is an apparent disconnect with the community that I can sense*".

Project leaders attended existing community meetings across the city to listen to people's questions and perspectives on their own turf and in their own terms. They made a concerted effort to reach out to people and groups that may have been overlooked in prior planning processes. Project staff members were also trained in responsive listening skills and on how to conduct productive one-on-one meetings. In addition, transparency and accountability were crucial to building trust. Thus, project leaders developed an ongoing feedback loop—meaning both delivering frequent and timely updates and notifying stakeholders of upcoming engagement opportunities, as well as soliciting community input in targeted ways. Communications methods (also known as information out) included a regular e-newsletter, project website [21] and social media posts, periodic canvassing and distribution of materials, and coverage in print, radio, and television outlets. Community input and perspectives (also known as information in) were sought through dialogue at meetings, events, one-on-one conversations, and story gathering, among other methods. These concurrent and complementary feedback mechanisms encouraged greater levels of engagement, trust, constituency building, and support from community stakeholders; contributed to the credibility of the process; and ameliorated the aforementioned challenges of planning fatigue, weak civic infrastructure, and negative racial and cultural dynamics.

Certainly, cultivating trust and enthusiasm through relationship-building and real-time feedback loops took time and resources. In fact, some of the project's Steering Committee members were hesitant at first to make the significant investment of time and resources in engagement activities that would not inform the content of the strategic framework plan directly. However, in the end, building trust proved a worthwhile investment. More than 60% of DWPLTP survey respondents indicated that they felt "more hopeful" after participation in the initiative [22]. In the words of one participant, "*I see how feedback has been used, I see that the team is thinking about implementation*". Not only did the trust-building strategy improve the transparency and authenticity of the process, it also better positioned the Framework Plan for sustained demand and buy-in when the time came for implementation.

3.2.2. Strategy #2: Designing a Process that Elevates Community Expertise

The civic culture of Detroit has worked frequently in an "us *versus* them" environment [23]. For example, local community groups have often organized for dialogue and change out of necessity because of the absence and/or distrust of institutional leadership, including government, philanthropy, and business. "Community" in this case refers to the "grassroots", defined as Detroit residents and those organized to work on behalf of local Detroiters. The anger expressed by the audience at the first public meetings in September 2010 had built from a desire to have Mayor Bing address them directly, particularly after having emphasized the importance of community engagement in his first State of the City address just a few months earlier. There was also a common feeling that once again, the usual public officials ("grass tops") and "outsider" consultants already had a plan. Evidence of how the grassroots would be meaningfully a part of the process was not presented or represented clearly in the initial engagements.

Often in planning processes, the tension between adopting a "community-driven" approach *versus* the perceptions of a "stakeholder-led" approach surfaces early, and usually a choice is made to adopt one or the other. It became clear that DWPLTP's goals of robust and authentic engagement required not settling for a "grassroots" *versus* a "grass tops" approach to engagement, but rather creating a new paradigm of engagement that expanded the definition of community to include multiple sectors in dialogue together to share expertise and inform the final planning outcome. In other words, project leaders opted to avoid the dichotomy of either a "top-down" or "bottom-up" approach. Instead, they explored how to involve community members deeply in participating in the process and influencing its outcomes without misleading the public into thinking that all decisions would stem from the ground up. In this way, civic engagement for the project was reconceived to be *citizen-shaped*, not *citizen-driven*. This was an important distinction that helped clarify expectations and roles, although it proved a challenge at times not to revert back to one end of the spectrum or the other.

While the initial five town hall meetings were difficult to manage logistically, and, in some cases, became a forum for residents' anger and frustration, the meetings surfaced ultimately an enormous amount of data about what Detroiters witnessed firsthand about the condition of their neighborhoods. In the early meetings, many people spoke in great detail about their communities, often citing problem addresses, blocks and intersections, as well as about the resources and local activism already working to remedy the problems. Pivoting from the early large-scale meetings, which represented primarily residents as just one segment of Detroit's larger community, project leaders sought to adopt a more expansive and inclusive view of community. This more expansive definition of community included large and small business owners, the financial and philanthropic sectors, health and educational leaders, and academics. This local knowledge about current conditions and expertise about urban issues were highly valued by project leaders. Since many local leaders wore multiple hats as different segments of community crossed over, DWPLTP sought to weave these segments together at meetings, events, and activities where possible, rather than approaching them in isolation from one another. Indeed, early project planning by the Detroit Collaborative Design Center often featured a graphic image of the interconnected segments of community (depicted in Figure 1).

Figure 1. Segments of community. (Image credit: Detroit Collaborative Design Center).

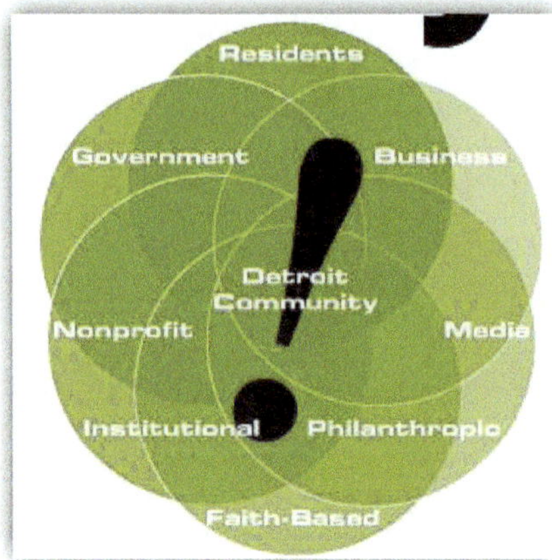

Local expertise was elevated by both a more expansive definition of community and by the careful consideration of who would be at the table to shape and execute the engagement. Another key engagement strategy that the DWPLTP developed and executed was the intentional strengthening of local capacity for ongoing and future engagement efforts. An essential part of this, and one that grew directly out of the early investment in trust building and relationship building, was the creation of a new group called the Process Leaders.

The more than a dozen local community leaders who comprised this group were selected for their expertise in civic engagement and local community knowledge among different racial, cultural, and age groups, as well as geographic areas of the city. The Process Leaders were charged with meeting weekly to advise and guide the engagement team and to help design and execute particular engagement activities and events. This role went beyond the typical "community advisory group" to active participation in decisions about where, when, and how engagement would roll out. This ameliorated the widespread perception of "us *versus* them" that had plagued previous projects. The Process Leaders helped to signal to Detroiters that engagement would go beyond the "usual suspects", and this generated broad excitement.

Examples of Community "Fingerprints" in DWPLTP

One of the richest areas of community input dealt with the Neighborhoods planning element—mentioned most frequently out of more than 180 DWPLTP planning topics. Participant comments and reflections about the most critical quality of life indicators shaped this aspect of the strategic framework. Such comments and ideas included:

- ✓ Reduce blight by making properties cleaner and safer;
- ✓ Encourage people to stay in Detroit, contribute to community, be neighborly, and maintain property;
- ✓ Get together with a group of neighbors and fellow citizens as a way of improving the neighborhood;
- ✓ Encourage community-based and youth-focused programs to improve safety in moderate vacancy areas.

The Process Leaders group was an innovative way to respond to the sense of weakened civic infrastructure and the sense of us *versus* them or insiders *versus* outsiders that often permeated previous planning projects in Detroit. It was a method used not only to cultivate broad participation and collaborative ownership, but also to build the local capacity for ongoing and future engagement efforts. Process Leaders strengthened relationships with each other, and in some cases, sought opportunities to work together even beyond DWPLTP. Engagement capacity was also built by training project ambassadors and street team canvassers, connecting people across silos and sectors where possible, and exposing the local civic leaders to new engagement tools and methods. The Process Leaders group also benefitted the engagement process by helping to build trust and getting resources to flow back into the city. In sum, the principle of developing and supporting local leadership helped address the capacity and infrastructure challenges and take the effort to scale. As one Process Leader reflected, "*I think over the next three to five years, you're going to see radical change in Detroit. It's already underway* [24]".

3.2.3. Strategy #3: Creating an Intentional Approach for Blending Technical and Community Expertise at a Citywide Scale

The first two strategies described how the engagement process was improved through methods that helped build trust and develop local capacity. In addition to these process design elements, an optimal outcome—the actual content of the Framework Plan—was also extremely important to all involved. Given the community's concerns about a "top-down" planning approach, where city leaders and their planners were perceived to already have the solution to an already framed set of problems, there was early push back that the process should be "community-driven". A community-driven process was understood as a process where the knowledge of local residents, both subjective and spatial, reflects their common experience and knowledge of everyday life, therefore better equipping them as the "experts" at identifying technical solutions that meet their realities and needs [25]. It was clear that this either/or proposition would not yield the best solution for addressing Detroit's severe economic, social, and spatial challenges that operated at both the citywide and neighborhood scale. Instead, an interdisciplinary approach was needed, integrating the best practices of planning and design, social science, and community-based problem solving. Values around immediate needs combined with innovative spatial solutions allowed both the practical and the aspirational. The technical solutions reflected the complexity of city dwellers' everyday lives.

Therefore, one of the unique features of the DWPLTP process was the aspiration to truly blend technical expertise with community expertise. Community expertise was defined as the inputs collected by community participants that included experiential knowledge and observations about current conditions and activities on the ground. That is, Detroiters were seen as the data "recorders" or experts of their own communities. Including their observations and ideas in the work would ensure that they could see their "fingerprints" on the resulting strategic framework. Technical and planning experts accepted the premise that these types of experiential and anecdotal inputs are valid.

Towards this end, project leaders designed ways to gather and record significant and wide-ranging input and perspectives, incorporate them into the technical analysis in real time, and demonstrate to the community how their inputs were incorporated. This process was referred to as a "civic engagement feedback loop" (see diagram in Figure 2). In terms of the precise methodology for gathering and analyzing community input, extensive methods were built into the process and were used in a variety of face-to-face and virtual engagement efforts (see Figure 3 for types and locations of face-to-face engagements). These included participant comment forms (largely handwritten responses which were then transcribed), electronic and phone surveys and snap polls, Facebook surveys, comment forms on the project website, Keypoint®/"clicker" polls, interactive mapping exercises, geographically-linked responses through the Detroit 24/7 online game, dot voting exercises, extensive meeting notes, live court reporter scribes and transcriptions, video interviews, small group participant note taking, one-on-one interview notes, and evaluation forms. Both open-ended and close-ended questions were asked. All of these data inputs were analyzed using Microsoft Access and Microsoft Excel.

Figure 2. The civic engagement feedback loop. (Image credit: Detroit Future City: 2012 Detroit Strategic Framework Plan).

Figure 3. Map of all face-to-face civic engagement activities executed by the Detroit Works Project throughout the city. (Image credit: Detroit Future City: Civic Engagement Appendix).

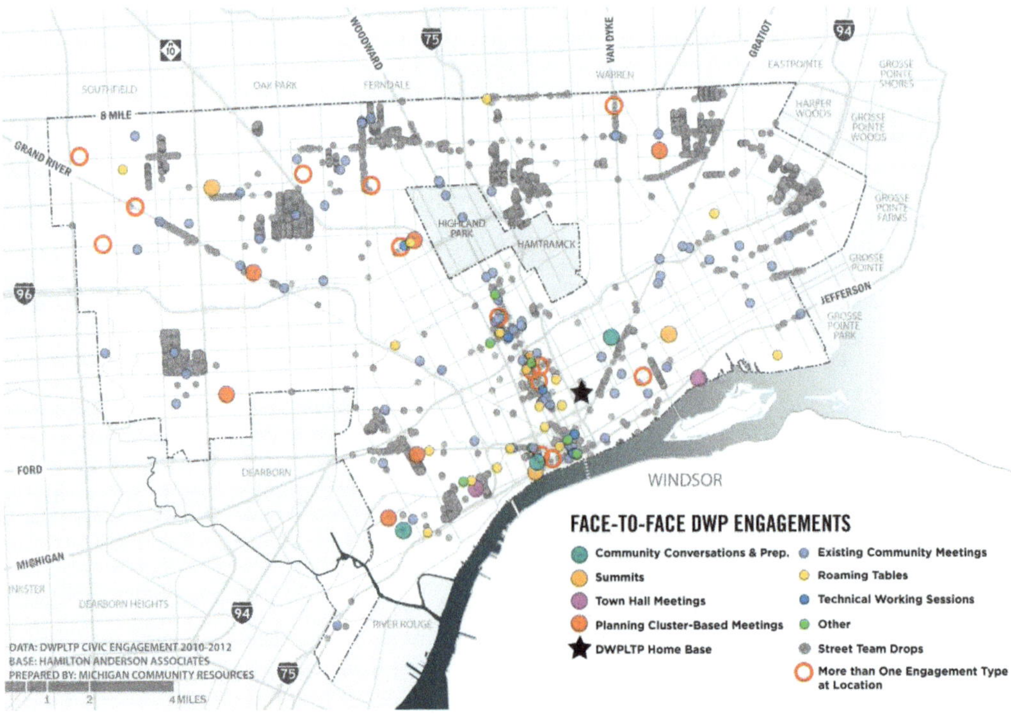

Project leaders strove to make engagement authentic, meaningful, and ongoing in the creation of a strategic framework that would be reflective of community perspectives and priorities, combined with technical data research and analysis. The goal of blending technical and community expertise was seen not only to benefit the final plan itself, but also to prime all community sectors to sustain its demand and implementation after the planning process was completed.

However, this ambitious aspiration of blending community and technical expertise came with pitfalls as well. For one, it was messy. The observations and priorities articulated by residents did not always translate laterally to the levels and types of analysis on the technical side. For example, parents shared specific input about the quality and characteristics of particular Detroit public schools, with comments such as: "*Safety at a school is the number one thing that determines whether or not a parent will send their child to that school—is my child safe and do you care about my child* [26]?" While valid and deeply important to residents' quality of life in neighborhoods, such input sometimes did not translate easily to technical assessments and recommendations at the systems level.

In addition, the integration of community expertise proved logistically and intellectually challenging at times. It proved enormously difficult to incorporate feedback from community sectors in real time (and to circulate out what was heard, what changed, and why) in the midst of also developing and advancing the rigorous, multifaceted technical work. Several reasons explain this difficulty. For one, resource and timing constraints—and sometimes unjustified deadlines—competed constantly with the need for depth and assessment of the community inputs. Certainly, this inhibited the ability to incorporate community input, but project leaders and advisors strove for the ideal of blending as much as possible. A second reason for the difficulty in blending inputs in real time was imbedded in the very nature of the "wicked problem [27]" of the planning scope. The large amount of content coming from both technical and community data created a constant cycle of discovering the nature of the problems while at the same time also investigating their dimensions and solutions in order to make timely decisions to move the process forward. And finally, even with a robust community engagement process, the thousands of people who participated and provided significant input still represented only 15% to16% of Detroit's total population. Still fewer of this number were signed up to receive ongoing project communications and updates via email, so the project was unable to stay connected over time with most participants.

Perhaps the greatest degree of successful technical and community blending came within the three months following the initial meetings in 2010 through smaller organized follow-up conversations designed to gather more community intelligence about the challenges and opportunities for Detroit's neighborhoods. These smaller meetings engaged not only residents, but also attracted other segments of the community to surface insights and data. For instance, many participants (32.8%) felt that access to high quality education was "the most critical" to their quality of life [28]. A careful recording of these inputs represented the initial set of "community expertise data" incorporated into the early phase of analysis. These dialogues ran parallel to an intensive technical assessment of several land use, economic, and social indicators of Detroit's current conditions. When the technical policy audits were blended with the community data, the technical team was able to develop a more accurate and inclusive "case for change"—a comprehensive set of data, both quantitative and qualitative, about issues most critical for the city to address in order to transition towards a more sustainable future.

The results of these integrated discussions between the community and the technical team produced twelve Imperative Actions—the important core values, goals, and quality of life/quality of business elements necessary to move Detroit into a more sustainable and affordable future. The Imperatives were set during the initial six months of the project and established the direction for subsequent phases and the final recommendations of the Framework.

12 Imperative Actions

1. We must re-energize Detroit's economy to increase job opportunities within the city for Detroiters and strengthen the tax base;
2. We must support our current residents and attract new residents;
3. We must use innovative approaches to transform our vacant land in ways that increase its value and productivity and promote long-term sustainability;
4. We must use our open space to improve the health of all Detroit's residents;
5. We must promote a range of sustainable residential densities;
6. We must focus on sizing the networks for a smaller population, making them more efficient, more affordable, and better performing;
7. We must realign city systems in ways that promote areas of economic potential, encourage thriving communities, and improve environmental and human health conditions;
8. We must be strategic and coordinated in our use of land;
9. We must promote stewardship for all areas of the city by implementing short- and long-term strategies;
10. We must provide residents with meaningful ways to make change in their communities and the city at large;
11. We must pursue a collaborative regional agenda that recognizes Detroit's strengths and our region's shared destiny;
12. We must dedicate ourselves to implementing this framework for our future.

Moving into the strategy development phase, blending community and technical expertise became more difficult for the DWPLTP team mainly because: (1) the technical work was at least two to three months ahead of the engagement due to gaps in the civic engagement process to redesign more effective approaches; and (2) the accelerated project schedule with deadlines driven by the political calendar—local elections, legislative conferences, *et cetera*—constrained greatly the amount of time that the team had for multiple rounds of data synthesis, confirmation, and integration with community sectors. And certainly, the skepticism and planning fatigue mentioned earlier still arose throughout the process, as participants voiced concern about the timeframe for implementation. "*You all have collected great ideas for improving the city, but I fear it will go into a report and no action will be taken. I would like to see how these things can become a reality and soon* [29]", cautioned one participant.

In hindsight, working groups, inclusive of all community sectors, working across the various planning elements of land use, economic growth, neighborhoods, city systems, public assets, and civic engagement should have been put in place earlier in the process to ensure deeper integration of the blended expertise. That said, the attempt to blend community and technical expertise was successful in the following ways. First, there were many instances where strategies developed by the technical team and community expertise were compatible and well aligned. For example, one common idea that surfaced repeatedly in the civic engagement process was to "*target different industries and businesses (for growth and development) in different areas of the city*". This idea took shape on the technical side as well, as the technical team examined how best to connect the strategies of economic development and land use. Second, the technical team included several Detroit residents who were able to add to as well as validate community expertise coming from the engagement process. Community perspectives about blue-green infrastructure, public safety, and environmental health, in particular, surfaced many times—as one person put it, "*We need to transform the city of Detroit into a cleaner, healthier, safer environment* [30]." Third, a constant dialogue between the technical team, civic engagement team, and process leaders through weekly coordination meetings helped ensure the most critical and commonly expressed issues and ideas from the community sectors were directed back to the technical team.

3.2.4. Strategy #4: Viewing Civic Engagement as an Ongoing Two-Way Conversation Rather Than a Series of Large-Scale Episodic Events

Oftentimes, civic engagement equates to large-scale events open to the public at large. Routinely a passive experience for attendees, the series of events tend to culminate in a tally of the total number of attendees as proof that the community was, in fact, engaged [31]. Through the DWPLTP process, particularly after its first set of town hall meetings, we learned that the approach needed to be more interactive and responsive to the readiness of the civic community to discuss and participate in designing change for the future.

Authentic civic engagement for this effort demanded that a relationship be built across all community sectors and neighborhoods around the city, not only to complete the planning process successfully, but also to strengthen the capacity to implement the strategies for change (see Figure 4 for the distribution of participant engagement by city zip codes). The process had to involve local participants actively and more visibly in leadership positions. Also paramount was creating a constant demonstration of transparency to maintain engagement, enthusiasm, and trust. In part, this meant placing a high value on the inputs of the community sectors equal to that of the technical data and expertise brought in to provide the best practices from around the world. Small and big creative ideas were shared throughout the civic engagement process, from raising people's awareness of the "*LightsOn! Detroit online tracking system allowing residents to communicate where street lights need to be fixed* [32]" to discussions of "*how we retrofit our built environment, [which] has implications for future costs and usage. You may be paying a higher rate, but with energy efficiency, you are using less* [33]".

While project leaders recognized that face-to-face community meetings and events would provide ways for people to connect, learn, and engage, DWPLTP also layered on a host of virtual and in-person engagement methods for engaging a wider and more diverse range of community participants to learn about, interact with, and provide feedback to shape the strategic framework. These innovative, fun, and interactive tactics were meant to engender an open and ongoing two-way dialogue among all community participants.

Figure 4. Map of registered participants by zip code. (Image credit: Detroit Future City: Civic Engagement Appendix).

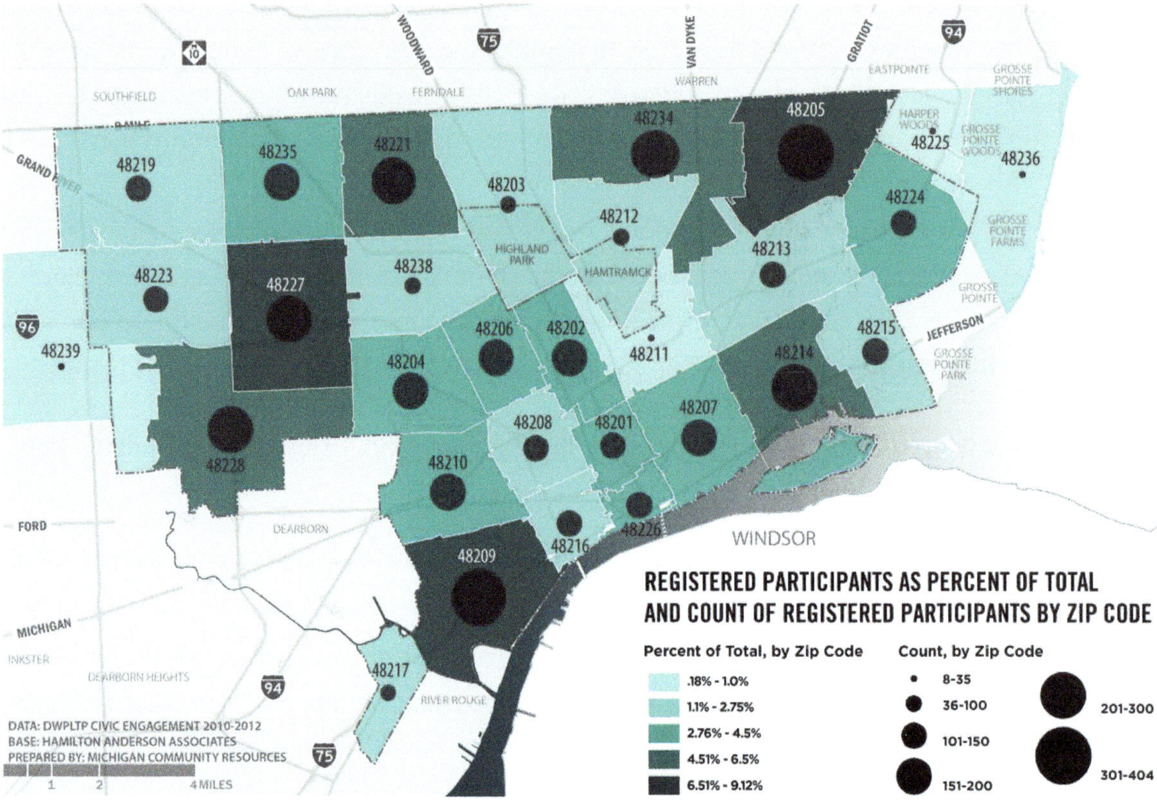

In addition to the more typical meeting-style formats of focus groups, charettes, and town hall forums, the following four innovative tactics are worth highlighting here:

- One of the tactics that helped to facilitate two-way conversation and learning was the "Roaming Table", a small travelling conversation booth that was staffed and stationed at various community locations. Tabling at events and meetings is a common engagement tactic for many efforts. However, the Roaming Table was unique because it could be set up anywhere rather than reaching only event attendees. And rather than being staffed by a project volunteer sitting passively behind a table full of flyers and materials it was designed to be approachable and interactive. The booth was designed inexpensively and could be collapsed to fit into the trunk of a car, so that organizers could transport it easily from place to place. The Roaming Table was ideal for engaging people in brief, one-on-one interactions about DWPLTP as they walked by, both to expose new people to the project as well as to solicit input on a specific planning question. This tactic facilitated more than 600 one-on-one conversations about the project (See Figure 5);
- The project also hosted a series of hour-long telephone town hall events, which reached more than 6000 people and engaged approximately 1300 participants. These were distinct from the five town hall meetings held at the beginning of the project, which were face-to-face facilitated exchanges between technical experts and a larger than expected crowd of participants. A telephone town hall, by contrast, is a virtual meeting that takes place via phone and can be joined from any telephone, such as from the convenience of a person's home. It is a cost-effective

and accessible technique for those for whom face-to-face meetings are challenging or unappealing. The technology uses an auto-dialer to call a specified list of phone numbers, offering a way to reach a large group of people instantly. People can listen to a speaker or a panel, ask questions in real time, or record messages expressing their perspectives on an issue. The telephone town halls promoted the two-way conversation between the speakers and the callers. These virtual town hall meetings allowed callers to hear project updates from Detroit community leaders, and to join the conversation live by posing their questions and comments. This tactic was particularly effective in engaging those who did not have the interest or ability to attend an in-person meeting, such as those with limited accessibility or who worked at times when community meetings took place. Many seniors participated in the telephone town halls.

Figure 5. The project's collapsible "roaming table" allowed the outreach team to engage over 6000 residents at transit stops, grocery stores, and community events throughout Detroit. (Photo credit: Detroit Collaborative Design Center).

A third tactic that attracted a good diversity of participants, particularly youth, was an online planning game called Detroit 24/7 [34]. The game was intended to gather input about elements of the strategic framework using a fun and interactive game platform. The idea to use game design was rooted in the theory that online social games can promote an environment for civic learning where engagement is combined with active participation and reflection [35]. More specifically, the tactic offered a new and innovative venue for trust building within and across the communities, as well as fostering recognition of alternative perspectives and ideas through reflection and discussion [35]. The platform for DWPLTP was designed as part of a Knight Foundation grant to Community PlanIt, a project of the Engagement Game Lab of Emerson College. The engagement and technical teams worked closely with Community PlanIt to determine the platform's content, approach, and language. The game featured challenges, a point system and leaderboard, prizes, and awards—all of which encouraged friendly competition and active game play. The game was open to people of all ages, but over the course of three weeks, DWPLTP partnered with Detroit Public Schools and youth organizations, setting up computer kiosks in schools and libraries to allow easy access to the game to ensure young people had a voice in the process (see Figure 6). Partner outreach was a key facet of this engagement tactic, as it allowed the project to achieve scale. Detroit 24/7 [36] succeeded in helping to engage young people in the planning process, with 8400 comments being submitted to inform the strategic framework. The game culminated in a face-to-face event, yet another opportunity to foster two-way conversation between participants and members of the project team.

Figure 6. A high percentage of young people between the ages of 10 to 24 participated in the project's online planning game "Detroit 24/7". (Photo credit: Detroit Collaborative Design Center).

One participant, George, wrote "*I think it's great when artists are able to harness their visions and engage community in meaningful ways.*" And this translated into several quality of life recommendations within the new neighborhood designs for the "live + make" neighborhoods featured in the strategic framework. Another game participant suggested, "*Projects like the Riverwalk's wetlands filter clean the water and bring back animals.*" This suggestion and several like it influenced the team's approach to blue and green landscape infrastructure strategies, which can be found in the land use and city systems chapters of the Framework.

Lastly, an oral history project called Detroit Stories [36] aimed to solicit, video record, and share everyday stories of Detroiters. This tactic was an example of an effective way to increase awareness and enthusiasm about the project and celebrate people as one of Detroit's greatest assets. Through Detroit Stories, dozens of these video stories were gathered to learn about the personal perspectives of Detroiters and the things that mattered most in their lives. Since real Detroiters shared stories in their own words, the oral history project offered a highly engaging way to highlight conversation and stories and to inspire further reflections and input from participants following the video viewings. Watched by hundreds of people throughout the process, the Detroit Stories videos also provided an archive of rich, personal stories that could be preserved and shared even beyond the scope of the project.

Each of these tactics was designed to enable ongoing, two-way conversations among participants and project team members. These methods successfully engaged a combination of different demographics, sectors, constituencies, and geographic areas throughout the city. Some of the tactics allowed for more in-depth conversation and input, while others allowed wider reach and raised broad awareness of the project. By contrast, other tactics were designed to inform more directly the technical aspects of the strategic framework. While the tactics varied in terms of the depth of input and feedback received, together they advanced the three objectives of building enthusiasm and trust, building engagement capacity, and fostering community learning and dialogue. In short, the engagement tactics enabled the planning conversation amongst Detroiters to extend beyond the start and end of a meeting.

However, executing such a diverse menu of ways to engage the community also required investing time, money, and attention to advance planning. Careful thought to logistics and event details was vital. On the front end, project leaders had to reimagine the purpose and structure of each meeting and how the series of meetings and events fit together. The deliberate planning also included identifying the target participants for a particular meeting as well as promotion and outreach to ensure robust attendance. Attention was given to designing the agenda and flow of a meeting, securing skilled and trusted facilitators, and identifying the format and key discussion questions. In addition, project organizers devoted significant time to important details such as optimal timing, venue, accessibility, childcare, language translation, creating a welcoming atmosphere, capturing ideas and input accurately, and other factors to maximize inclusivity and participation. These are considered some of the most effective and necessary practices for inclusive and participatory engagement [37]. Finally, organizers debriefed after the meetings and followed up to thank and update participants—a step commonly forgotten in event planning. All of these steps made project events and meetings more effective and successful.

4. Conclusions

The DWPLTP planning process culminated in January 2013, as *Detroit Future City: 2012 Detroit Strategic Framework Plan* was released to significant acclaim across sectors. After nearly three years of dialogue, analysis, planning, and collaborative work by a variety of community sectors, consultants, and civic leaders, this comprehensive strategic framework represented a shared vision for the future of the city's economic health, land use reform, neighborhood revitalization, public service delivery, management of abandoned lands, and elevation of civic capacity to implement and sustain change.

All in all, the scope and scale of civic participation in the Detroit Works Project: Long-Term Planning process was unprecedented in Detroit. The project conducted hundreds of meetings, had 30,000 conversations, connected with people over 163,000 times, and incorporated more than 70,000 surveyed responses and comments from participants [38]. Relationships, trust, buy-in, and momentum were invested in and built. The participation was inclusive in terms of race, age, gender, and neighborhood/geographic area, as shown in Figures 7–10.

Figure 7. Map of all registered participants of DWPLTP within Detroit who provided race and ethnicity. (Image credit: Detroit Future City: 2012 Detroit Strategic Framework Plan).

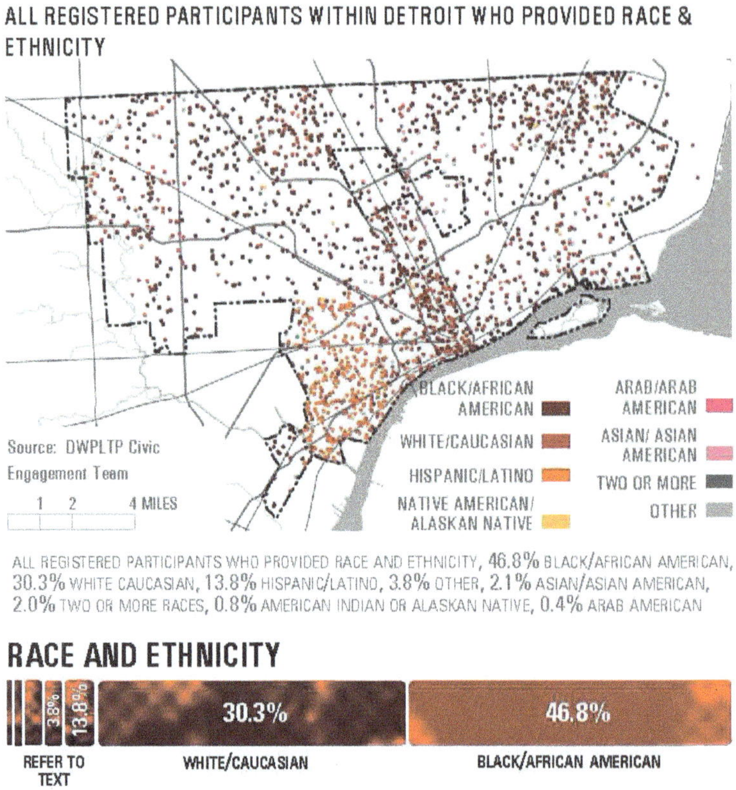

Figure 8. Map of all registered participants of DWPLTP within Detroit who provided gender. (Image credit: Detroit Future City: 2012 Detroit Strategic Framework Plan).

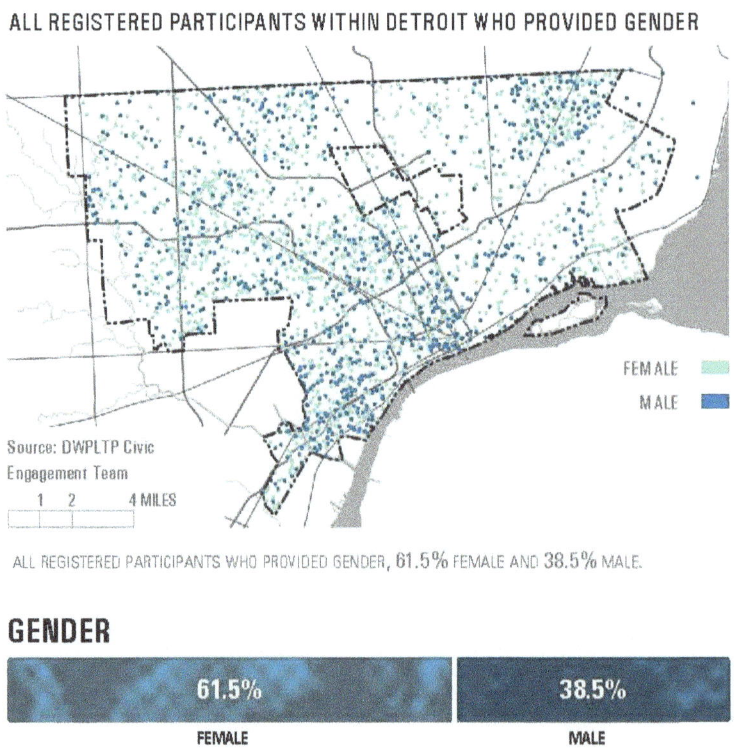

Figure 9. Map of all registered participants of DWPLTP within Detroit who provided age. (Image credit: Detroit Future City: 2012 Detroit Strategic Framework Plan).

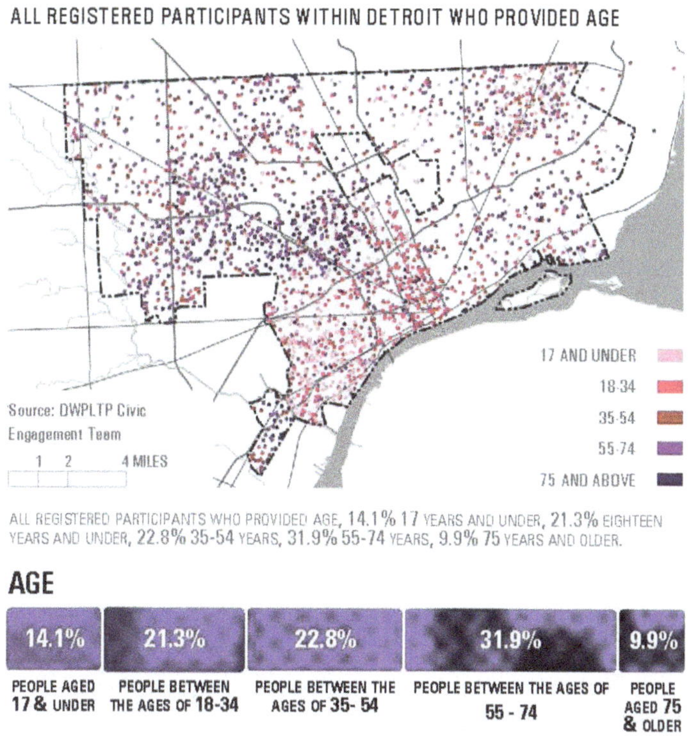

Figure 10. Map of all registered participants of DWPLTP within Detroit who provided geography. (Image credit: Detroit Future City: 2012 Detroit Strategic Framework Plan).

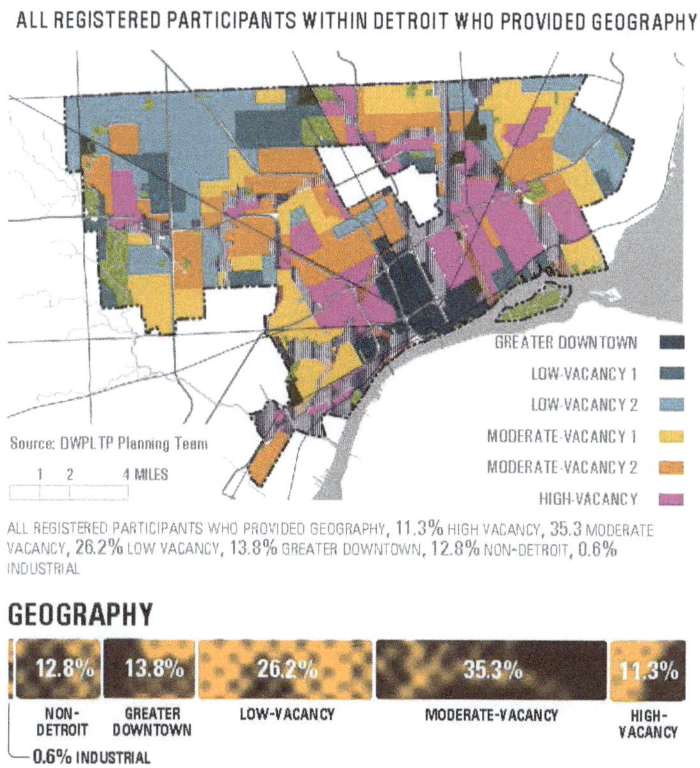

Perhaps most notably, most of the Process Leaders who were involved deeply in the civic engagement process have continued meeting regularly on a voluntary basis, sharing learning, collaborating, and continuing to champion the strategic framework within their own networks. Each Process Leader helps raise the visibility of different aspects of the strategic framework. For instance, one member of the group highlights the need to "*repurpose land to provide one of our most basic needs—healthy food*" [39], while others champion aspects such as community development, blight removal, the activation of block clubs, and small business incubation. These community leaders continue to shine a light on the need for civic engagement as Detroit Future City begins to be implemented across the city, and they raise both critiques and ideas in order for the project to remain accountable to all segments of community.

The success of the strategic framework plan can be measured in at least four distinct ways. First, the extraordinary amount of multi-sector involvement in the planning process was unprecedented. Over 200 local business, nonprofit, faith-based, institutional and grassroots organizations, together with governmental working groups, a Mayor's Advisory Task Force participated actively in the creation of the Framework. Second, government, private, nonprofit and philanthropic organizations have adopted all or parts of the Framework into their strategic investment plans for Detroit. Third, the strategic framework and the civic engagement process received over one million largely positive media impressions within the first month of its release and three professional recognitions including the American Planning Association's Daniel Burnham Award for a Comprehensive Plan, the Michigan Association of Planning's Planning Excellence Award, and the Michigan Chapter of the American Society of Landscape Architects' Honor Award for Landscape Planning and Analysis. The Detroit Collaborative Design Center was also given a SEED Award for Excellence in Community Interest Design in 2014, by Design Corps and the Social Economic Environmental Design® (SEED) Network, for several of its engagement approaches and materials.

The fourth, and perhaps most important, measurement of success is the ongoing commitment of Detroiters and the enhanced civic capacity across sectors to implement change through collaboration, partnerships and strategic coordination. Because the civic engagement process was designed around four key principles—overcoming cultural and racial dynamics, elevating local knowledge, blending community and technical expertise, and providing a wide range of opportunities to engage in two-way conversation—the strategic framework was solidified as the definitive way forward for reinventing the city's future [40]. The investment in building trust, fostering community learning, and strengthening local engagement capacity paid off in the end, as community members were able to see themselves reflected in the resulting strategic framework. Though the process was far from perfect, the willingness of project leaders to be responsive and make adjustments along the way was critical to the largely successful outcome.

To be sure, significant concern remains about whether this enhanced community capacity will continue to be meaningfully included in the process of implementation. The composition of newly elected and first-time city government officials and the uncertainty of the emergency manager and bankruptcy resolution at the time this article was completed leave many unsure about how well local government will partner with all community sectors going forward. In fact, the media often criticizes Detroit Future City for not being able to confirm the roles of community members in implementation, citing "*we engage our citizens for feedback but leave them on the sidelines when it comes time for deployment*" [41].

Some of the enduring capacity that was developed through the process included exposure to new engagement tools and tactics, the expansion of local leadership capability, and an increased ability to communicate and collaborate across siloes. Examples of this sustained capacity include the creation and funding of the Detroit Future City Project Implementation Office, which is staffed by a diverse group of multi-generational Detroit leaders; the participation of several young planning activists who ran for city council seats in the 2013 elections, all of whom were new to politics; and the incorporation of Detroit Future City's goals and strategies into the implementation agendas of several public and philanthropic sector organizations, including the Kresge Foundation and the Detroit Economic Growth Corporation. Over the last eighteen months since the plan's release, Detroit Future City has become a relevant framework for guiding collaboration, investment, and action, as Detroiters across all community sectors, including new government leadership at the state and local levels, continue to access, learn about, and integrate the framework into their efforts on the ground.

Acknowledgments

The authors would like to acknowledge the hard work of all who were instrumental in the success of Detroit Works Project: Long-Term Planning, particularly the members of the Steering Committee, the Office of the Mayor of Detroit, the City Council, the planning team, civic engagement and communications team, the Process Leaders, the Mayor's Advisory Task Force, partner organizations, and the various working groups and peer review group that lent their time and expertise to the process. We also extend special thanks to the project's funders and their staff: the Kresge Foundation, Ford Foundation, W.K. Kellogg Foundation, John S. and James L. Knight Foundation, Hudson Webber Foundation, Erb Family Foundation, and the Community Foundation for Southeast Michigan.

Authors Contribution

Planning Team: Toni L. Griffin, Hamilton Anderson Associates, Stoss Landscape Urbanism, Initiative for a Competitive Inner City, Mass Economics, Interface Studio, Happold Consulting, Center for Community Progress, and Carlisle Wortman. Civic Engagement Team: Detroit Collaborative Design Center, Michigan Community Resources, and Grassroots Solutions.

Conflicts of Interest

The authors declare no conflict of interest.

References

1. Leighninger, M. *Creating Spaces for Change: Working Towards a "Story of Now" in Civic Engagement*; Deliberative Democracy Consortium and W.K. Kellogg Foundation: Battle Creek, MI, USA, 2010.
2. Pastor, M.; Ito, J.; Rosner, R. *Transactions, Transformations, Translations: Metrics That Matter for Building, Scaling, and Funding Social Movements*; USC Program for Environmental and Regional Equity (PERE): Los Angeles, CA, USA, 2011.

3. Powell, J.R.J.; Hardaway, P. *Recommendations for Assuring Robust Civic Engagement & Equity in Detroit's Shrinking City Planning Effort*; Metropolitan Organizing Strategy Enabling Strength (MOSES) and The Kirwan Institute for the Study of Race and Ethnicity; Ohio State University: Columbus, OH, USA, 2010.
4. Barnes, W.; Mann, B. *Making Local Democracy Work: Municipal Officials' Views about Public Engagement*; Center for Research and Innovation, National League of Cities: Washington, DC, USA, 2010; pp. 13–18.
5. Detroit Works Project: Long-Term Planning. *Civic Engagement Appendix: Supporting Lasting Civic Capacity in Detroit*; Detroit Economic Growth Corporation: Detroit, MI, USA, January 2013. Available online: http://detroitworksproject.com/wp-content/uploads/2013/01/05-Civic-Engagement.pdf (accessed on 31 January 2014).
6. Cramer, C. *Detroit Engagement Audit Report*; Grassroots Solutions: Minneapolis, MN, USA, 2010.
7. WCHB News Detroit. Mayor Dave Bing Gives State of the City Address. 24 March 2010. Available online: http://wchbnewsdetroit.newsone.com/246091/mayor-dave-bing-gives-state-of-the-city-address-full-text/ (accessed on 19 August 2014).
8. Hoene, C.; Kingsley, C.; Leighninger, M. *Bright Spots in Community Engagement: Case Studies of U.S. Communities Creating Greater Civic Participation from the Bottom Up*; National League of Cities: Washington, DC, USA, 2013; pp. 8–9.
9. *Detroit Future City: 2012 Detroit Strategic Framework Plan*; Project of Detroit Future City (DFC) Strategic Framework; DFC Implementation Office: Detroit, MI, USA, December 2012. Available online: http://detroitworksproject.com/wp-content/uploads/2013/01/The-DFC-Plan.pdf (accessed on 7 October 2013).
10. Clark, A. Dave Bing's Detroit. *The American Prospect*, 2 October 2013. Available online: http://prospect.org/article/dave-bing%E2%80%99s-detroit (accessed on 21 April 2014).
11. Davey, M. The Odd Challenge for Detroit's Planners. *New York Times*, 5 April 2011. Available online: http://www.nytimes.com/2011/04/06/us/06detroit.html?pagewanted=1&_r=2&sq=detroit&st=cse&scp=1 (accessed on 21 April 2014).
12. Cramer, C. *Detroit Engagement Audit Report*; Grassroots Solutions: Minneapolis, MN, USA, 2010; p. 30.
13. Cramer, C. *Detroit Engagement Audit Report*; Grassroots Solutions: Minneapolis, MN, USA, 2010; p. 51.
14. Chavez, C.; Matsumura, E.; Bacon, D. *Mo(vement) Town: Building Civic Engagement in the Detroit Region*; Working Partnerships USA; Detroit Collaborative Design Center; the Ford Foundation: San Jose, CA, USA, 2011; pp. 6–10.
15. Oakley, D.; Tsao, H. A new way of revitalizing distressed urban communities? assessing the impact of the federal empowerment zone program. *J. Urban Aff.* **2006**, *28*, 443–471.
16. Cramer, C. *Detroit Engagement Audit Report*; Grassroots Solutions: Minneapolis, MN, USA, 2010; pp. 13–15.
17. *State and County Quick Facts*; U.S. Department of Commerce, United States Census Bureau: Allegany, NY, USA. Available online: http://quickfacts.census.gov/qfd/states/26/2622000.html (accessed on 7 January 2014).

18. Cramer, C. *Detroit Engagement Audit Report*; Grassroots Solutions: Minneapolis, MN, USA, 2010; pp. 26–28.
19. Ishem, L.H. Beyond Insiders and Outsides: Conceptualizing Multiple Dimensions of Community Development Stakeholders. In *The Paradox of Urban Space*; Sutton, S.E., Kemp, S.P., Eds.; Palgrave Macmillan: Hampshire, UK, 2011; p. 95.
20. Manzo, L.C.; Perkins, D.D. Finding common ground: The importance of place attachment to community participation and planning. *J. Plan. Lit.* **2006**, *20*, 340.
21. Detroit Works Project: Long-Term Planning Project Website. Available online: http://www.detroitworksproject.com (accessed on 31 January 2014).
22. Detroit Works Project: Long-Term Planning. In *Civic Engagement Appendix: Supporting Lasting Civic Capacity in Detroit*; Detroit Economic Growth Corporation: Detroit, MI, USA, January 2013; p. 91. Available online: http://detroitworksproject.com/wp-content/uploads/2013/01/05-Civic-Engagement.pdf (accessed on 31 January 2014).
23. City *versus* Suburb: A Longstanding Divide in Detroit. *Weekend Edition Sunday, National Public Radio*, 9 March 2014. Available online: http://www.npr.org/2014/03/09/287877060/city-versus-suburb-a-longstanding-divide-in-detroit (accessed on 21 April 2014).
24. ARISE Detroit! Unites Groups across the City. *The Kresge Foundation*, 4 February 2014. Available online: http://kresge.org/news/news-extra/arise-detroit-unites-groups-acrosscity#sthash.28AyDIQ6.dpuf (accessed on 22 April 2014).
25. Kemp, S.P.; Sutton, S.E. Integrating social science and design inquiry through interdisciplinary design charettes: An approach to participatory community problem solving. *Am. J. Commun. Psychol.* **2006**, *38*, 125–139.
26. Detroit Works Project: Long-Term Planning. In *Civic Engagement Appendix: Supporting Lasting Civic Capacity in Detroit*; Detroit Economic Growth Corporation: Detroit, MI, USA, January 2013; p. 227. Available online: http://detroitworksproject.com/wp-content/uploads/2013/01/05-Civic-Engagement.pdf (accessed on 31 January 2014).
27. Mugerauer, R.; Manzo, L. *Environmental Dilemmas: Ethical Decision Making (Toposophia: Sustainability, Dwelling, Design)*; Lexington Books: Lanham, Maryland, USA,, 2008; pp. 24–27.
28. Detroit Works Project: Long-Term Planning. In *Civic Engagement Appendix: Supporting Lasting Civic Capacity in Detroit*; Detroit Economic Growth Corporation: Detroit, MI, USA, January 2013; p. 49. Available online: http://detroitworksproject.com/wp-content/uploads/2013/01/05-Civic-Engagement.pdf (accessed on 31 January 2014).
29. Detroit Works Project: Long-Term Planning. In *Civic Engagement Appendix: Supporting Lasting Civic Capacity in Detroit*; Detroit Economic Growth Corporation: Detroit, MI, USA, January 2013; p. 175. Available online: http://detroitworksproject.com/wp-content/uploads/2013/01/05-Civic-Engagement.pdf (accessed on 31 January 2014).
30. Detroit Works Project: Long-Term Planning. In *Civic Engagement Appendix: Supporting Lasting Civic Capacity in Detroit*; Detroit Economic Growth Corporation: Detroit, MI, USA, January 2013; p. 145. Available online: http://detroitworksproject.com/wp-content/uploads/2013/01/05-Civic-Engagement.pdf (accessed on 31 January 2014).

31. Lukensmeyer, C.; Goldman, J.; Brigham, S. A Town Meeting for the Twenty-First Century. In *The Deliberative Democracy Handbook: Strategies for Effective Engagement in the 21st Century*, 1st ed.; Gastil, J., Levine, P., Eds.; Jossey-Bass: San Francisco, CA, USA, 2005; pp. 154–163.
32. Detroit Works Project: Long-Term Planning. In *Civic Engagement Appendix: Supporting Lasting Civic Capacity in Detroit*; Detroit Economic Growth Corporation: Detroit, MI, USA, January 2013; p. 158. Available online: http://detroitworksproject.com/wp-content/uploads/2013/01/05-Civic-Engagement.pdf (accessed on 31 January 2014).
33. Detroit Works Project: Long-Term Planning. In *Civic Engagement Appendix: Supporting Lasting Civic Capacity in Detroit*; Detroit Economic Growth Corporation: Detroit, MI, USA, January 2013; p. 160. Available online: http://detroitworksproject.com/wp-content/uploads/2013/01/05-Civic-Engagement.pdf (accessed on 31 January 2014).
34. Detroit 24/7 Online Game; Community PlanIt. Available online: http://detroitworksproject.com/engagement/detroit247 (accessed on 31 January 2014).
35. Gordon, E.; Baldwin-Philippi, J. Playful civic learning: Enabling reflection and lateral trust in game-based public participation. *Int. J. Commun.* **2014**, *8*, 759–786.
36. Detroit Stories Oral History Project Website. Available online: http://www.detroitstoriesproject.com (accessed on 31 January 2014).
37. Creighton, J.L. *The Public Participation Handbook: Making Better Decisions Through Citizen Involvement*; Wiley: Jossey-Bass: San Francisco, CA, USA, 2005; pp. 85–179.
38. Detroit Works Project: Long-Term Planning. In *Civic Engagement Appendix: Supporting Lasting Civic Capacity in Detroit*; Detroit Economic Growth Corporation: Detroit, MI, USA, January 2013; pp. 322–333. Available online: http://detroitworksproject.com/wp-content/uploads/2013/01/05-Civic-Engagement.pdf (accessed on 31 January 2014).
39. Turner-Handy, S. A Common Theme throughout the Detroit Future City Framework. *Michigan Citizen*, 28 March 2013. Available online: http://michigancitizen.com/urban-agriculture-as-a-tool-to-help-revitalize-the-city/ (accessed on 22 April 2014).
40. Thompson, A.; Wheeler, H. Commentary: No Debate, Detroit Future City Is the People's Plan. *MLive*, 3 April 2014. Available online: http://blog.mlive.com/news/detroit_impact/print.html?entry=/2014/04/commentary_no_debate_detroit_f.html. (accessed on 21 April 2014).
41. Lorimer, J. Detroit Future City Is Full of Small Mistakes. *HuffPost Detroit*, 22 January 2013. Available online: http://www.huffingtonpost.com/jason-lorimer/full-of-small-mistakes_b_2517018.html (accessed on 5 May 2014).

The Design and Simulation of Natural Personalised Ventilation (NPV) System for Multi-Bed Hospital Wards

Zulfikar A. Adamu [†,*] and Andrew Price [†]

School of Civil and Building Engineering, Loughborough University, LE11 3TU Leicestershire, UK; E-Mail: a.d.f.price@lboro.ac.uk

† These authors contributed equally to this work.

* Author to whom correspondence should be addressed; E-Mail: z.a.adamu@lboro.ac.uk

Academic Editor: Adrian Pitts

Abstract: Adequate ventilation is necessary for thermal comfort and reducing risks from infectious bio-aerosols in hospital wards, but achieving this with mechanical ventilation has carbon and energy implications. Natural ventilation is often limited to window-based designs whose dilution/mixing effectiveness are subject to constraints of wind speed, cross ventilation, and in the case of hospital wards, proximity of patients to external walls. A buoyancy-driven natural ventilation system capable of achieving dilution/mixing was shown to be feasible in a preceding study of novel system called natural personalised ventilation (NPV). This system combined both architecture and airflow engineering principles of space design and buoyancy and was tested and validated (salt-bath experiment) for a single bed ward. This research extends the previous work and is proof-of-concept on the feasibility of NPV system for multi-bed wards. Two different four-bed ward types were investigated of using computational fluid dynamics (CFD) simulations under wind-neutral conditions. Results predict that NPV system could deliver fresh air to multiple patients, including those located 10 m away from external wall, with absolute flow rates of between 32 $L \cdot s^{-1}$ and 54 $L \cdot s^{-1}$ for each patient/bed. Compared to same wards simulated using window design, ingress of airborne contaminants into patients' breathing zone and summer overheating potential were minimised, while overall ward dilution was maximised. Findings suggest the NPV has potentials for enabling architects and building service engineers to decouple airflow delivery from the visualisation and illumination responsibilities placed upon windows.

Keywords: natural personalised ventilation (NPV); multi-bed hospital ward; airborne contaminants; overheating; mixing ventilation; computational fluid dynamics (CFD)

---

1. Introduction

In the United Kingdom (UK), mechanical ventilation contributes over 40% of energy allocated by typical hospitals for space conditioning [1]. Low energy alternatives have therefore been explored as intervention measures for existing stock of medium rise hospitals [2]. High ventilation rates from natural ventilation [3,4] make it an attractive option in the absence of cold temperate weather. Traditional window-based simple natural ventilation (SNV) systems used in UK hospitals are challenged by limitations of guidelines of the Department of Health [5] which specify that the width of any opened window must not exceed 100 mm. The recent Health Building Notes [6] also emphasises the role of window restrictors in maintaining this maximum width. This 100 mm restriction is largely for security and safety reasons [5,6]. It is not evident from historical or contemporary literature that this area of opening has been fully studied in terms of its impact on adequate ventilation rates, specifically for dilution or airborne pollutants of pathogens. There is nevertheless research evidence pointing towards substantial increase in risk of cross-infection when such windows in multi-bed wards are closed in winter to save heating energy [7]. In summer, the current and long-term potential for overheating of naturally ventilated wards have also been shown [8] implying that better natural ventilation is desirable. There is also the issue of room air distribution, through displacement or mixing strategies and how this can affect clinical outcomes in multi-bed wards.

1.1. Performance of Displacement Ventilation in Clinical Spaces

The relatively new advanced natural ventilation (ANV) system uses buoyancy-driven airflow through displacement of stale air by cooler air supplied from low-level inlets instead of windows [9]. The ANV system (a wind-neutral system) is interesting because it circumvents the 100mm restriction since air supply is not through windows. The windows remain as fixed glazed components for visual and illumination purposes. However, buoyancy-driven gravity current airflows are inseparably linked to displacement ventilation strategy in natural ventilation [10,11]. Recent findings about displacement ventilation in hospital wards however suggest a split in research opinion about the use of this approach for aseptic or clinical purposes. For example, a study of person-to-person cross-infection in mechanically ventilated offices revealed that displacement and under-floor air distribution (UFAD) strategies performed better overall than mixing strategy but the distance from infected to susceptible person was more important than airflow rate or mode of ventilation [12]. However, results from a study on displacement ventilation by Gao *et al.* [13] led to conclusions that with high temperature gradients, exhaled droplets are trapped at the breathing height of an adult. Likewise, it has been found [14] that mechanical displacement strategy was clinically inferior to mixing strategy because emissions from human pulmonary events (e.g., sneezing and coughing) occur in the lower zone of stratification whereas the mixing desired for dilution happens in the upper strata. They [14] established that under displacement ventilation driven by gravity currents, the momentum from emitted puffs in the occupied zone would be too strong, relative to the momentum of displacement air, thereby not allowing mixing and subsequent dilution. Other related studies on displacement ventilation [15,16], add weight to this argument, including a study [17] where experimental manikins were subjected to displacement, mixing strategies as well as in unventilated scenarios. In this case, it was also concluded that exhaled air travelled further at higher concentration in displacement ventilation, than when mixing is used. The balance of evidence from contemporary research, therefore, seems to favour mixing over displacement as far as dilution of airborne contaminants is concerned in clinical spaces. The implications of these studies are an important motivation for investigating alternative ways of achieving mixing ventilation in clinical spaces.

Mechanised mixing ventilation systems could provide mixing, but the use of electric energy to power fans makes it unattractive and unsustainable for wide scale adoption. Modern aseptic ventilation of hospital wards is usually achieved through personalised ventilation (PV) systems which provide an individual with dedicated air supply and control of the direction and flow rate of the air [18]. Yang *et al.* [19] have established that the integration of ceiling mounted (mechanised) PV systems with background mixing regime would lead to lower energy consumption at various room temperatures, than when mixing ventilation systems alone are used. Other relevant studies include: seat headrest personalized ventilation (SHPV) [20]; the performances of underfloor and PV systems [21]; as well as the use of PV under mixing and displacement strategies [22]. Yet, the electricity required to operate mechanical PV systems could be a drawback for their sustainable and wide scale adoption in hospitals. This downside of mechanised PV system inspired the development of a zero-energy version called natural personalised ventilation (NPV) [23]. The NPV system is designed to be used for wind-neutral (buoyancy-driven) conditions-where the air could come from, shielded points of air intake, or linked to underground labyrinths as done in advanced natural ventilation (ANV) [9].

1.2. Personalised Ventilation (PV): Mechanical and Passive Options for Mixing Regime

In the original NPV system, an elevated duct delivered fresh outdoor air directly over a patient in a single-bed ward (Figure 1a). In the process, the ingress of contaminated air into the patient's breathing zone and overall bed area is minimised. The NPV also achieved a mixing regime for the entire ward because of the subsequent rising of warmed (buoyant) air and its escape via a stack whose point of discharge was higher than the duct's inlet (Figure 1b). Follow-up investigation was carried out to validate the performance of the NPV using a scaled (1:20) Perspex model of the single-bed ward using salt-bath experimental modelling [24]. The flow regime was confirmed for interface layers (Figure 2) including the formation of "eddies" around the air supply duct. The scaled experimental validation of the NPV confirmed the descent of fresh air over the bed, the mixing regime as well as the tendency of fresh air to be "thrown" horizontally in a forward direction, which is a known characteristic of cold jet of air entering a warm space [10,25].

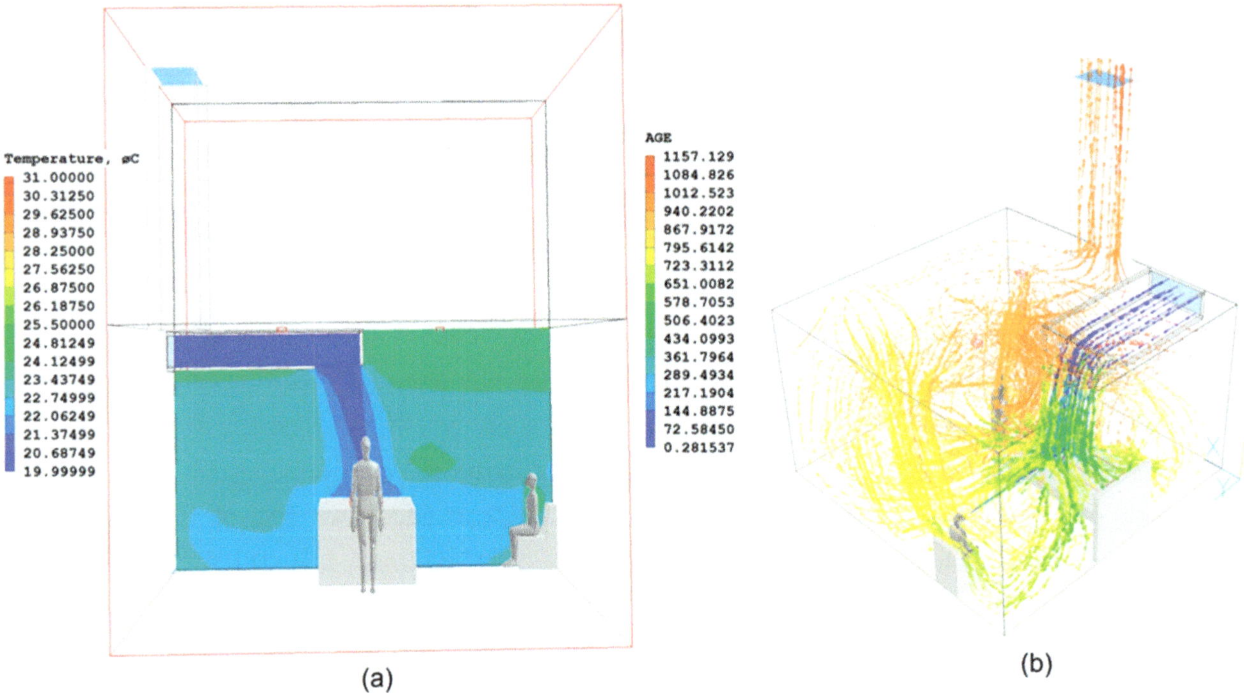

Figure 1. The natural personalised ventilation (NPV) system showing: (**a**) Fresh air delivered over patient's bed; and (**b**) The overall mixing regime achieved in the ambient space.

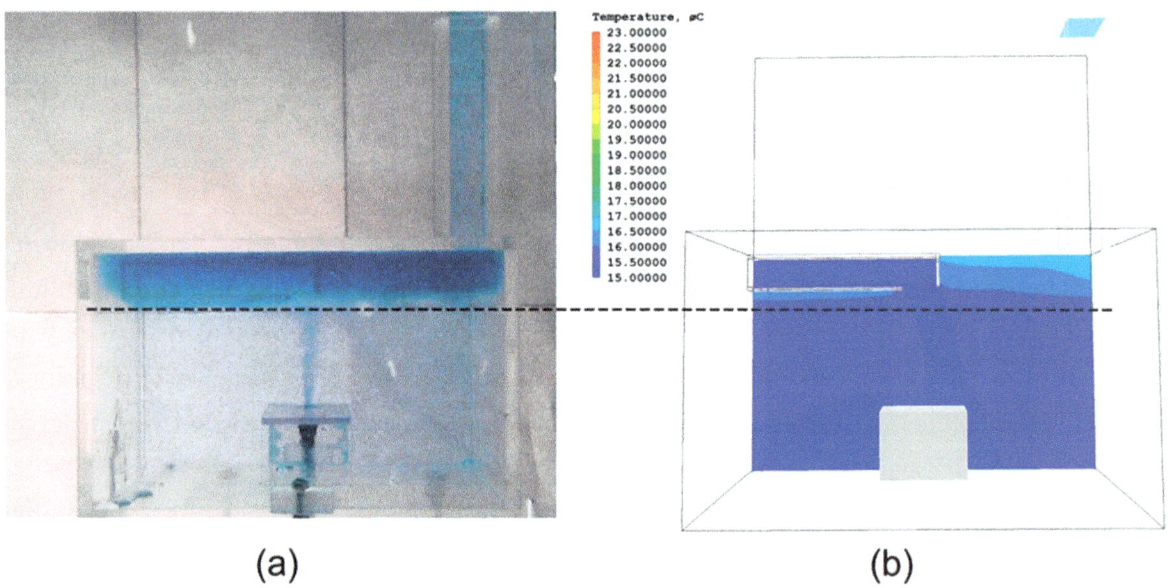

Figure 2. Compared interface layers in: (**a**) Salt-bath model; and (**b**) Computational fluid dynamics (CFD) model [24].

In summary, achieving dedicated personalised ventilation and mixing regime in occupied clinical spaces using the buoyancy-driven NPV system is an example of feasible low energy innovation. However, this approach requires integrating the architectural enclosure and bed layout with the design principles of buoyancy-driven natural ventilation system including internal heat loads and temperature differentials.

The objective of this research is therefore, to scale the NPV system by customising its design for multi-patient wards. Salt-bath validation of the NPV [24] has provided the confidence needed to further investigate the feasibility of NPV system for multi-bed hospital wards using steady-state CFD. The choice of CFD for this investigation is that it has been shown [26] to be helpful and growing in reliability and wide spread adoption for ventilation-related research since 2002. Other studies [27,28] that have adopted CFD for modelling indoor air quality also support this notion. Subject to validation, CFD is in fact cheaper for studying person-to-person exposure to airborne contaminants [12].

2. Materials and Methods

2.1. The CFD Application and Its Governing Equations

The Parabolic Hyperbolic or Elliptic Numerical Integration Code Series (PHOENICS) is one of the earliest commercially available CFD code developed in 1980 [29]. It is used to quantitatively simulate the flow of fluids (e.g., air and water). PHOENICS utilizes the fundamental conservation model (Equation (1)) for balancing quantities (e.g., mass, energy, momentum) where:

$$[outflow\ from\ a\ cell] - [inflow\ to\ a\ cell] = [net\ flow\ in\ cell] \tag{1}$$

The single phase equation for conservation is expressed as shown in Equation (2):

$$\frac{\partial(\rho\emptyset)}{\partial t} + \frac{\partial}{\partial X_k}\left(\rho U_k \emptyset - \Gamma_\Phi \frac{\partial_\emptyset}{\partial X_k}\right) = S_\emptyset \tag{2}$$

where $\emptyset$ is a given variable; $S_\emptyset$ is the conventional source term; Γ_Φ is the diffusive exchange coefficient and U is the vector velocity. These are expressed through a relationship where the boundary conditions (e.g., S_{bc1}) and sources are on the right hand side of Equation (3):

$$\frac{\partial(\rho\emptyset)}{\partial t} + \frac{\partial}{\partial X_k}\left(\rho U_k \emptyset - \Gamma_\Phi \frac{\partial_\emptyset}{\partial X_k}\right) = S_\emptyset + S_{bc1} + S_{bc2} + \ldots + S_{bcn} \tag{3}$$

For modelling turbulence, PHOENICS has the option to implement the k-ε renormalisation group (RNG) model [30] for continuity (Equation (4)), momentum (Equation (5)), as well as the transport models for turbulence kinetic energy, k, (Equation (6)) and rate of dissipation, ε, (Equation (7)) respectively as follows:

$$\frac{\partial u_i}{\partial x_i} = 0 \tag{4}$$

$$\frac{\partial u_i}{\partial x_i}(u_i u_j) = \frac{\partial}{\partial x_j}\left[(v + v_t)\frac{\partial u_i}{\partial x_i}\right] - \frac{1}{\rho}\frac{\partial_p}{\partial x_i} \tag{5}$$

$$\frac{\partial}{\partial x_j}(k u_j) = \frac{\partial}{\partial x_j}\left[\left(v + \frac{v_t}{\sigma_k}\right)\frac{\partial k}{\partial x_j}\right] + v_t S_{ij}\frac{\partial U_i}{\partial x_j} - \varepsilon \tag{6}$$

$$\frac{\partial}{\partial x_j}(\varepsilon u_j) = \frac{\partial}{\partial x_j}\left[\left(v + \frac{v_t}{\sigma_\varepsilon}\right)\frac{\partial \varepsilon}{\partial x_j}\right] + C_{\varepsilon 1}\frac{\varepsilon}{k} v_t S_{ij}\frac{\delta U_i}{\delta x_j} - C_{\varepsilon 2}\frac{\varepsilon^2}{k} - \frac{C_\mu \eta^3\left(1 - \frac{\eta}{\eta_0}\right)\varepsilon^2}{1 + \beta\eta^3}\frac{\varepsilon^2}{k} \tag{7}$$

where u_i represents mean velocity in x_i direction, u_j represents fluctuating velocity in x_j direction; p is mean pressure of fluid; ρ is density of fluid; while k is turbulence kinetic energy; and ε is rate of dissipation. The turbulence constants are represented by: σ_k, σ_ε, C_μ, $C_{\varepsilon 1}$, $C_{\varepsilon 2}$, η_0 and β.

2.2. Indoor Air Quality Metrics

One method used to measure the effectiveness of ventilation systems is the contaminant removal efficiency (CRE) as proposed by Mundt *et al.* [31] and later by Cheong and Phua [32]. CRE is used to determine the level of exposure to contaminants for a person (or manikin) at the point or zone of inhalation. It is non-dimensional and typically expressed as a percentage. Given the concentration of a pollutant at return/exhaust point, C_R, and the concentration of same pollutant at a location of interest (e.g., breathing point) C_{exp}, then CRE can be calculated as shown in Equation (8):

$$CRE = \frac{C_R}{C_{exp}} \tag{8}$$

Age of air (τ) is another metric, used to evaluate the quality of air because it indicates the time spent by air particles since entering a space [10]. In simple terms, Age of air determines the relative freshness or staleness of air, making it helpful for this kind of study. The predictions of percentage of people dissatisfied due to draught (PPDR) is based on PHOENICS' implementation of ISO 7730 for Class I buildings [33] also described in Awbi [10].

Two types of ward designs available in the UK were selected for the CFD investigation: the four-bed activity database (ADB) ward and the four-bed Nightingale Cruciform ward. Subsequent sections describe the studies carried out on these ward types. The specific boundary conditions for CFD modelling that were applied to these wards are described in the next section.

2.3. Boundary Conditions for CFD Models

The four-bed ADB ward was re-created in PHOENICS, to serve as the Base Case upon which the performance of the NPV system would be appraised as an alternative. The computational domain of the four-bed ADB ward was modelled without the toilet or store areas but including the lobby which separates them, giving a net area of 62.25 m^2. This domain was meshed with 90 × 100 × 45 (405,000) cells. This fine mesh was arrived at after carrying out tests (coarse, medium and fine) on the best grid resolution that would produce reliable results while being computationally affordable. All envelope surfaces were assumed to be adiabatic. The internal heat load utilised for the CFD simulation was based on assumed peak occupancy and equipment utilisation. For lighting, using an 80:20 radiation-to-convection ratio for ceiling mounted fluorescent lighting as contained in CIBSE Guide A [34], the floor re-radiated 127 W while the lighting fixtures themselves produced a total of 28 W, giving a combined lighting load of 155 W. Heat load from miscellaneous medical equipment was assumed to contribute a total of 300 W. In the CFD model, each patient emitted 80W of heat while standing healthcare workers and seated visitors were allocated 100 W. A peak-time total of eight persons (healthcare workers and visitors) were assumed in addition to four patients. This assumption would represent a fully occupied ward, allowing the resilience of the window openings to cope with high internal heat loads to be evaluated.

The *k*-ε renormalisation group (RNG) model and the boussinesq approximation model were used for turbulence and buoyancy, respectively. The effect of wind was kept minimal at a fixed inlet velocity of 0.2 m/s for the window cases: (in order to make the airflow through the window case as close as possible to the NPV, this low wind speed was allowed since windows cannot constantly guarantee zero wind conditions like the NPV). For the NPV model, airflow was driven by buoyancy forces only. A steady-state ambient temperature of 18 °C was used. For simulating released airborne pathogens, a surrogate (passive scalar) airborne contaminant of dimensionless value and initial concentration of 100% was assumed. Passive scalar contaminants have been shown [35,36], to be adequate in representing airborne contaminants (droplet nuclei) whose size does not exceed 1μm in CFD modelling studies. Cox and Wathes [37] also reported that a contaminant density of 1.1 kg/m^3 (similar to that of air) is widely accepted for computational applications. The contaminant was assumed to be produced via a cough at a mass flow rate of 0.005 kg/s and temperature of 30 °C, from an adult male. The mouth area was taken as 0.0049 m^2 which is within the range found by Gupta *et al.* [38]. The cough model was assumed because it is not only representative of a possible source of a single, forceful emission, but also for the presumption that five minutes of talking could generate as much bio-aerosols as a single cough [39]. The contaminant source is assumed to be a sitting visitor, whose back is to the external wall (position "*X*" in Figure 3) with the mouth at an elevation of 1.05 m, for a seated person.

The second type of ward simulated is the Nightingale Cruciform Ward. As in the ADB ward, a mild non-winter scenario was assumed, when outdoor temperatures could be 18 °C. Similar boundary conditions used in the ADB ward (*i.e.*, internal heat load, occupancy and airborne contaminants) were retained, with the exception of the cell density in computational domain which was now meshed using 105 × 110 × 50 (577,500) cells. The same pollutant removal metric (*i.e.*, CRE) applied in the ADB ward was also used to further predict the contaminant removal/dilution capacity of the Nightingale Cruciform ward.

3. Study 1: The ADB Ward

The design of the NPV system integrates architectural layout (e.g., bed locations and height of headroom) with natural ventilation principles as earlier demonstrated in the original study of the NPV [23]. For this study, these integrated design factors are described in the next sub-sections.

3.1. Architectural Considerations, Materials and Methods

The standard four-bed ADB ward (Base Case) has a combined bed area of 56.25 m^2 and a floor to ceiling height of 3.6 m gives it a volume of 202.5 m^3 (Figure 3a). The en-suite shared bathroom found in this type of ADB ward (with mechanical extract fan) was ignored in order to focus on the natural (buoyancy-induced) airflow regime. Each bed is within an area approximately 3.3 m × 3.9 m (Figure 3a). The natural ventilation openings of the Base Case were side-hung windows, modelled as five bays of side-hung casement windows. Given the 100 mm maximum restrictions [5], the total effective area of each opening was found to be 0.15 m^2. This area of opening was modelled as a vertical orifice measuring 0.125 m by 1.2 m on the external wall. The 125 mm opening marginally exceeds the 100 mm requirements because the upper and lower "triangular" portions of the open window were added to the horizontal 100 mm width. The total area of window opening is 0.6 m^2.

For the NPV system in the ADB ward, four ducts (each measuring 0.3 m × 0.6 m in cross section) and two supplementary ducts (each measuring 0.4 m × 0.7 m) were used for fresh air supply. The total area of air inlets was therefore 1.28 m^2 which is 53% more than total area of window openings. Due to the need to ensure fresh air was supplied to the exact location on each of the four beds, two possible designs of the NPV system that were considered and modelled are: the *Side–Side Duct Arrangement*; and the *Top–Bottom Duct Arrangement* (Figure 3). The three-dimensional positions of the Side-Side and Top–Bottom NPV ducts along (Figure 3b) represent the distinct feature of the NPV cases being investigated for the ADB ward. The clear headroom and height of opening for each duct in the Side–Side arrangement was 2.45 m while for the Top–Bottom arrangement, the headroom was 2.15 m. The exhaust stack was proportionately sized in cross section as 1.28 m^2, representing 1% of the net floor area as done in Lomas and Ji [40] for sizing advanced natural ventilation systems. The central exhaust stack was located internally along the external wall (Figure 3a) to ensure balanced access of stale to the exterior. The stacks were assumed to rise up to 2.5 m above the roof level. The longest ducts meant for the furthest beds were required to span a distance of 5.4 m (Figure 3a).

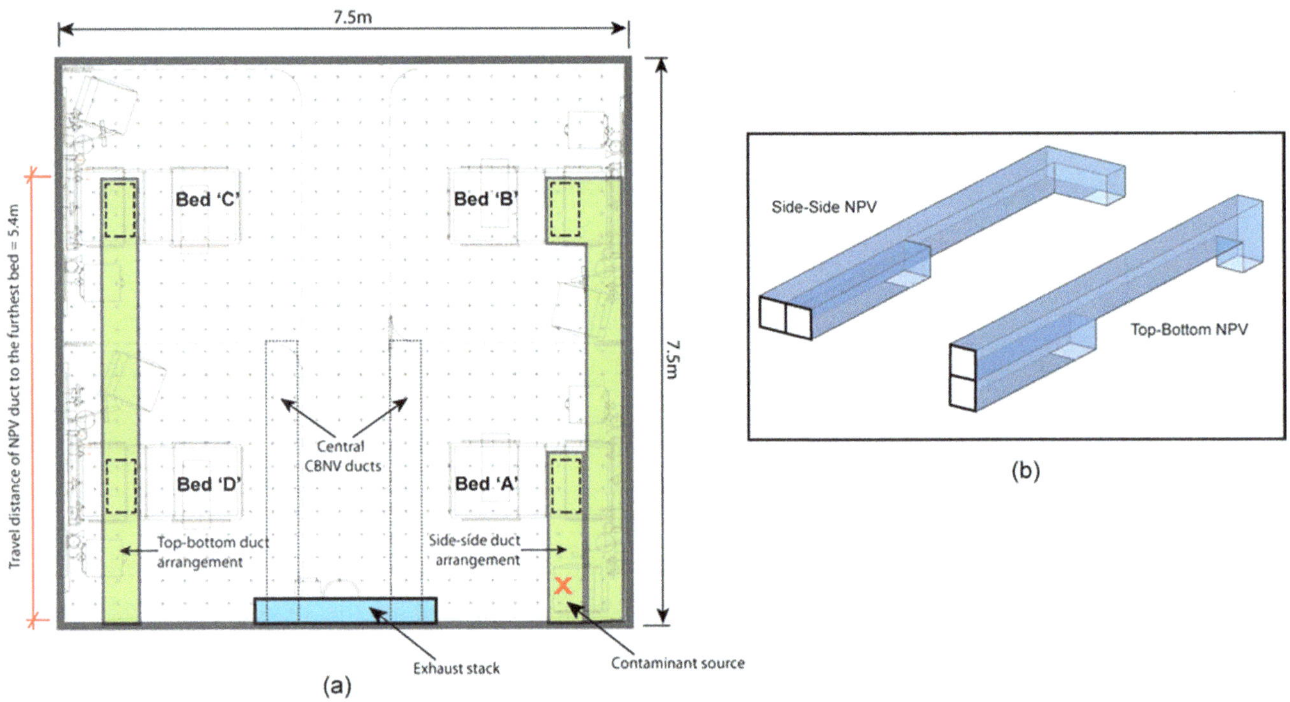

Figure 3. Layout of activity database (ADB) ward showing: (**a**) Plan view of Side–Side and Top–Bottom NPV duct locations, central CBNV ducts and two possible locations of exhaust stacks; as well as (**b**) The NPV duct designs in 3D.

3.2. Results of the ADB Ward

The predictions from CFD are in agreement with findings from literature that with displacement strategy (*i.e.*, Base Case windows), fresh air from the side-hung openings flood the space at floor level (Figure 4). As the same orifice serves as air outlet, the mean age of air of at the floor is 2100s, before the currents rise upwards upon reaching the opposite wall (Figure 4d). Upon reaching the opposite wall, the air currents begin a return sweep with a mean Age of air at bed height of between 2730s and 2940s. Thermal stratification is clear from the predictions of ward temperature (Figure 4a), PMV (Figure 4b) and Age of air (Figure 4c). The temperature at 1 m (bed heights) is generally around 27 °C, indicating a rise of around 9 °C from ambient (outdoor) temperature. The overall mean ward temperature at head height is 29 °C. There would be pockets of stale warm air collecting at the celling with values approaching 36 °C. The relative air change rate predicted from the CFD simulations for the Base Case is 1.8 ACH.

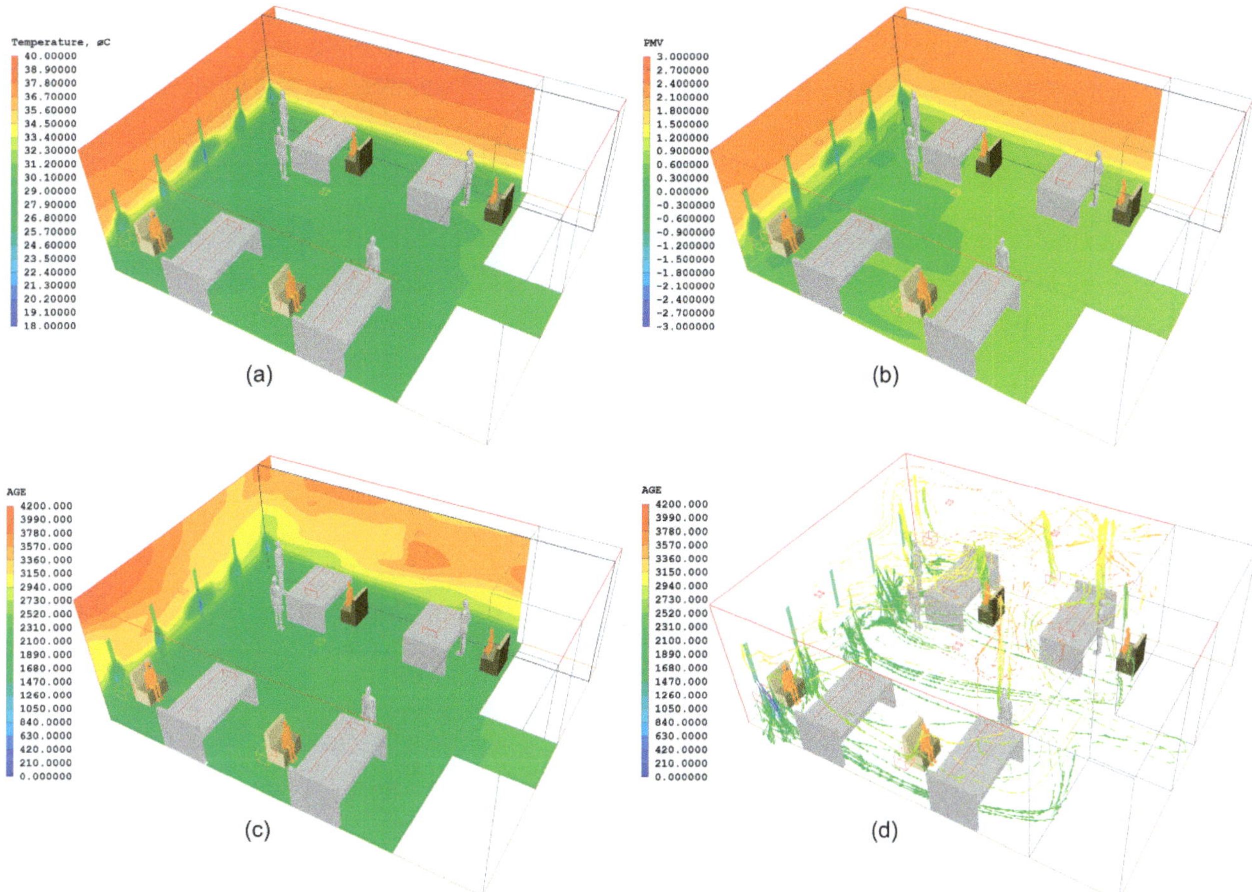

Figure 4. Airflow characteristics of the base case design showing contours of: (**a**) Temperature; (**b**) PMV; (**c**) Age of air; and (**d**) Streamlines of displaced air particles as they flood the space.

When the ADB ward is fitted with the NPV system, the shorter ducts of both Case 1 (Side–Side NPV) and Case 2 (Top–Bottom NPV) bring in fresh air with a horizontal throw as expected (Figure 5). The Top–Bottom design delivers greater volume of fresh air for the longer ducts, while for the Side–Side design, the elbow (corner) in the longer duct leads to a reduction in flow velocity and subsequently air flow rates (Figure 5a). In both cases, airflow within the ducts can be described as steady laminar (Rayleigh number, $Ra < 10^5$). In the Top–Bottom duct arrangement (Figure 5b) the presence of canopy is useful re-directing air from the top duct unto the bed, whereas air from the bottom (shorter) duct experiences a horizontal throw.

In all NPV cases, the mean Age of air at bed height is ≈240 s and temperature stratification gradient is much less evident. There is potential for draught to be experienced by the patients, due to the instant discharge of fresh air over the beds with minor trubulence. The predictions of PPDR (for both types of NPV designs) show that for an outdoor simulation temperature of 18 °C, between 11% and 25% of patients may experience draught (Figure 5c,d). This proportion of dissatisfied persons would expectedly be higher at a lower ambient temperature. The relative air change rates predicted from the CFD simulations are: 4.8 ACH for Side–Side NPV and 5.2 ACH for Top–Bottom NPV.

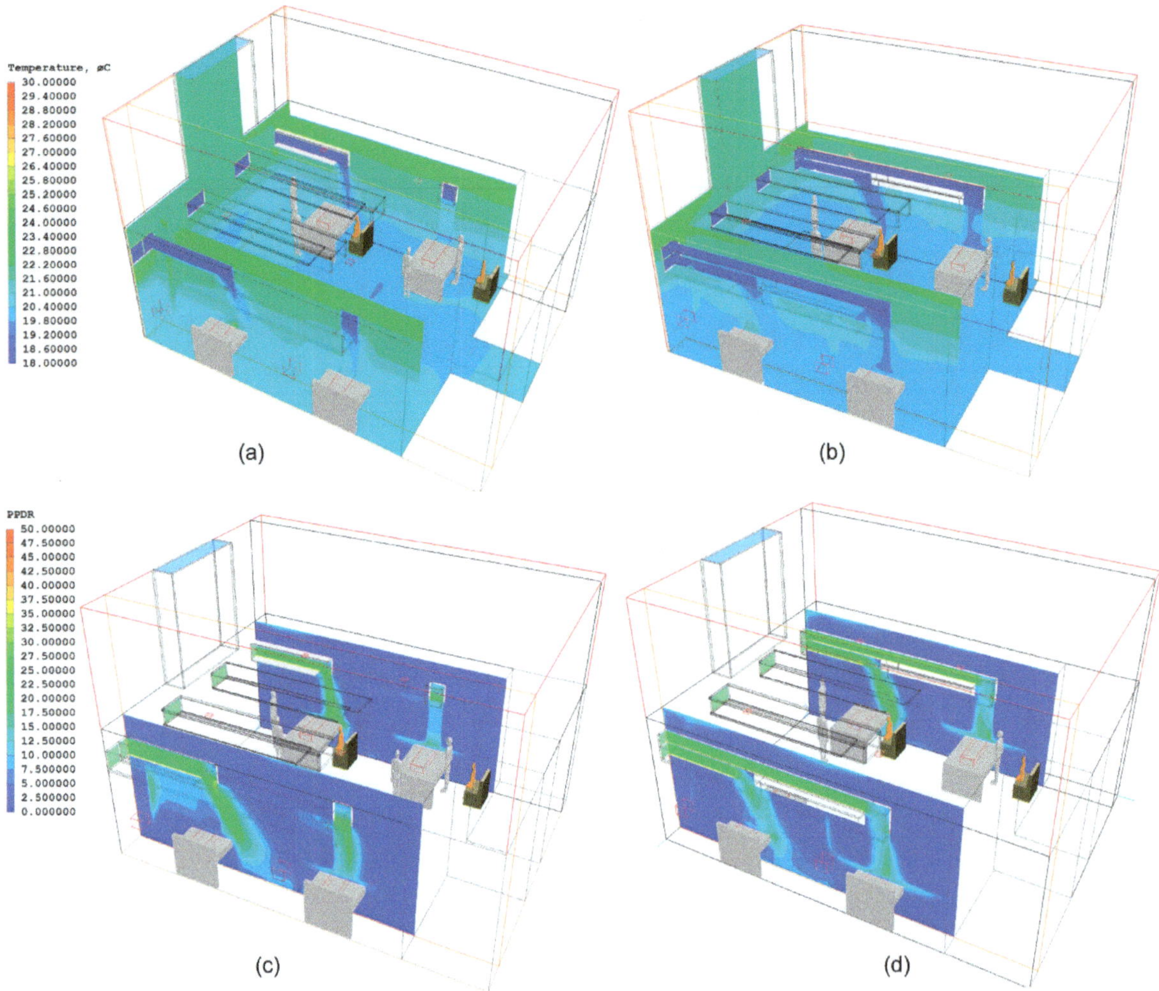

Figure 5. Temperature contours for: (**a**) NPV Case 1; and (**b**) NPV Case 2 showing horizontal throw of fresh air from shorter ducts; with draught potential for (**c**) NPV Case 1; and (**d**) NPV Case 2.

The distinct designs of the NPV ducts in Case 1 and Case 2 also have marginal implications on the speed of incoming air within the ducts and hence on airflow rates. From the steady-state CFD results of the Side–Side design (Case 1), the computed mean velocity of air at the discharge orifice is 0.31 m/s for the shorter duct and 0.26 m/s for the longer duct. The reduced velocity in the longer duct is due to the "L" shape bend which leads to pressure loss, and not particularly due to distance. This was confirmed in the Top–Bottom design, where the mean velocity of ≈0.3 m/s is maintained for both long (top) and short (bottom) ducts. Relative to the cross-sectional area of duct, this air speed gives each patient an absolute mean flow rate of 0.0324 m^3/s or 32.4 $L \cdot s^{-1}$.

The airborne contaminants originate from source location (Figure 3a, Position "*X*") in all Cases. The concentrations were calculated at the assumed (horizontal and vertical) position of a sleeping head for two bed positions (Figure 6). In all Cases, the airborne contaminants are able to travel further across the ward with a relatively high concentration at a distance of up to 8 m from the source. However, in the Base Case, the concentration across the width of Bed "A" builds up from 7% up to 13.9% before it attenuates to ≈8% across Bed "B" (Figure 6a). In the NPV Cases, however, the concentration across Bed "A" drops from ≈12% down to ≈3% in Case 1 (Side–Side) and to ≈2% in Case 2 (Top–Bottom). Across

Bed "B" the concentration is ≈2% for Case 1 and ≈1% for Case 2 (Figure 6a). The marked differences in the horizontal spread of the contaminants in the three Cases can be appreciated from the contour plots for the Base Case (Figure 6b), Case 1 (Figure 6c) as well as Case 2 (Figure 6d).

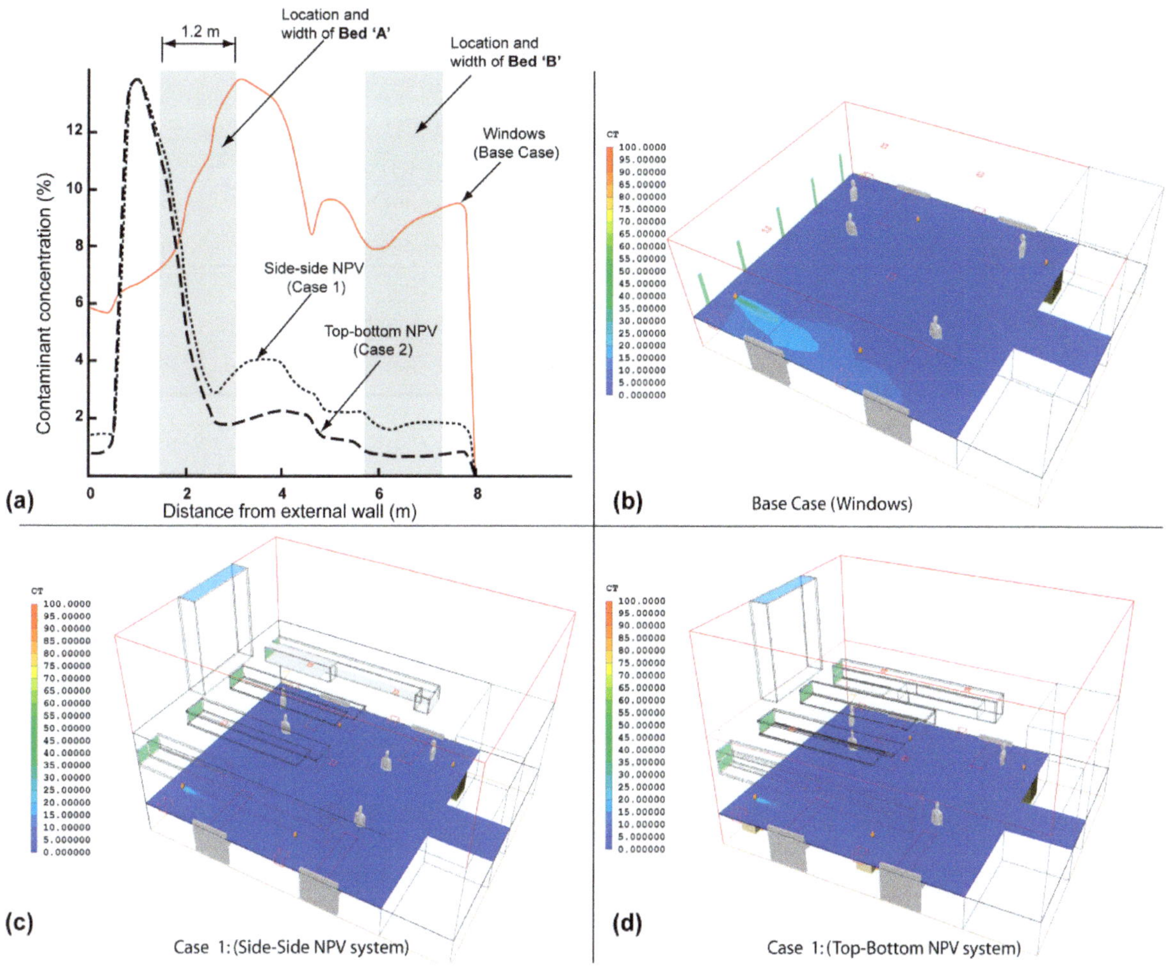

Figure 6. Predicted contaminant (CT) concentration results for: (**a**) All three cases at height of 1.05 m along trajectory of release; and 2D horizontal axis contours for (**b**); Base Case (**c**) Case 1; and (**d**) Case 2.

Contaminant removal efficiency (CRE) was estimated at assumed head position on all four beds as well as at standing height in the centre of the ward. The CRE for each position in the Base Case is significantly higher than at similar positions for Case 1 and Case 2 (Table 1). Although the NPV in Case 2 (Top–Bottom) offers relatively better CRE values than Case 1 (Side–Side), both systems are significantly superior to the CRE of Base Case, by a minimum order of 34.4% (Bed C, in Case 1) and a maximum order of 80% (Bed D in Case 2).

The airflow characteristics the Top–Bottom and Side–Side duct arrangements differ marginally in airflow rates, but the differences in the direction of airflow are significant. The short ducts of both cases are observed to deliver fresh outdoor air into the space with an expected lateral throw in the same direction as the duct. In previous NPV studies [23], such horizontal throw led to the ducts being offset backwards by 0.3 m from the desired point of discharge. This horizontal throw is also present in the multi-bed use of NPV where the discharged air from shorter ducts partially overshoots the bed (Figure 7). An improvement was needed to channel the discharged air directly over the bed.

Table 1. The contaminant removal efficiency (CRE) values computed at five locations for all three Cases in ADB ward.

Cases	Bed-A	Bed-B	Bed-C	Bed-D	Ward Centre
Windows (Base Case)	0.73	0.57	0.32	0.30	0.40
Side–Side NPV (Case 1)	0.34	0.24	0.21	0.09	0.20
Top–Bottom NPV (Case 2)	0.30	0.18	0.13	0.06	0.16
Difference between Base Case and Case 1	53.4%	57.9%	34.4%	70.0%	50.0%
Difference between Base Case and Case 2	58.9%	68.4%	59.4%	80.0%	60.0%

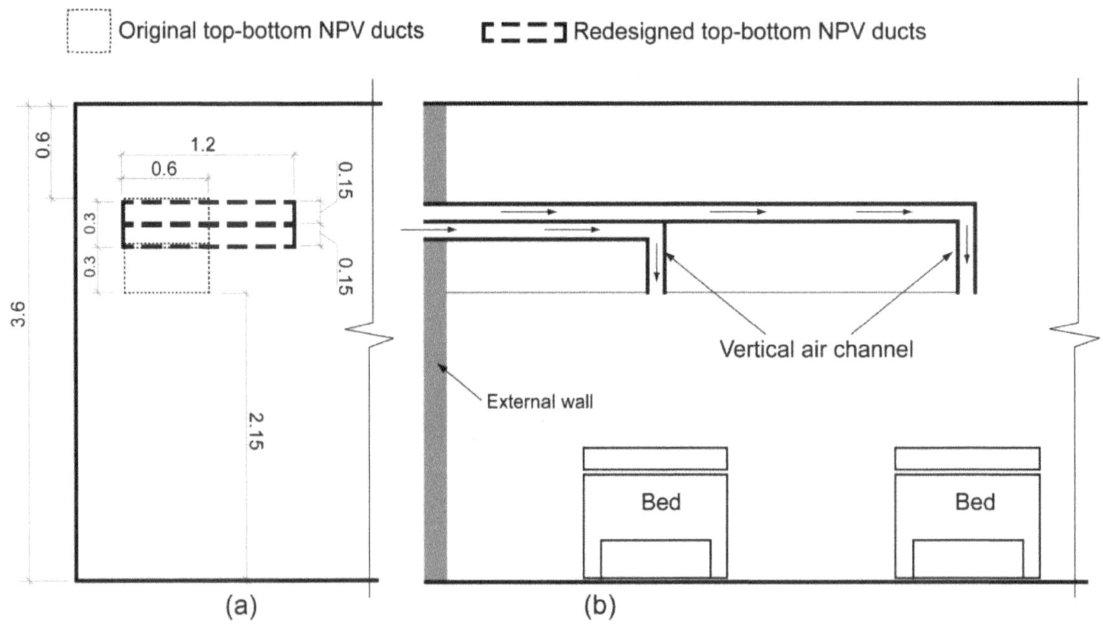

Figure 7. Re-design of the Top–Bottom NPV duct to create vertical air discharge channels, shown as: (**a**) External facade view; and (**b**) Cross-sectional view.

3.3. Improved Design of NPV Duct Using Vertical Air Channels

To constrain the horizontal throw of air, a channel was created to direct the flow vertically and downwards, unto the middle of the bed. It was necessary to work around the architectural features of the ward (e.g., headroom). The vertical channel in the Top–Bottom concept was adopted for the shorter duct but to avoid encroaching into the headroom (currently 2.15 m), both ducts were compressed from a cross-sectional area of 0.6 m × 0.6 m into 0.15 m × 1.2 m. The same area of opening was achieved with different (flattened) geometry (see Figure 7a). This compression freed up a height 0.3 m which was used to elevate the lower duct. The new headroom and height of opening for the lowest Top–Bottom duct (Figure 7b) became 2.45 m.

For Case 2 (Top–bottom NPV) the new design simulated in CFD produced desired results (Figure 8b) without much horizontal throw (Figure 8a). The horizontal throw was largely eliminated by the use of vertical channels, as observed from the beds with shorter ducts. This vertical channel helps with discharge of fresh air directly over the patient, without any significant loss of head room.

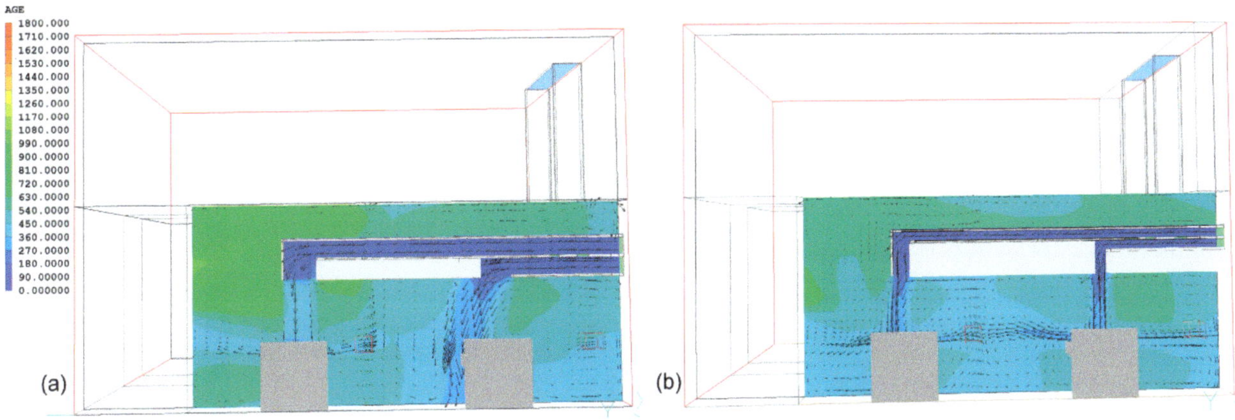

Figure 8. 2D contours and vectors indicating directionality of air in: (**a**) The original Top–Bottom design; (**b**) The narrow vertical air discharge channels.

4. Study 2: The Nightingale Cruciform Ward

To further test the scalability and adaptability of the NPV system to different design scenarios, an alternative four-bed ward (Nightingale Cruciform ward) developed by an architectural practice was modelled and simulated. This ward had significantly different spatial and bed layout requirements. The next subsections describe the NPV design considerations and the results obtained from the simulations carried out.

4.1. Architectural Considerations, Materials and Methods

The floor area of the Nightingale Cruciform ward [41] is approximately 107.5 m^2, with an en-suite bathroom measuring 9.5 m^2 (Figure 9). Disregarding the bathroom space which is often fitted with extract fans, the effective volume of the ward space to be naturally ventilated is 352.8 m^3, assuming a uniform floor to ceiling height of 3.6 m. The ward has two window bays, located at two corners of the ward. These corners were designated as waiting and dining areas (Figure 9a,b) and for the purpose of this investigation, their windows were assumed to be top-hung. Following the DH regulations that restrict vents to 100 mm width, the effective area of ventilation were taken as three (lower, middle and upper) equidistant vents measuring 0.18 m^2 each, with a distance of 1.2 m between the lower and upper vents. The total area of window openings is 1.08 m^2.

The collapsible partitions which demarcate each bed space were presumed to be fully stretched out, to represent a scenario that might impede natural airflow. This is because other than airflow rates, two other parameters: airflow pattern and airflow direction, are known to affect the migration of airborne contaminants in ventilated spaces [42,43].

For testing the NPV system in the Nightingale Cruciform ward, four ducts—A, B, C and D (Figure 9b) each with cross-sectional flow area of 0.18 m × 1.2 m were created at the top for supplying fresh air over corresponding patient beds (Figure 9c). The total area of NPV inlets is 0.86 m^2, which is 80% of the total area of the window openings. Based on learning from the improved Top–Bottom duct design of ADB ward (*i.e.*, use of flattened ducts and narrow air discharge channels), all ducts in the Nightingale Cruciform Ward were also modified to achieve similar vertical descent of cool fresh air directly over each patient. The air discharge channels had to be oriented along the length of each bed.

The longest NPV duct (Duct C) was required to transport fresh air up to 10 m to Bed C. Four exhaust stacks (each measuring 0.5 m × 0.65 m) were created at each corner of the ward to remove stale air. These stacks represent located at the four corners represent edge-out and centre-out strategies as defined in [40]. A plenum-like space 0.6 m high was left above the ducts for buoyant stale air to collect and make its way towards the exhaust stack. The lowest natural ventilation inlets in waiting and dining areas were retained in order to supply air to these spaces.

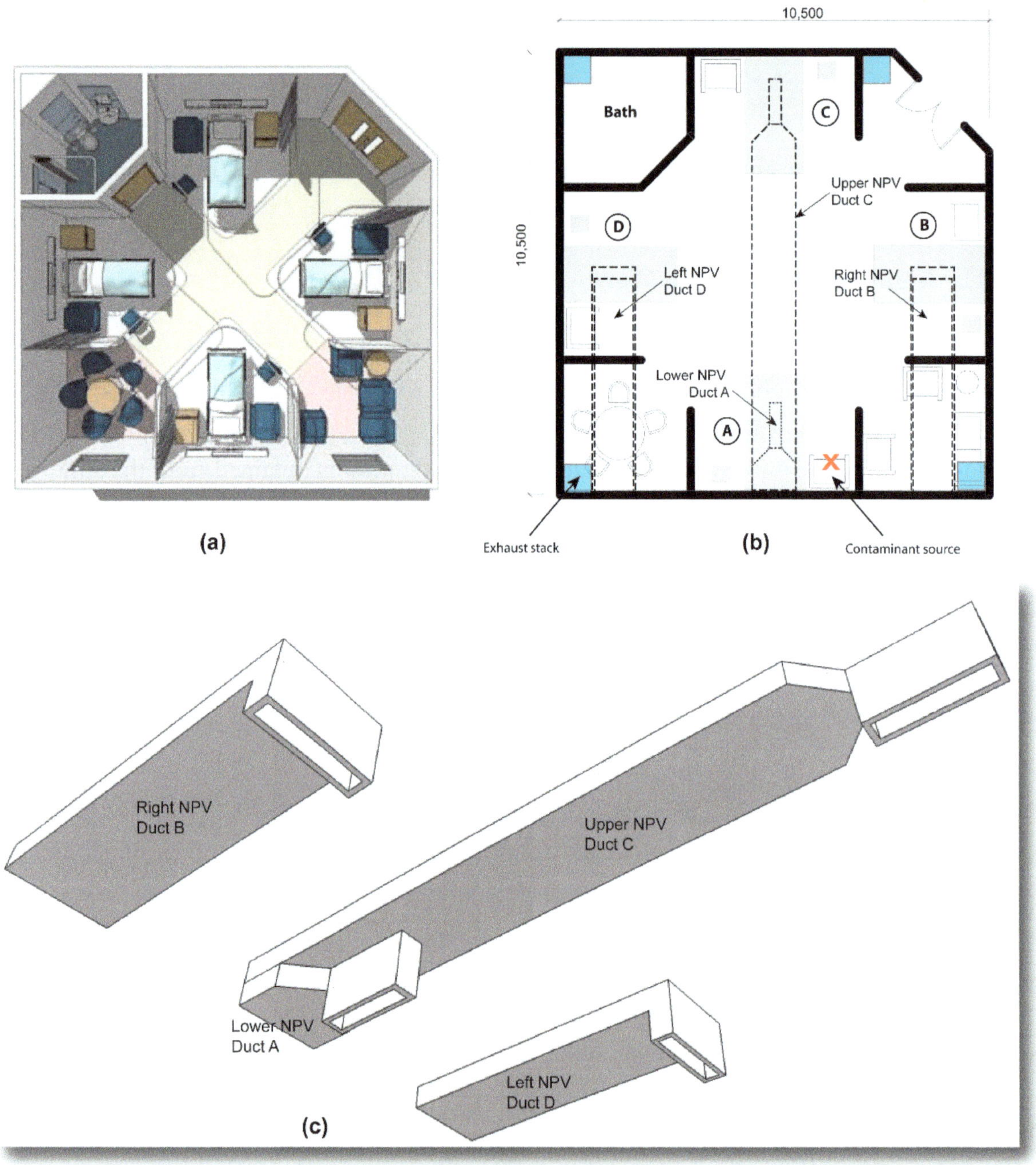

Figure 9. Design of original Cruciform ward © Nightingale Associates showing: (**a**) Top view; (**b**) NPV ducts fitted in plan; (**c**) Vertical air channel for NPV ducts in 3D orientation over four beds.

4.2. Results of the Nightingale Cruciform Ward

In the base case of the Nightingale Cruciform ward, the waiting and dining areas are flooded by cascading cooler outdoor air from their windows. The air currents are then displaced towards other parts of the ward, but presence of partitions in both areas constrains the free flow of fresh air (Figure 10a,b). The top-hung windows acted like dual openings, with lower vents bringing in fresh air and upper vents taking out stale air (Figure 10c).

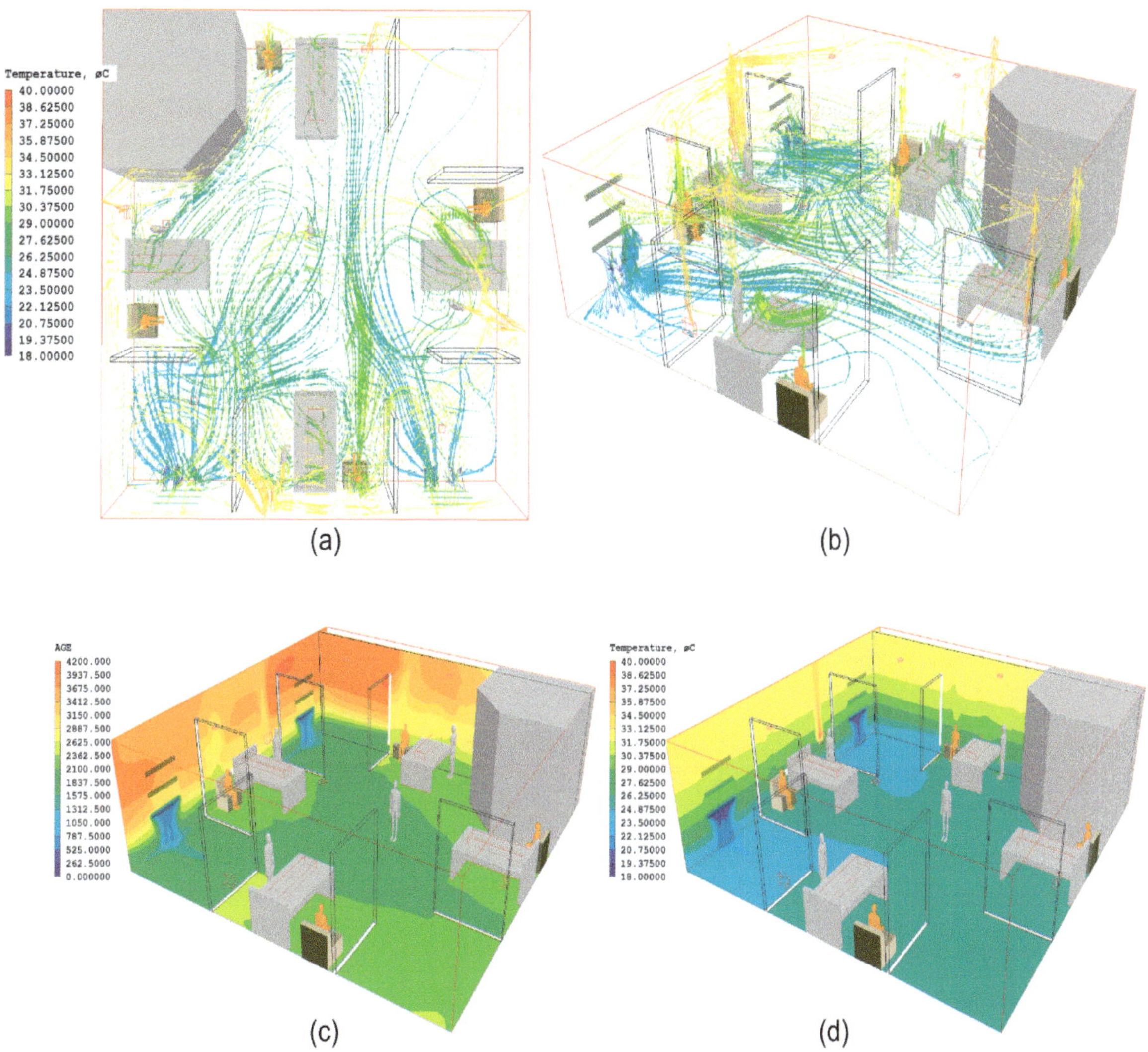

Figure 10. Airflow characteristics in Cruciform Ward showing pattern and direction of air movement in: (**a**) Top view; and (**b**) Isometric view; with contours for (**c**) Age of air; (**d**) Stratified temperature.

Deterioration in Age of air (τ) at floor level (Figure 11c) means fresh air creeps towards the interior, from the central part of the ward (τ = 1575 s) to the partial detriment of Bed A (see Figure 10a), which is actually nearer to the facade. The side beds (*i.e.*, Beds B and D) are also partially bypassed by the gravity currents of fresh air. As the air currents make a return sweep after encountering the opposite wall, the Age of air at bed level is up to 2625 s. In other words, the partitions separating each bed area in the Nightingale Cruciform ward are seemingly causing a delayed entrance of fresh air into the bed zones. This is with the exception of Bed C, which is furthest from the window but is directly in the path of creeping and rising gravity currents (Figure 10b). Overall, the Age of air at the floor of furthest end of the ward is similar to the Age predicted for the ADB ward (≈2100 s).

The 18 °C temperature of outdoor air rapidly increases towards the inner part of the ward where, even at floor level, it is already up to 24 °C around the furthermost bed (Figure 10d). Like with the ADB ward, the temperature of the Nightingale Cruciform ward is stratified, as expected in natural displacement ventilation systems (Figure 10d). The stratification means that at a bed height of 1.0 m, temperatures could rise up to 24 °C, which is 3 °C lower than at similar positions in the ADB ward. The mean temperature at head height is found to be 28 °C, while towards the ceiling, the temperature of collected pockets of stale air collect is up to 31 °C. The relative air change rates predicted from the CFD simulations are in the Nightingale Cruciform ward are 2.1 ACH for default windows Case and 5.5 ACH for NPV Case.

When the Nightingale Cruciform ward is fitted with NPV ducts, the most obvious implication is relative low Age of air (≈180 s) at the air discharge zone of each patient's bed (Figure 11a,b). In addition, there is a lack of significant stratification in temperature throughout the ward (Figure 11c,d). This time, the mean temperature gradient between floor and head level is 2 °C, suggesting a reduction of 4 °C from the mean temperature obtained in the window case.

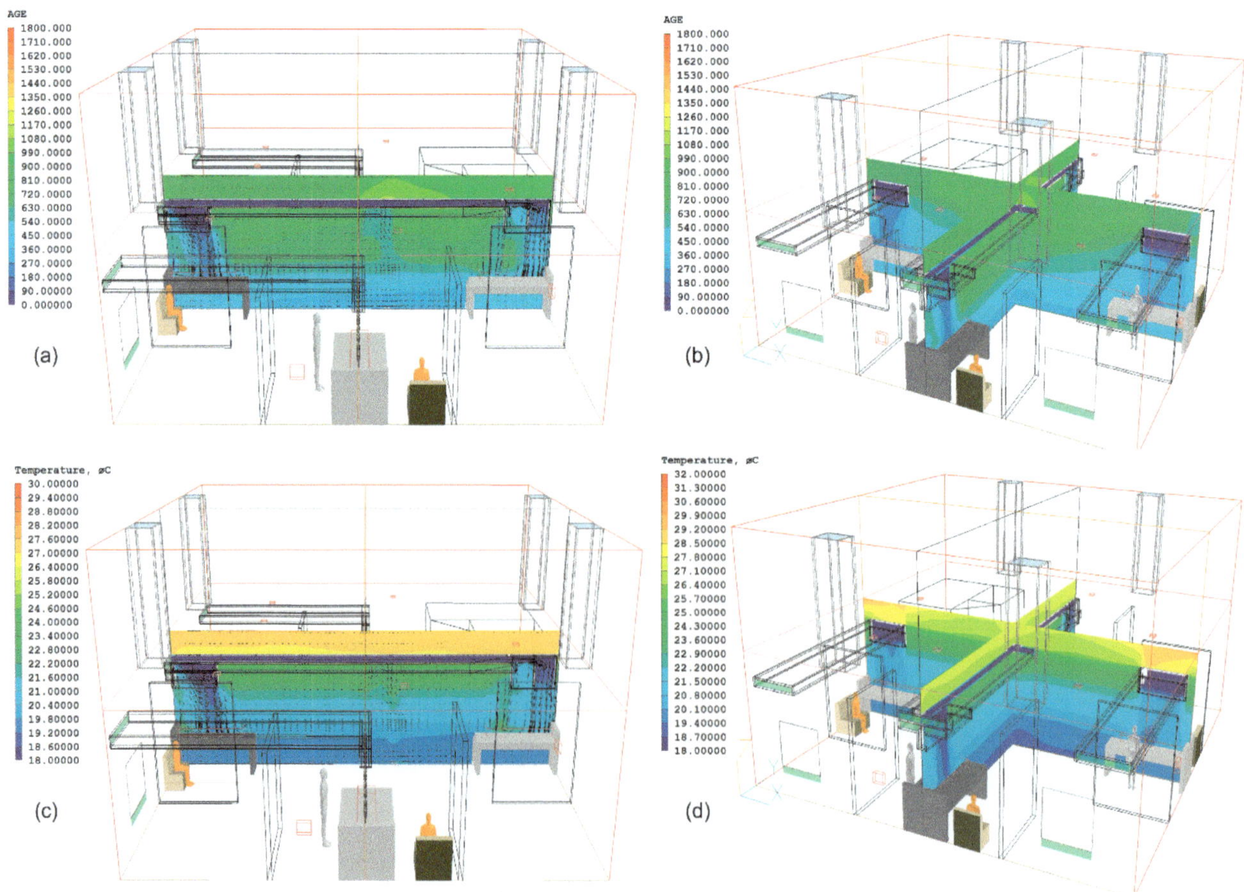

Figure 11. The Cruciform Ward: contours and vectors for age of air shown as: (**a**) Side view; (**b**) Isometric view; and for temperature (**c**) Side view; and (**d**) Isometric view.

The NPV ducts (A, B and D) have mean velocities of ≈0.32 m/s, indicating a flow rate of 0.069 m^3/s or 69 $L \cdot s^{-1}$, relative to their cross-sectional areas (*i.e.*, 0.18 m × 1.2 m). The NPV duct C which is longest, spanning up to 10 m has a mean velocity of 0.27 m/s, delivering a mean absolute flow rate of 0.054 m^3/s or 54 $L \cdot s^{-1}$.

The source of scalar airborne contaminant in the Nightingale Cruciform Ward is assumed to be a visitor sitting by the side of Bed A. As the emitted pathogens are discharged towards the ward centre, the lateral spread is restricted to one side due to the presence of partitions (Figure 12a). The central area of the ward receives a relatively significant discharge where the contaminants collect in a pocket. With the NPV system in place, the contaminants discharged are not able to collect in the ward centre. Relative to the window case, dilution at the source location and centre of ward is evident (Figure 12b).

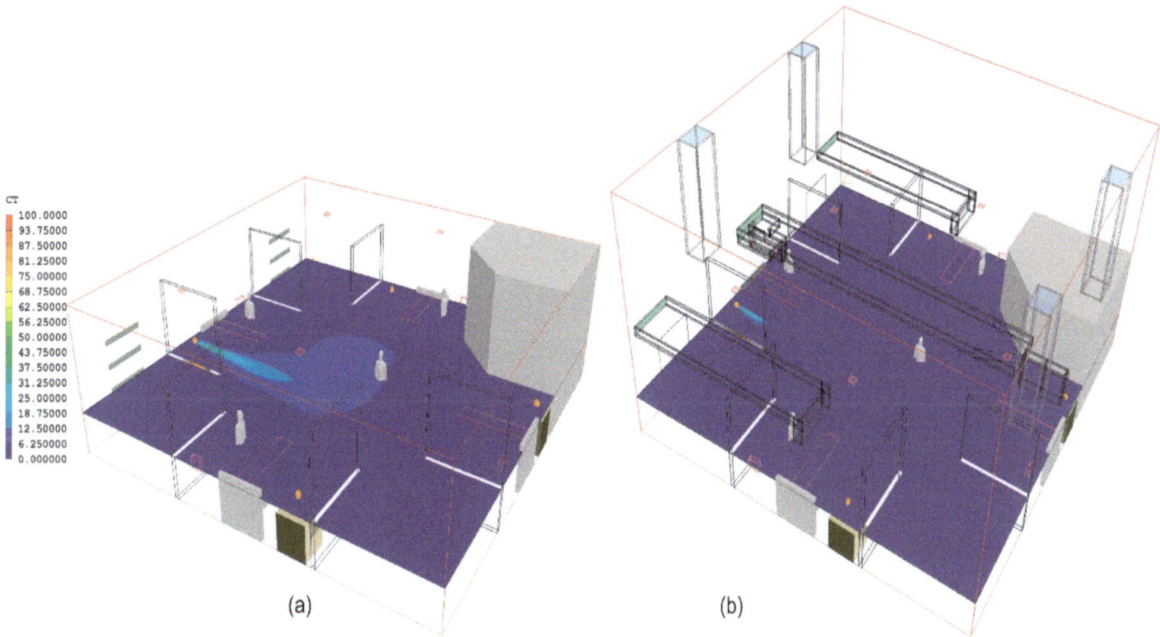

Figure 12. Differences in distribution of passive scalar airborne contaminants in Cruciform Ward based on: (**a**) Window design; (**b**) NPV system design.

Both ventilation systems were compared in terms of the differences in efficiency of contaminant removal. It is predicted that the Nightingale Cruciform ward fitted with NPV system would have better CRE in all five key locations (Table 2). Specifically, under the NPV regime, Bed A would have a 71.6% reduction in CRE compared to Bed C which has up to 93% reduction. For the window case, both beds represent the source location and the most ventilated locations, respectively.

Table 2. The CRE values computed at five locations for all three Cases in Cruciform ward.

Cases	Bed A	Bed B	Bed C	Bed D	Ward Centre
Cruciform Windows (Case A)	0.81	0.35	0.43	0.37	0.65
Cruciform NPV (Case B)	0.23	0.06	0.03	0.05	0.10
Difference between Window and NPV Cases	71.6%	82.9%	93%	86.5%	84.6%

5. Assessment of Thermal Comfort Levels in All Wards

Thermal comfort evaluation of these wards was done using temperature, air velocity, predicted mean vote (PMV) and percentage of people dissatisfied due to draught (PPDR) as the metrics. These parameters were calculated (*i.e.*, measured virtually in CFD model space) directly beneath the point of fresh air discharge, at a height of 1 m for each bed, using the Probe Object in PHOENICS. In addition, other determinants of thermal comfort, which are similar for all wards include: initial RH of 50% for each ward space; Clothing level (Clo)-assumed to be 0.6 for all occupants; and Metabolic rate (Met) assumed to be "Seated and relaxed" for patients. The results for temperature, air velocity, PMV and PPDR for all the ward designs are summarized in Table 3.

Table 3. Thermal comfort variables, showing values at four bed locations at centre of bed (height = 1.0 m).

Cases	Temperature				Air Velocity (m/s)				PMV *				PPDR (%)			
	Bed A	Bed B	Bed C	Bed D	Bed A	Bed B	Bed C	Bed D	Bed A	Bed B	Bed C	Bed D	Bed A	Bed B	Bed C	Bed D
ADB: Windows	27.4	27.8	27.2	27.1	0.02	0.01	0.01	0.02	1.22	1.41	1.44	1.27	0.0	0.0	0.0	0.0
ADB: Side–Side NPV	20.3	22.1	22.3	20.1	0.13	0.07	0.08	0.12	−1.72	−1.06	−1.1	−1.81	24.0	15.4	15.8	24.2
ADB: Top–Bottom NPV	21.9	20.3	20.3	21.8	0.18	0.21	0.23	0.18	−1.03	−1.34	−1.23	−1.05	11.3	15.8	15.9	11.8
ADB: Modified Top–Bottom	19.4	19.5	19.4	19.4	0.3	0.32	0.31	0.3	−1.32	−1.35	−1.31	−1.33	24.8	24.8	24.8	24.9
Cruciform: Windows	24.4	24.7	25.4	24.6	0.02	0.01	0.01	0.02	0.62	0.67	0.71	0.65	0.0	0.0	0.0	0.0
Cruciform: NPV	18.5	19.2	19.4	19.1	0.1	0.09	0.06	0.09	−2.25	−1.68	−1.63	−1.71	35.9	22.5	21.2	22.3

Notes: * PMV scale: +3 (hot); +2 (warm); +1 (slightly warm); 0 (neutral); −1 (slightly cool); −2 (cool) and −3 (cold).

Based on these thermal comfort results, it can be generalized that the window ventilated (ADB and Cruciform) wards do not have problems with PPDR. In fact, air velocity is hardly noticeable at the points of probe measurement. However, the PMV values of window ventilated wards are relatively poorer than the PMV of NPV ventilated wards (Table 3). The main noticeable thermal comfort concern for the NPV ventilated wards is their PPDR values over the bed.

For the Side–Side NPV design in ADB ward, the predicted PPDR values are between 15% and 24% (Table 3). The lower PPDR values belong to the longer ducts which have elbows, leading to lower air velocity for Beds B and C. Due to pressure loss at these bends, these beds (which are furthest from external wall) are expected to experience lesser draught (Figure 5c) than Beds A and D. The less draughty beds (B and C) are therefore warmer (*i.e.*, based on PMV values), than Beds A and D.

With the Top–Bottom NPV design in the ADB ward, the horizontal throw of air (occurring in on Beds A and D) means such patients would not receive much cooler air directly (Figure 5a). This is because a lot of the fresh air is discharged outside the bed (see Figure 8a) leading to slightly higher temperatures at the points of measurement. The furthest beds (B and C) get direct air discharge (and more draught) due to channel effect. Such draught-related direct discharge of air on Beds B and C is evident in their PMV and PPDR values (Table 1). Similar observations about relatively higher PPDR due to NPV air discharge can be made for the Cruciform ward.

6. Discussion

It should be noted that if occupancy and miscellaneous internal heat loads are lower than assumed for this study, then it is likely that the windows would perform better in achieving thermal comfort. Their airborne contaminant dilution capacity could increase as well, but modelling a worse case internal heat load and occupancy scenario is a better risk management strategy. For instance, the overheating potential predicted by this study, the mean ward temperature at head height is 29 °C and 28 °C for the ADB and Nightingale Cruciform wards respectively are within the range found [8] concerning overheating in Glenfield Hospital, UK. In future warmer climates, it is likely that such predicted overheating potential as well as low relative ventilation rates could pose serious risks to susceptible occupants due to constricted (maximum of 100 mm width) window-based natural ventilation. With the NPV system however, the temperature is not stratified and the mixing regime ensures that at head height, the temperature is much lower: 21 °C for Side–Side NPV and 20 °C for Top–Bottom NPV (Figure 5a,b). The thermal comfort levels are obviously a function of both outdoor temperature (18 °C) and the internal heat loads and a change in either could lead to different results. However, the main benefit of the NPV system is the delivery of fresh air into relatively deep plans as demonstrated by the two wards studied. The capacity of NPV to induce mixing regime due to the descent of fresh air and its consequent rise and escape as stale air through the stacks. This mixing largely eliminates thermal stratification as observed, ensuring that temperature remains relatively uniform in the areas surrounding the point of fresh air discharge. As observed, the mean Age of air at bed height is significantly lower with the NPV system than window cases, implying the level of air freshness to be expected by patients.

Although, the total area of inlets for the NPV openings is 53% more than total area of window openings in the ADB case. The ventilation rate obtained from the NPV system in ADB ward depends on the design/arrangement used, but the rate of 4.8 ACH (Side–Side) or 5.2 ACH (Top–Bottom) are significantly higher than the air change rates obtained from window designs (1.8 ACH). In the Nightingale Cruciform ward, the total area of window vents is 20% more than the total area of NPV openings, yet the ventilation rate obtained for NPV is 5.5 ACH compared to 2.1 ACH for windows. This suggests that appropriate "design" of a natural ventilation system (in this case the duct and stack-based NPV system) can lead to superior airflow rates than merely depending on area of opening (in this case traditional top-hung window system). The role of vertical air channels in restricting the horizontal throw of air fresh air can be appreciated with the improved NPV in the ADB ward as well as in the Cruciform ward. However use of such air channels could mean loss of head room and with the ADB ward, the effective headroom above the bed was 2.45 m, but the headroom in other parts of the ward remained unchanged (3.6 m). In retrospect, the NPV ducts could have been elevated higher towards the ceiling, but this may mean increasing the length of stack, because distance between openings and stack outlets are important for driving airflow.

For both types of wards, the Top–Bottom arrangement is found to be a better approach to multi-bed application of the NPV system. The longer (upper) ducts which take fresh air to the furthest beds needed a downward channel to enforce the perpendicular descent of air unto the bed. However, a similar air channel for the shorter (lower) ducts might encroach into the clear head room above the bed. The Top–Bottom design also has better consistency in airflow rates with speeds of 0.3 m/s over a distance of 5 m in the ADB ward and a speed of 0.27 m/s over 10 m in the longest duct of the Nightingale Cruciform ward. Such consistency in airflow rates would be important for system reliability, particularly for predicting long-term performance and for operation and control purposes. The computed absolute airflow rates for these ducts are 32.4 $L \cdot s^{-1}$ and 54 $L \cdot s^{-1}$ for ADB and Nightingale Cruciform wards respectively. The airflow rate into via NPV system into Nightingale Cruciform ward approaches the 60 L/s/person recommended by the WHO [42] for controlling airborne contaminants in naturally ventilated hospitals. This WHO requirement puts the NPV system into proper context since it can guarantee delivery of fresh air to each patient for effective ventilation. It is also found that the use of vertical air channel is beneficial in redirecting the flow of air downwards. This helps to eliminate the tendency for cooler air to be "thrown" horizontally when entering a warm space [10,25], a phenomenon that was also observed in the experimental validation done of NPV [24] for a single bed ward.

Results obtained for all NPV concepts investigated in this study, suggest that draught is a risk to be taken into account due to the direct discharge of air over the beds. It is observed that the NPV ventilated wards have better PMV values, relative to the windows, but their PPDR values are high. The channel effect, though desirable/intended for re-directing fresh air to where it is most needed (*i.e.*, over the bed) appears to increase PPDR (with knock-on effect on PMV), whereas the elbows in the side–side NPV design actually improve PMV and PPDR due to reduced air velocity. Nevertheless, the actual thermal comfort experiences of occupants (especially patients) will be a function of their clothing and physical condition of health. These are parameters that cannot be modelled easily.

The low CRE from windows in both ADB and Nightingale Cruciform wards cases confirm previous research in which airborne contaminants where found to travel far under displacement [14] as well as their tendency to get trapped in pockets [14,17], needing more time to be diluted. The concentration of contaminants simulated in these cases were relatively high after travelling up to 8 m (ADB ward) and up 5 m (Nightingale Cruciform ward). The high CRE values predicted for the NPV in both ward types are consistent with the original findings [23] and the expected behaviour of such a personalised ventilation system.

7. Conclusions

The objective of this research was to build on previous work which introduced the NPV system by scaling its application to a multi-bed hospital ward. This required the design of a system for two different kinds of four-bed wards found in the UK. Steady-state CFD was beneficial for predicting and comparing detailed space-based variations in both airflow and contaminant distribution between windows (SNV) and two kinds of NPV systems. The Side–Side and Top–Bottom NPV designs were first modelled using the four-bed ADB ward. Practical design considerations for occupant head room as well as superior airflow characteristics led to the Top–Bottom NPV design being improved by flattening the supply ducts. This flattened duct design was then adopted for studying the four-bed Nightingale Cruciform ward, where the NPV system was able to deliver dedicated fresh air to a patient located up to 10 m away from the external wall. The room air distribution of the NPV system in both types of four-bed wards was a mixing regime, offering better, non-stratified internal temperature as well as improved dilution of passive scalar airborne contaminants. The results support the expected protection from cross-infection, which could be provided to each patient at risk from passive scalar airborne contaminants.

It is however not feasible to use the NPV in its current form during winter in cold climates, unless there is pre-heating of fresh air either within the duct or at point of air discharge (*i.e.*, diffusers). The airflow and mixing/dilution performance of the NPV system therefore, seems a feasible proposition that could improve airflow compared to traditional window-based systems which are restricted by current UK legislation to 100 mm opening. It is also plausible that in other countries where some form of natural ventilation is practicable all year round, the NPV system can be a practical alternative to traditional window-based ventilation of hospital wards.

7.1. Limitations of Study

Exhaled contaminants are often comprised of several droplet sizes of which only a fraction are droplet nuclei that remain airborne. So in its present form, the NPV may not protect against contaminated large droplets or large particulates. This study does not consider the impacts of deposition of contaminated droplets on surfaces and human body. Although posture of coughing patients could also affect exposure to coughed airborne pathogens [44] only the uni-directional emission from a seated visitor was considered. We also acknowledge that modelling the NPV system under a wider temperature range would be beneficial, but this has not been done, given the conceptual scope of this work. Finally, in reality, PV systems would have control mechanisms allowing each user to regulate the flow of air including its speed, direction and possibly temperature, but for the scope of this paper and the exploratory nature of this CFD-based investigation, such control mechanisms were not modelled. Due to limitation of the PHOENICS simulation application, RH could not be measured at specific points inside the ward. Hence, it is unclear at this stage, the extent to which humidity varies as a result of direct discharge of fresh air over the beds through the NPV system. Many of these limitations will be tackled in subsequent work.

7.2. Further Work

Full-scaled experiments are planned for the NPV system in either a real hospital ward or in a mock-up ward. In such experiments, a wider range of outdoor temperature would be used to test whole year performance of the system, but obviously, pre-heating of fresh air would be needed in winter. The airflow performance of NPV ducts should also improve using circular ducts but in the studies carried out so far, software constraints have limited the duct design to rectangular shapes. For future full-scale experiments, the desirability of flattened ducts in Top–Bottom arrangement might rule out circular ducts, however, a series of small radius ducts could be arranged in parallel. This circular duct concept could offer further opportunity for users to control airflow delivery by switching some ducts on or off.

Author Contributions

The first author was responsible for carrying out simulations and putting together the results/analysis, while the second author provided guidance/direction on the research, as well as proof reading and editing of the paper.

Conflicts of Interest

The authors declare no conflict of interest.

References and Notes

1. Department of Health (NHS Estates). *Health Technical Memorandum 07-02: EnCO2de-, Making Energy Work in Healthcare, Environment and Sustainability*; Department of Health Estates and Facilities Division, The Stationary Office: London, UK, 2006.
2. Short, C.A.; Cook, M.; Cropper, P.C.; Al-Maiyah, S. Low-energy refurbishment strategies for health buildings. *J. Build. Perform. Simul.* **2010**, *3*, 197–216.
3. Qian, H.; Li, Y.; Seto, W.H.; Ching, P.; Ching, W.H.; Sun, H.Q. Natural ventilation for reducing airborne infection in hospitals. *Build. Environ.* **2010**, *45*, 559–565.
4. Escombe, A.R.; Oeser, C.C.; Gilman, R.H.; Navincopa, M.; Ticona, E.; Pan, W.; Martínez, C.; Chacaltana, J.; Rodríguez, R.; Moore, D.A.J.; *et al.* Natural ventilation for the prevention of airborne contagion. *PLoS Med.* **2007**, *4*, 309–317.
5. Department of Health (NHS Estates). *Health Technical Memorandum HTM 55 (Building Components) Windows*; The Stationery Office: London, UK, 1998.
6. Department of Health (NHS Estates). *Health Building Note 00-10 Part D: Windows and Associated Hardware*; Department of Health: London, UK, 2013. Available online: http://www.dhsspsni.gov.uk/hbn00-01-partd.pdf (accessed on 12 January 2014).
7. Gilkeson, C.A.; Camargo-Valero, M.A.; Pickin, L.E.; Noakes, C.J. Measurement of ventilation and airborne infection risk in large naturally ventilated hospital wards. *Build. Environ.* **2013**, *65*, 35–48.
8. Giridharan, R.; Lomas, K.J.; Short, C.A.; Fair, A.J. Performance of hospital spaces in summer: A case study of a "Nucleus"-type hospital in the UK Midlands. *Energy Build.* **2013**, *66*, 315–328.
9. Lomas, K.J. Architectural design of an advanced naturally ventilated building form. *Energy Build.* **2007**, *39*, 166–181.
10. Awbi, H.B. *Ventilation of Buildings*, 2nd ed.; Spon Press: London, UK, 2003.
11. Moodie, T.B. Gravity currents. *J. Comput. Appl. Math.* **2002**, *144*, 49–83.
12. Chen, C.; Zhu, J.; Qu, Z.; Lin, C.-H.; Jiang, Z.; Chen, Q. Systematic study of person-to-person contaminant transport in mechanically ventilated spaces. *HVAC&R Res.* **2014**, *20*, 80–91.
13. Gao, N.; He, Q.; Niu, J. Numerical study of the lock-up phenomenon of human exhaled droplets under a displacement ventilated room. *Build. Simul.* **2012**, *5*, 51–60.
14. Li, Y.; Nielsen, P.V.; Sandberg, M. Displacement ventilation in hospital environments. *ASHRAE J.* **2011**, *53*, 86–88.
15. Beggs, C.B.; Kerr, K.G.; Noakes, C.J.; Hathway, E.A.; Sleigh, P.A. The ventilation of multiple-bed hospital wards: Review and analysis. *Am. J. Infect. Control* **2008**, *36*, 250–259.

16. Qian, H.; Li, Y.; Nielsen, P.V.; Hyldgaard, C.E.; Wong, T.W.; Chwang, A.T.Y. Dispersion of exhaled droplet nuclei in a two-bed hospital ward with three different ventilation systems. *Indoor Air* **2006**, *16*, 111–128.
17. Olmedo, I.; Nielsen, P.V.; Ruiz de Adana, M.; Jensen, R.L.; Grzelecki, P. Distribution of exhaled contaminants and personal exposure in a room using three different air distribution strategies. *Indoor Air* **2012**, *22*, 64–76.
18. Melikov, A. Personalized ventilation. *Indoor Air* **2004**, *14*, 157–167.
19. Yang, B.; Sekhar, C.; Melikov, A.K. Ceiling mounted personalized ventilation system in hot and humid climate—An energy analysis. *Energy Build.* **2010**, *42*, 2304–2308.
20. Melikov, A.; Ivanova, T.; Stefanova, G. Seat headrest-incorporated personalized ventilation: Thermal comfort and inhaled air quality. *Build. Environ.* **2012**, *47*, 100–108.
21. Cermak, R.; Melikov, A. Protection of occupants from exhaled infectious agents and floor material emissions in rooms with personalized and underfloor ventilation. *HVAC&R Res.* **2007**, *13*, 23–38.
22. Cermak, R.; Melikov, A.K.; Forejt, L.; Kovar, O. Performance of personalized ventilation in conjunction with mixing and displacement ventilation. *HVAC&R Res.* **2006**, *12*, 295–311.
23. Adamu, Z.A.; Cook, M.J.; Price, A.D.F. Natural personalised ventilation: A novel approach. *Int. J. Ventil.* **2011**, *10*, 263–275.
24. Todd, S.P.; Adamu, Z.A.; Cook, M.J.; Price, A.D.F. Natural Personalised Ventilation for Hospital Wards: Experimental Validation. In Proceedings of the 4th CIBSE ASHRAE Technical Symposium, Dublin, Ireland, 3–4 April 2014.
25. Koestel, A. Paths of horizontally projected heated and chilled air jets. *Trans. ASHAF* **1955**, *61*, 231–232.
26. Li, Y.; Nielsen, P.V. CFD and ventilation research. *Indoor Air* **2011**, *21*, 442–453.
27. Chen, Q. Ventilation performance prediction for buildings: A method overview and recent applications. *Build. Environ.* **2009**, *44*, 848–858.
28. Chen, Q. Using computational tools to factor wind into architectural environment design. *Energy Build.* **2004**, *36*, 1197–1209.
29. CHAM. Phoenics version 2009. Cham Co. Ltd.: Wimbledon, UK, 2014. Available online: http://www.cham.co.uk/ (accessed on 5 July 2013).
30. Yakhot, V.; Orszag, S.A. Renormalization group analysis of turbulence. *J. Sci. Comput.* **1986**, *1*, 3–51.
31. Mundt, M.; Mathisen, H.M.; Moser, M.; Nielsen, P.V. *REHVA Ventilation Effectiveness Guide*; Federation of European Heating, Ventilation and Air-Conditioning Association: Brussels, Belgium, 2004. Available online: http://www.rehva.eu/publications-and-resources/guidebooks-shop/ (accessed on 8 August 2012).
32. Cheong, K.W.D.; Phua, S.Y. Development of ventilation design strategy for effective removal of pollutant in the isolation room of a hospital. *Build. Environ.* **2006**, *41*, 1161–1170.
33. Olesen, B.W. The philosophy behind EN 15251: Indoor environmental criteria for design and calculation of energy performance of buildings. *Energy Build.* **2007**, *39*, 740–749.
34. The Chartered Institution of Building Services Engineers (CIBSE). *Guide A: Environmental Design*; CIBSE: London, UK, 2006.

35. Jiang, Y.; Zhao, B.; Li, X.; Yang, X.; Zhang, Z.; Zhang, Y. Investigating a safe ventilation rate for the prevention of indoor SARS transmission: An attempt based on simulation approach. *Build. Simul.* **2009**, *2*, 281–289.
36. Morawska, L. Droplet fate in indoor environments, or can we prevent the spread of infection? *Indoor Air* **2006**, *16*, 335–347.
37. Cox, C.S.; Wathes, C.M. *Bioaerosols Handbook*; Lewis Publishers: Boca Raton, FL, USA, 1995; p. 17.
38. Gupta, J.; Lin, C.-H.; Chen, Q. Flow dynamics and characterization of a cough. *Indoor Air* **2009**, *19*, 517–525.
39. Des Prez, R.M.; Heim, C.R. Mycobacterium Tuberculosis. In *Principles and Practices of Infectious Diseases*; Mandell, G.L., Douglas, R.G., Jr., Bennett, J.E., Eds.; Churchill Livingston: New York, NY, USA, 1990; pp. 1877–1906.
40. Lomas, K.J.; Ji, Y. Resilience of naturally ventilated buildings to climate change: Advanced natural ventilation and hospital wards. *Energy Build.* **2009**, *41*, 629–653.
41. Paradise, C. Innovative Patient Accommodation. In Proceedings of International Academy for Design and Health 7th World Congress and Exhibition, Boston, MA, USA, 6–10 July 2011. Available online: http://www.designandhealth.com/uploaded/documents/Awards-and-events/WCDH2011/Presentations/Friday/Session-8/CarolineParadise.pdf (accessed on 5 January 2013).
42. *Natural Ventilation for Infection Control in Health-Care Settings*; Atkinson, J., Chartier, Y., Pessoa-Silva, C.L., Jensen, P., Li, Y., Seto, W.-H., Eds.; WHO: Geneva, Switzerland. 2009. Available online: http://www.who.int/water_sanitation_health/publications/natural_ventilation.pdf (accessed on 18 May 2012).
43. Li, Y.; Leung, M.K.H.; Seto, W.H.; Yeun, P.L.; Leung, J.; Kwan, J.K.; Yu, S.C.T. Factors affecting ventilation effectiveness in SARS wards. *Hong Kong Med. J.* **2008**, *14*, 33–36.
44. Kierat, W.; Bolashikov, Z.D.; Melikov, A.K.; Popiołek, Z.; Brand, M. Exposure to Coughed Airborne Pathogens in a Double Bed Hospital Patient Room: Impact of Posture of Coughing Patient and Location of Doctor. In Proceedings of the ASHRAE IAQ Conference, Kuala Lumpur, Malaysia, 10–12 November 2010; paper 128.

6

A Production Model for Construction: A Theoretical Framework

Ricardo Antunes * and Vicente Gonzalez

Department of Civil and Environmental Engineering, University of Auckland, 20 Symonds Street, Auckland 1142, New Zealand; E-Mail: v.gonzalez@auckland.ac.nz

* Author to whom correspondence should be addressed; E-Mail: rsan640@aucklanduni.ac.nz

Academic Editor: George Baird

Abstract: The building construction industry faces challenges, such as increasing project complexity and scope requirements, but shorter deadlines. Additionally, economic uncertainty and rising business competition with a subsequent decrease in profit margins for the industry demands the development of new approaches to construction management. However, the building construction sector relies on practices based on intuition and experience, overlooking the dynamics of its production system. Furthermore, researchers maintain that the construction industry has no history of the application of mathematical approaches to model and manage production. Much work has been carried out on how manufacturing practices apply to construction projects, mostly lean principles. Nevertheless, there has been little research to understand the fundamental mechanisms of production in construction. This study develops an in-depth literature review to examine the existing knowledge about production models and their characteristics in order to establish a foundation for dynamic production systems management in construction. As a result, a theoretical framework is proposed, which will be instrumental in the future development of mathematical production models aimed at predicting the performance and behaviour of dynamic project-based systems in construction.

Keywords: production model; system model; lean construction; process control; building construction; uncertainty; risk management; variability

---

1. Introduction

The construction industry faces ongoing challenges. From an external perspective, economic uncertainty, increased competition within the industry and the growing influence of regulatory agencies drive profit margins down. From an internal viewpoint, the increase in the numbers of features in a project's scope, against shorter construction deadlines and restricted budgets raise the complexity of building construction projects. Additionally, building construction often occupies the bottom of industrial productivity rank reports worldwide [1], even showing negative productivity rates [2]. The construction sector relies on management practices based on intuition, experience and poor risk management. These practices often preclude the appropriate level of ability to handle the uncertainty and complexity [3,4] involved in construction projects, resulting in project failures in terms of finishing projects within deadlines and budgets [5]. As a common practice, the building construction industry utilizes ordinary project management practices and frameworks. Practices are limited to assessing the consequences of deviations from the project plan rather than dealing with the causes of delays in the production system [6]. "Heisenberg's principle indicates that the conventional construction management practice of measuring the overall timeliness and budgetary status (position) of a project using schedule/cost variances will result in a less precise assessment of the throughput (momentum) of the project" [4]. At the tactical management level, little attention has been paid to production management in building and construction [7]. Traditional scheduling approaches in construction, such as the critical path method, have been used unrestrictedly, producing unfinished and erratic plans [4]. Even more, a developed estimation practice, such as the line of balance, fails to deliver a reliable outcome. Instead of being derived from manufacturing practices, the line of balance was purely copied regardless of the inherent production differences between manufacturing and construction projects. Both approaches assume that construction processes are constant with some degree of variability, setting constant estimation values for production and adding an extra measure to accommodate variations. In summary, the dynamics of the production system is overlooked [8]. In fact, the building construction industry has no history of applying mathematical approaches to model and manage production [9]. Mathematical models have enabled a comprehensive understanding of production mechanisms [10]. For instance, manufacturing has a record of successfully developing and using robust mathematical models to improve productivity, system comprehension and event forecasting [11,12]. Nevertheless, the manufacturing model does not directly apply to construction. Steady-state conditions, nonterminating events, long production runs, medium to low levels of uncertainty and permanent production facilities characterize production in manufacturing [12]. Production in construction takes place under a mix of steady- and unsteady-state conditions, with terminating events, short production runs and various levels of uncertainty, all conducted in temporary production facilities [13–15].

The impacts of World War II in Japan restricted market demand and the production model. Mass production changed to the production of a high variety of products in small quantities. Moreover, in order to adapt to the post-war scenario of the financial incapability of acquiring new machinery, the Japanese industry had to work with the existing production equipment in a more efficient manner. The new diversified small-batch production and post-war recession pushed the industry to re-evaluate the use of resources and production methods and resulted in a shift in production paradigms in the manufacturing industry. Consequently, a new production model should be developed to increase productivity while using existing resources. The production philosophy named just-in-time (JIT) together with automatization established the foundations of the Toyota Production System (TPS) in 1945. TPS is a compilation of observations from Taiichi Ohno and Shigeo Shingo and driven by Toyota's goal of "catching up" with the American automotive industry. TPS only received notable exposure during the oil crisis in the autumn of 1973, when the improvements brought by the relentless elimination of waste helped Toyota through the Japanese economic crisis [16]. In an attempt to reproduce the success achieved by Toyota, the automotive industry adopted TPS as a production model and then forged a new manufacturing philosophy, later entitled "lean manufacturing" [17]. The lean principles—waste eradication, flow, value, value-stream and zero-defect—crossed the boundaries of manufacturing. These principles permeated several productive systems and industries, such as software development [18], even reaching business models [19]. At the same time, manufacturing was formulating mathematical explanations of production management. Studies about the variables of the productive process—a process is a repetitive collection of interrelated tasks aimed at achieving a specific goal—provided a quantitative meaning to production management practices in manufacturing. Equations could now represent queueing systems [10], lead time [20], machine maintenance [21], Kanban [22], inventory management [23], throughput, cycle time and work in progress [11]. The set of equations enabled a deeper understanding of the current practices and the outcomes of their interaction.

The building construction industry has also demonstrated interest in lean principles [7]; later named lean construction [24]. The transformation, flow and value generation views associated with a set of principles constitute the transformation-flow-value (TFV) theory of production. However, the implementation of lean principles has been sporadic, rather than an industry standard [25]. Several contradictions to the lean values, such as excessive consumption of raw material [26], disconnection of activities, preventing the establishment of a flow [7], a focus on costs over value [27], an inefficient measurement system [28], high levels of rework due to production errors or changes in technical specifications [29] and worker safety [30], still populate the list of the most common problems in construction endeavours. Inefficiency at the tactical level is only one barrier to the adoption of manufacturing principles. Intrinsic project characteristics also increase the challenges to the industry. Uniqueness—each project produces a distinct product, service or result with individual features of the project work [31]—temporality—"indicates that a project has a definite beginning and end" [32]—and uncertainty [32–34] are some attributes that set apart projects and manufacturing. As a matter of fact, even repetitive construction projects—projects characterized by repetitive activities, as, for instance, pipelines, tunnels, mass housing projects and high-rise buildings [35,36]—are exposed to the effects of uniqueness, temporality and uncertainty. Despite an insufficient pace to "catching up" with market demand, construction has progressed towards the incorporation of TPS practices [7] and an understanding of its production.

Construction has successfully comprehended the effects of variability in production [37] and the influence of buffers on sequential activities [38]. Researchers consistently advance the design and management of work-in-progress buffers. Additionally, approximate models can be obtained by approaches, such as discrete simulation [39], fuzzy logic [40], neural networks [41], statistics [42] and control theory [43]. However, the results are still theoretical and restrictively applied [15,44,45]. Moreover, obtaining mathematical expressions that quantitatively explain the production mechanisms of different repetitive construction projects has been a challenge [15,46] and is still to be formulated.

Much work has been carried out on how manufacturing practices apply to construction projects, mostly lean principles. Nevertheless, there has been little research to understand the fundamental mechanisms of production in construction. The current knowledge is at a rudimentary stage and has not yet fully established the foundations of the production model in construction projects. This research entails an in-depth literature review to examine the existing knowledge about production models and their characteristics in order to establish a foundation for the production model in construction. This literature review does not intend to pursue lean construction or its production models. This study searches outside the construction scenario for best practices in production and then inside for the barriers to the implementation of these practices. The study retraces production characteristics from manufacturing that are applicable to building construction projects in order to obtain a similar system view that propitiates the development of manufacturing production laws, such as variability, conservation of material and capacity laws [11]. Furthermore, the concepts are reviewed from a microeconomic/strategic/tactical vantage point required for project management development, regardless of the project technical specificities. As a result, a theoretical framework, which may serve as a basis for the further development of a mathematical production model aimed at predicting the performance and behaviour of dynamic project-based production systems in construction is proposed.

2. A Theoretical Framework

2.1. A Brief Understanding of Construction

The understanding of what is construction extrapolates its technical characteristics. In this research, construction is not restricted to civil engineering and architecture, but comprehends a broader understanding of building, putting up, setting up, establishing and assembling. Construction is the materialization of a concept through design, taking into account functional requirements and technical specifications for a project product utilizing specialized labour. In other words, it is the creation of a product that will fulfil a strategic goal. This study excludes prototyping as the final product; however, a project scope may include prototyping as a deliverable or stage, as, for example, in the design phase. Prototype objectives test a process or concept in real situations and then provide information about what works and what needs improvement before the final product. In contrast to prototypes, the project product is final and definitive. Project products, according to the design specifications, have usability in a first and unique building. Project products of building construction are, for instance, software, pipelines, roads, bridges, tunnels, house building and oil well construction. In summary, project deliverables with different technical backlogs and fields of application are in accordance with the definition of construction stated in this study. Moreover, the existence of a technical backlog brings a dimension of repetitiveness. Because these projects require specialized labour, the project executive is likely to be involved in projects with a similar technical background. For instance, an offshore oil well construction company is more likely to have future endeavours in the same technical field rather than in a road building. In this case, the repetitiveness occurs horizontally across different projects with a similar scope. A project can also have repetitiveness within an endeavour: for instance, the construction of similar products, such as housing units, floors of a skyscraper or the installation of cases in an oil well. These processes illustrate a vertical dimension of repetitiveness, *i.e.*, process repetition in a unique project.

In order to understand how construction applies to different fields and where construction is placed in project life cycles, ten different project management frameworks were analysed. Figure 1 displays the organization of several project-driven approaches in four clustered stages grouped by fundamental characteristics. The initial stage (feasibility) consists of a group of processes aimed at identifying the best possible solution to achieve the strategic business goal intended by the project sponsor. Additionally, these processes should determine if the project product corresponding to the selected solution can actually be built or created (several applicable practices can be found in [47]). The second stage (design) presents a group of processes where the project and technical teams examine the outcome of the feasibility stage. The teams produce design specifications and project plans that contain the information about how to make a suitable project product based on the best possible solution. The next stage (construction) is to build the product following the technical instructions and executing the management plans from the previous stage. The final step (operation) is delivering the finished product to the user, who will use the project product to generate the benefits and consequently achieve the benefits of the business goal [34]. Operation is not considered a project stage, although the beginning of this stage commonly determines the termination of the project life cycle [31].

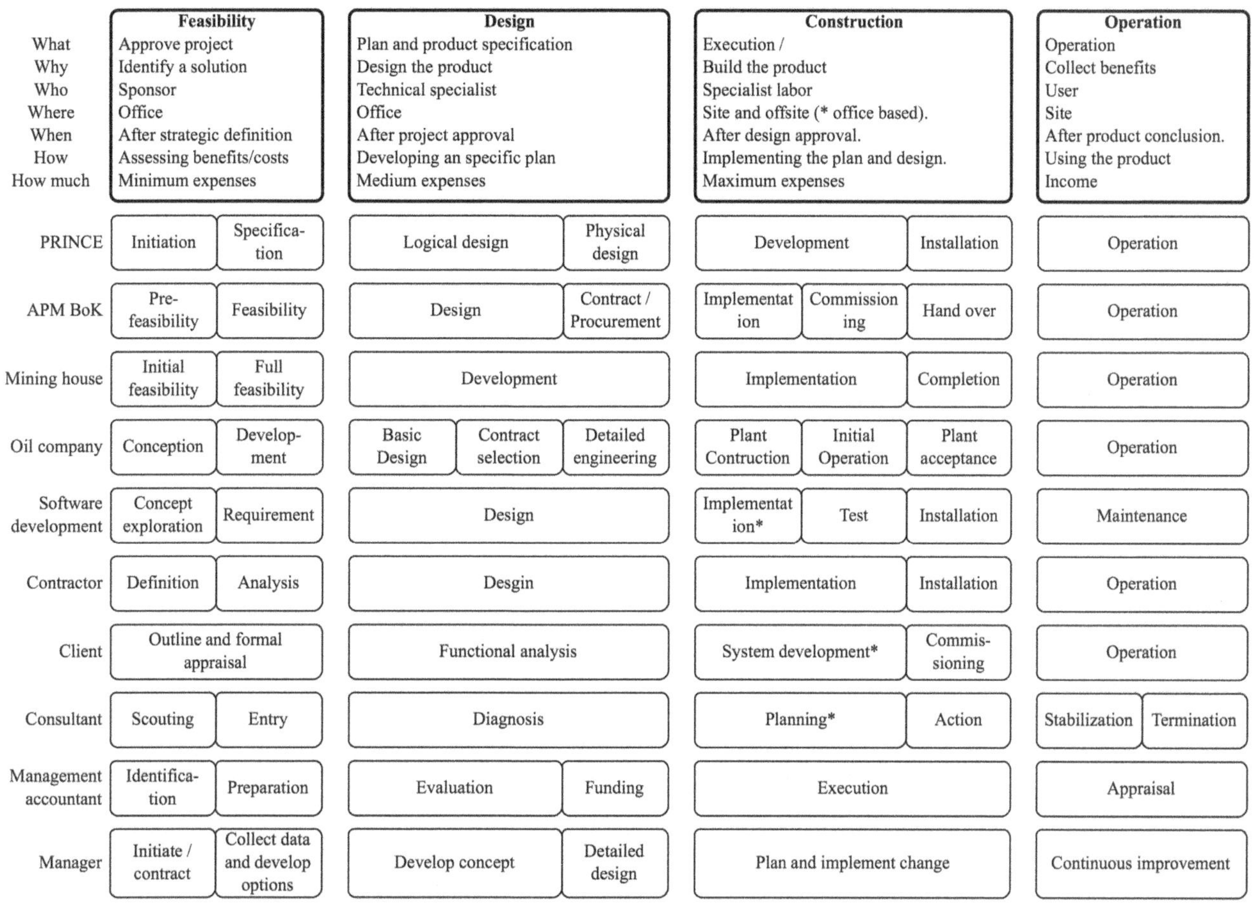

Figure 1. Characteristics of project stages (projects stages of various approaches adapted from Smith, Merna and Jobling [48], originally from McGetric, The Project Life Cycle, 1996). For additional information about the Project in Controlled Environment (PRINCE) and Association for Project Management Book of Knowledge (APM BoK), refer to [34] and [49], respectively.

2.2. Production Facilities

The project takes place in different locations through the three project stages (feasibility, design construction) and is frequently related to where the main phase stakeholders are located. First, it originates from strategic demand, and feasibility studies take place where the sponsor is located. At this stage, the management has recognized that normal operation is unable to address a particular business requirement [50]. Consequently, the management sponsors a unique endeavour to address this outstanding business requirement, initiating the project. Once initiated, it is up to the sponsor to choose the best solution and investment to solve the issue. Altogether, the feasibility stage is likely to occur as an internal business routine.

Project and product design require strong technical knowledge, which is usually not the sponsor's forte, since the initial motive for the project's existence is the inability of routine operations to accomplish a specific business requirement [50]. Hence, it may be necessary for a specialist third party to join the project and develop the technical features, collect all requirements, and plan and design the project product to execution. For this reason, the design stage tends to move from the project sponsor's office to third party offices with regular interaction between stakeholders. Thus, the influence of stakeholders is less than at the previous stage [51].

Similar reasoning about technical development is valid for the construction stage, where each task of the construction plan often requires either a specialist in the trade or a contractor. For this reason, work packages are broken down and distributed to several independent third parties. Frequently, these third parties are only connected to each other by the current project, without a prior collaboration background between the companies or teams involved. The location, where the final project product takes place, is undoubtedly a strategic decision. Nevertheless, the work packages of the project may be produced in different facilities, on-site or off-site. The work packages are spread and distributed to suppliers according to the project procurement management plan [25] and are, in general, produced in facilities subject to the supplier's business model, rather than the project owner's choice. Again, the influence of stakeholders is diminished [51]. The production facility is distinctively important, because this stage gathers most of the resources and labour needed for building the project product as designed and planned. This resource convergence is responsible for the majority of expenditures in the project [25,31], and it reveals any faults and inaccuracies of the plan developed in previous project stages.

Clients perceive several benefits of off-site production over on-site [52]. Benefits, such as production time, quality and safety, are often mentioned. Interestingly, the most important benefit of off-site production is not being executed on-site. In the rank of benefits, according to importance and likelihood, the top three answers refer directly or indirectly to the reduction of activity on-site. The benefits from the highest to lowest rating are: minimize on-site operation; reduce congested work areas; and multi-trade interfaces to minimize on-site duration [52,53]. Construction's on-site production implies challenging logistics. The location of a construction site solely depends on the final user needs, which are different from manufacturing, where the choice of factory location fulfils strategic logistic goals of product distribution and supplier connection. From a logistic perspective, "the delivery is to a temporary location, without permanent facilities for handling material" [14]. The provisional storehouse complicates the employment of optimal delivery routes and connections to preferred suppliers, further hindering the formation of an efficient supply chain. Location may severely interfere with the supply chain. Depending on the construction site location, several nontechnical restrictions may apply. For instance, building a skyscraper in the city centre may require obeying municipal restrictions. Timetables and regulations for moving load, material delivery and dispatch and also additional safety procedures due to population proximity [54] may be necessary. On the opposite side, construction in isolated locations endures materials transportation, because of the long distances from suppliers and improper or even inexistent transportation routes [55]. A long distance from preferred suppliers often raises costs to a level where the chain becomes unfeasible, forcing executors to work with alternatives sources. New parties are most likely to experience a slower project learning curve than the usual partners, because previous involvements support the creation of both product and process know-how via lessons learned.

For the building construction industry, the supply chain is critical due to the excessive amount of raw materials required for product assembly [56]. Expressly in the construction phase, the raw materials need to arrive on-site in the correct amount and sequence, since there is a minimal warehouse installed on-site, which is incapable of handling massive amounts of material [57]. Arriving and queueing coordination is essential to not overload the warehouse nor starve the production line. Material surplus may impact how material flows to the production line, because overstocks may delay picking activities, further reducing the necessary space for cargo handling inside the warehouse. Starving the production line is more critical, since it may switch off the production line or part of it.

2.3. Production Run

By definition, the short run is the window frame when steady factors drive production. Moreover, the addition of capital is not possible, so increased output requires more labour [58]. Scope, budget and schedule constraints are core characteristics of the production of projects [31], *i.e.*, steady factors that restrict the amount of work, the addition of capital and the project execution time frame. On a short production run, as in projects, the deadline, budget and total number of units to produce (scope) are clear from the beginning. Although clearly and explicitly stated in the project scope, the estimations of these constraints are, unfortunately, not always realistic. In contrast, a long production run is unclear about both the production end and total units to produce, thus relying on forecasts and market conditions [59]. For instance, upstream oil production rig expects, based on the oil prospect, to extract a set number of oil barrels in a set time. This notwithstanding, the extraction from the reservoir will continue until the extinction of oil or until its operation becomes unfeasible.

Consequently, the average cost (cost/units) [59] for a short production run is more accurate than for the long run, because the units to be produced can be considered constant. Initially, the long run (forecasted units) has a higher potential risk than the short run (exact units) for the same total cost. However, the long run may produce more units than the forecasted number, creating a more favourable cost/units relation [59]. Moreover, the short-run production aggregates higher costs per unit than long-run production, because the number of units to produce is much smaller than on that for long-run production. This value concentration makes short-run production more sensitive to risk in the case of unit variation. In other words, the expected value is higher because the risk impact is higher for the same risk probability (expected value = risk probability × impact) [60]. In summary, short-run production has a lower potential risk, but requires more accuracy, because value concentrates in the production units. Consequently, small variations in labour may affect production and the overall cost.

2.4. Event Termination

Projects by definition are finite terminating endeavours [31,34]. There are some divergences in the literature [31,33,34] about when exactly a project starts, although the termination is clear. The handover to the user sets the final milestone of product construction, marking the project's end. Any activities later on relate to project documentation and formal closure. Considering every stage, deliverable or activity startup and ending, the system (in this study, the project) receives several sequential on and off switches. These interruptions cause work fragmentation [42]. This fragmentation implies numerous transient states, *i.e.*, a timeframe in which the system is either adapting to a state change or responding to a disturbance, e.g., after the impact of a risk. Such variation may affect the state condition of production, changing the production state from a steady- to unsteady-state condition.

2.5. State Condition

Projects involve a unique state change [61]. During project execution, the amount of work is variable and concentrated on the "carrying out the work" stage [31]. Consequently, initial and final project stages are low-speed regions. The project velocity [31,62,63] increases from zero, at project start, accelerating to its maximum at the execution stage and then slowing down to a full stop at project end. However, the velocity variation is not the determinant of the state condition of the system. The variation, or not, over time of the relation of different properties that generate the system output from the input set the state condition of that system. Therefore, the system is in a steady-state when several properties are invariant over time [64]. Once these properties are invariant, the steady-state condition gives a means to the development of simple mathematical equations. For instance, in the steady-state, average work output is strictly less than average capacity, as claimed by the law (capacity) in manufacturing [11]. The law (capacity) does not directly apply to projects. Projects have a distinctive outcome, involving different teams with different technical knowledge and learning curves [65]. Regarding the production system of the project, these properties alter the output/input relation over time, which is the main characteristic of unsteady-state systems [15]. As a consequence, projects assume the unsteady-state condition. However, parts of the project—stages, deliverables, activities or processes—might assume a steady- or unsteady-state depending on process repetitiveness [13], variability and the conservation of process properties over time.

2.6. Complexity

Vastly present in the literature, the word "complex" seems to stand for a supernatural force supposedly responsible for disturbances, a scary ghost haunting projects. With no absolute definition of what complexity means, the only consensus among researchers is that there is no agreement about the specific definition of complexity [66]. However, a characterization of what is complex is possible. A structure is complex; if composed of several interconnected pieces [67], with dynamic networks of interactions, and their relationships are not aggregations of the individual static entities [68]. A theory derived from chaos theory follows a similar principle: "The complexity theory states that critically interacting components self-organize to form potentially evolving structures exhibiting a hierarchy of emergent system properties" [69]. Both statements point to the interaction between the internal structures as complexity characteristics.

Projects are complex. Projects involve several interrelated stakeholders, deliverables [70] and activities [71] in order to deliver a strategic state change [61,72]. Given that, projects, in a broad definition, show the characteristics of what is complex. Methods to assess, classify and measure project complexity have been discussed in the literature [69,73–75]. Most methods are restricted to computations involving scheduling and activity relationships [73,76,77]. Although these models provide a relative comparison of the complexity of similar projects in the same industry, they fail on a broader scale. The analytical hierarchy process (AHP) [78] provides a more inclusive and simple approach to complexity evaluation. AHP can assess a set of pre-chosen project criteria of a variety of projects, providing a calculated numerical grade from zero to one for complexity comparison [74]. Despite the fact that it is possible to use AHP to congregate projects into clusters, such a study does not exist. As a consequence, no explicit information comparing the complexity of projects in building construction with other industries could be traced.

2.7. Uncertainty

Similarly to complexity, there is no consensus about a pure definition of uncertainty. Uncertainty is an unintelligible expression without a straightforward description. Keynes was the first economist to incorporate uncertainty as a theory pillar, adding a factor of distrust to economic science. Keynes' studies explored how to be rational in an uncertain world, in what degree and how to specify uncertainty (refer to Mello [79] about Keynes' life and work). Later, Keynes studied the nature and effect of uncertainty on statistics. "Human decisions affecting the future, whether personal or political or economic, cannot depend on strict mathematical expectation since the basis for making such calculations does not exist" [80]. In the final analysis, uncertainty is beyond any prediction, forecast, calculation or measurement notably when considering human behaviour [49]. The main reason for the existence of a project is to achieve the value perceived by the stakeholders [33,34,49,61,72]. Nevertheless, value is relative [65]. Therefore, internal and external, direct and indirect stakeholders may have a different perception of the achieved project business outcome depending on their vantage points and individual backgrounds [17,65,81]. That, according to Keynes, is a fertile soil for uncertainty. Inevitably, projects are conducted in an uncertain scenario.

2.8. Risk

The viewpoint of uncertainty and risk vary from different disciplines [82]. However, uncertainty is not risk [83]. While uncertainty is a potential, unpredictable, unmeasurable and uncontrollable outcome [83,84], risk is a consequence of action taken in spite of uncertainty [85]. As an illustration, two men are about to skydive from an airplane, but there is only one parachute on-board. At this point, the level of uncertainty experienced by the men is equal, as to whether or not the parachute will open. The moment that one man chooses jumping, the uncertainty is the same to both, and he bears all the risk. The uncertainty ceases when the parachute opens during the skydiving, through time passing, event and action. Nevertheless, there is still risk regarding a safe landing even after uncertainty about the parachute opening has come to an end [85]. Similarly, in projects, the decrease of uncertainty and risk through time passing applies [86].

Projects are risky. The action of conducting an endeavour upon uncertainty is the intrinsic nature of risk [83,84]. Hence, dealing with risk is crucial in project management to achieve successful business results [31]. Controlling uncertainty is impossible [80,83]. However, it is possible to identify, measure, and formulate plans to transfer, provoke, mitigate, avoid or embrace the impact of risk [84], *i.e.*, manager risk. Risk management's objective is to assure uncertainty does not deviate the project from the business goals [84]. In detail, risk managers triumph by maximizing opportunities, minimizing threads and hedging against the risk of a contingent, uncertain loss [87]. The primary manoeuvring instrument against uncertainty is the contingency plan [88], whose determination counts on the summation of expected value [60]. The contingency plan consists of the probability of risk occurring multiplied by the financial impact if it occurs [60] plus an amount judged sufficient to handle unforeseen risks. In addition to the contingency plan, calculating how the contingency could spread over the project lifespan is paramount, especially in projects with a high concentration of events, such as construction projects. Value-at-risk (VaR) "describes the quantile of the projected distribution of gains and losses over the target horizon" [87]. In other words, VaR forecasts the expected value distributed across the project life, so the expected value can be used as a buffer to possible variations, positive or negative, in the project plan.

2.9. Variability

Variability is a measurement of variation, *i.e.*, the difference between results. Variability expresses a dataset that is clustered [89]. Risk impact causes variability, as more risk motivates a broader span of "potential scenarios of outcomes" [85]. The impacts of variability in construction production processes are well known and proven as a decrease in productivity [37], scheduled delays and cost overrun [38]. Considering short runs, the manipulation of the production output requires the administration of labour as an input. Hence, the understanding of the relation between labour and productivity potentially enables the control of production. However, correlating variability to labour is challenging. In statistics, the productivity of labour in building construction does not fit a normal distribution; provided that the variance is undefined, the curvature is leptokurtic [90]. Leptokurtosis may influence the productivity of labour analysis, while estimations made using a normal distribution overestimate at low levels of significance and underestimate at high levels of significance. The wider end of a leptokurtic distribution means risk is coming from outlier events, and extreme observations are likely to occur.

Projects are extremely exposed to variability caused by risk. On the one hand, the short-run production nature of projects requires the manipulation of labour input to control the output [91]. Labour is a prominent human factor in construction projects [42] and a primary source of uncertainty [80], which translates into an increase of risk probability. On other hand, the short-run production concentration of the average cost contributes to an increasing of risk impact. Consequently, both factors—probability and impact—of the risk expected value are affected. In conclusion, at the occurrence of a risk event, a higher expected risk value implies a higher difference between the plan and the actual result, *i.e.*, higher variability.

2.10. Cone of Uncertainty

Uncertainty not only decreases over time passing, but it also diminishes its impact by risk management, specifically by decision-making. In other words, the impact of uncertainty strongly depends on the decision of the risks taken. Researchers' findings indicate that the project estimation builds on foreseen quantities of uncertain events distributed at several stages, resulting in the process of resolving decisions [92]. Figure 2 shows a decrease in variability and the consequent increase in the accuracy of estimates over project progression. Still, the cone does not form itself naturally. It is a product of various decisions made as the project progresses concerning the plan and product. Narrowing the cone implies removing sources of variability sooner. On the contrary, a wider cone results from later decisions [92]. Figure 3 presents a parallel analysis illustrating the impact of decisions made in early and late stages on possible cost reduction in construction projects. Hence, to protect subsequent events from variability and its effects, decisions should be made in early stages to narrow the range of possible later events that will impact the project and cause variability.

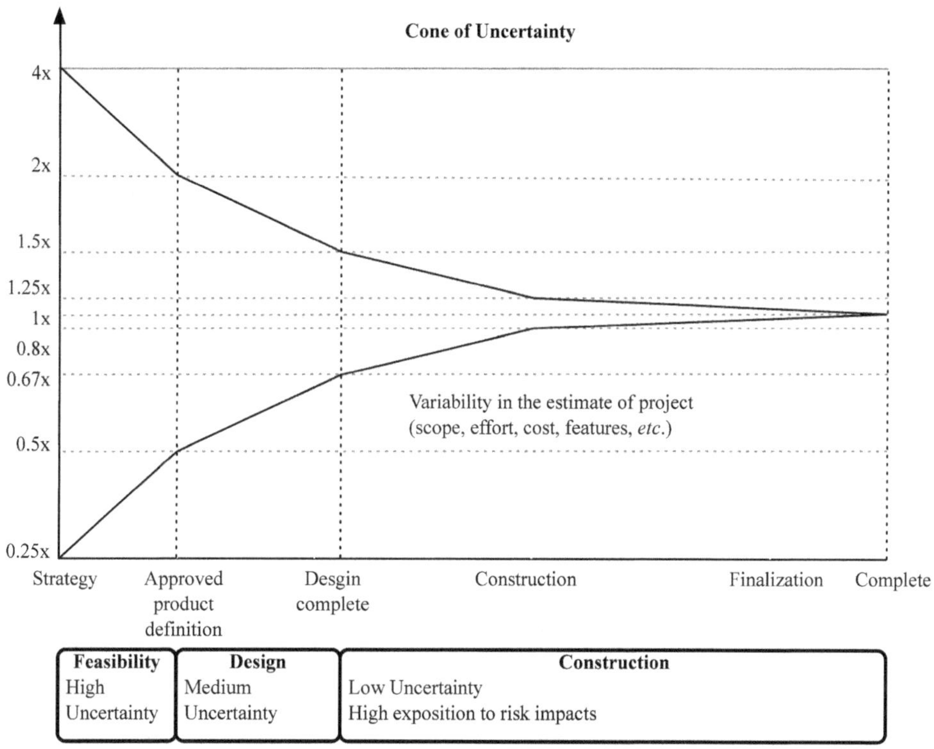

Figure 2. The cone of uncertainty for sequential project development (adapted from McConnell [92]).

Experience from past projects supports decision-making. Commonly called "lessons learned" by project literature and practitioners, the term defines the practice of learning from successful and unsuccessful past events. Insights from assessing captured outstanding project events in post-project reviews can dramatically narrow the cone of future projects. Despite having a remarkable potential to reduce variability in a project, in practice the "lessons learned" process frequently ends with capturing the outstanding events [91], dismissing further steps. The likelihood of reusing lessons learned depends on their applicability to future projects and the similarity between projects, in other words, project receptiveness.

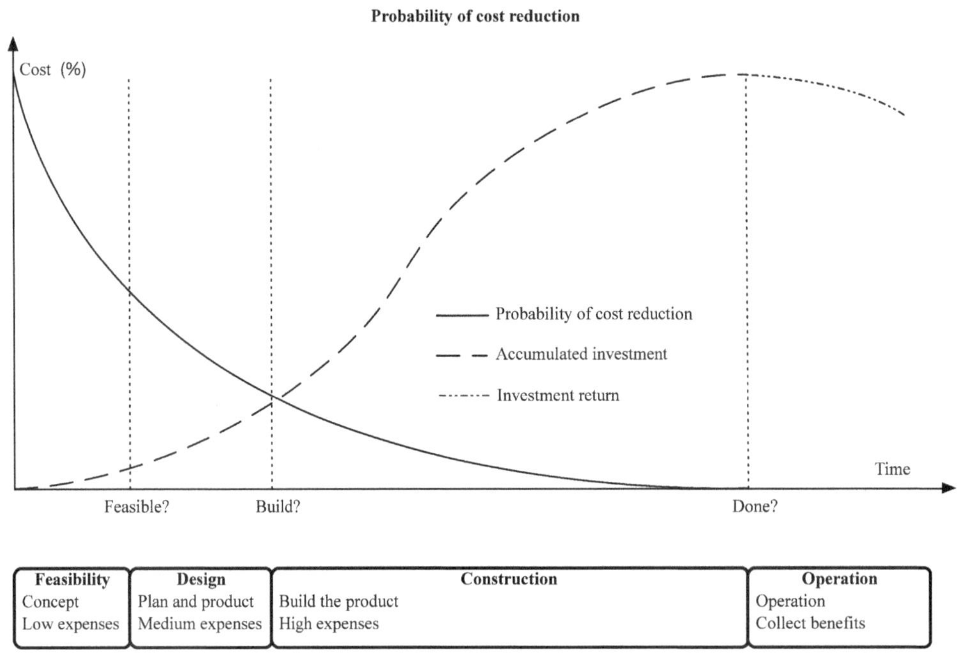

Figure 3. Graph of percentage cost over time demonstrating how early correct decisions may impact cost reduction (adapted from Smith *et al.*, [50]).

2.11. Repetitiveness

"Repetitive manufacturing is the production of discrete units in a high volume concentration of available capacity using fixed routings. Products may be standard or be assembled from standard modules. Production management is usually based on the production rate" [93]. Construction produces multiple products in low volume coupled with the process structured in the midway of the batch, matching the major stages of product and process life cycles from Spencer [93]. A construction production run is short and easily countable considering the final project product. Nonetheless, the project product or a deliverable may be broken into several units, creating an artificial repeatability. For instance, the installation of the windows on the World Trade Center One (or Freedom Tower) is unique and not repeatable. However, this deliverable can be subdivided into floors. Therefore, instead of one unique product, there are 94 repetitions of the same product of "windows installed by floor". The deliverable can be subdivided even more. If each of the 21,800 windows to install is considered, the window installation process can be interpreted as very repetitive [94].

Repetitiveness provides the opportunity for greater use of lessons learned, inevitably proportionating improvement in processes and products. A high level of repetitiveness may propitiate continuous enhancements due to the product likeness, ease of measurement and comparison between construction cycles [35,93]. Formed by four basic sequential activities, plan-do-check-act (PDCA)—also known as Shewhart's cycle—founded an unceasing circle of quality enhancement [31]. Likewise, Six Sigma, a set of techniques and tools for process improvement, aims for defect riddance and variability minimization [95]. At the sixth sigma level, processes in a state of statistical control expect no more than 3.4 defects per million opportunities. Additionally, the short-term process capability index (Cpk) is equal to two. These values typify a stable process almost free of variability. Consequently, processes at the sixth sigma level are steady state [96]. Measurement of Six Sigma levels can provide a useful form to qualify the state condition of the process. Higher Six Sigma levels correlate with a process at steady state (low variability), and lower six sigma levels correlate with the unsteady state (high variability). The levels of Six Sigma can be accurately calculated [96], as can the process variability. It is possible to establish a mathematical correlation between these variables, quantifying the state condition of a process through its variability. In summary, repetitiveness creates the opportunity for improvement based on the use of lessons learned, PDCA or Six Sigma and variability reduction through decreasing uncertainty and narrowing the cone of uncertainty earlier. In terms of system description, repetitiveness creates a feedback connection with activities or processes [11]. Furthermore, repetitiveness defines a closed loop system where deviations of past events may be used as input to correct and control [11] current and future system output.

3. Construction Project System

Several elements found in this literature review connect the characteristics of construction projects to the characteristics of a dynamic system. As shown in Figure 4, the interconnectivity is explicit between project stages, in the event that subsequent phases rely on the accomplishment and performance of previous ones. This dependent connection remains valid for divided *n*-substages or *n*-activities and also applies to the proposed framework. The dependence of processes and/or activities is well documented in the literature and well known by practitioners. An activity or stage may impair or favour a successive action depending on the level of correlation and dependence. The interdependence of activities forms a conduit to the propagation of unsure events. Potential risks captured through the entire project life may impact project execution whenever not properly treated, resulting in project deviations. This sequence of events is represented in the system by the flow of uncertainty to risk and the occurrence of risk events, through risk management filtering actions—avoidance, acceptance, sharing, transference, mitigation, motivation—and, finally, to variability. This flow resembles an intrinsic characteristic of systems in the presence of disturbance or noise. Control systems may transmit unfiltered noise across connections affecting vulnerable components and causing disturbances or unpredicted behaviour. Although the level of influence in this flow of sequential, parallel or overlapping relationships in the process or activity network have not been investigated at this point, understanding how risk transforms into variability, and especially how variability affects networked activities, propitiates an opportunity to develop methods aimed at avoiding and mitigating (filtering) the propagation of risk (noise). Regarding risk materialization in variability, different outcomes build on how concentrated or distributed the risk impact was. This scenario requires a function capable of scale variation and energy conservation (impact) when calculating the functional energy. The wavelet network evolved from the Fourier transformation: "wavelet network is a type of building block for approximation of unknown functions based on the concept of the multi-resolution approximation. The building block is formed by shifting and dilating the basis functions, the mother wavelet and father wavelet" [97]. A wavelet network may be used as universal function approximator ("a universal function approximator is a system that, given a set of predictor variables, can output an accurate estimate of some predicted variable" [97]) to estimate unknown nonlinear functions and to attain a required control performance. A new concept in the control area, wavelets have been successfully used in several applications, such as physics, signal processing and statistics, where small complicated details matter [98]. Operating on possibly the same conditions of wave theory—

linear/nonlinear, deterministic/stochastic, time-domain/frequency domain, direct/inverse problems, discrete/continuous models [99]—control theory may create a proxy theory to explain the effects of variability in construction projects by extending the elements of the dynamic systems.

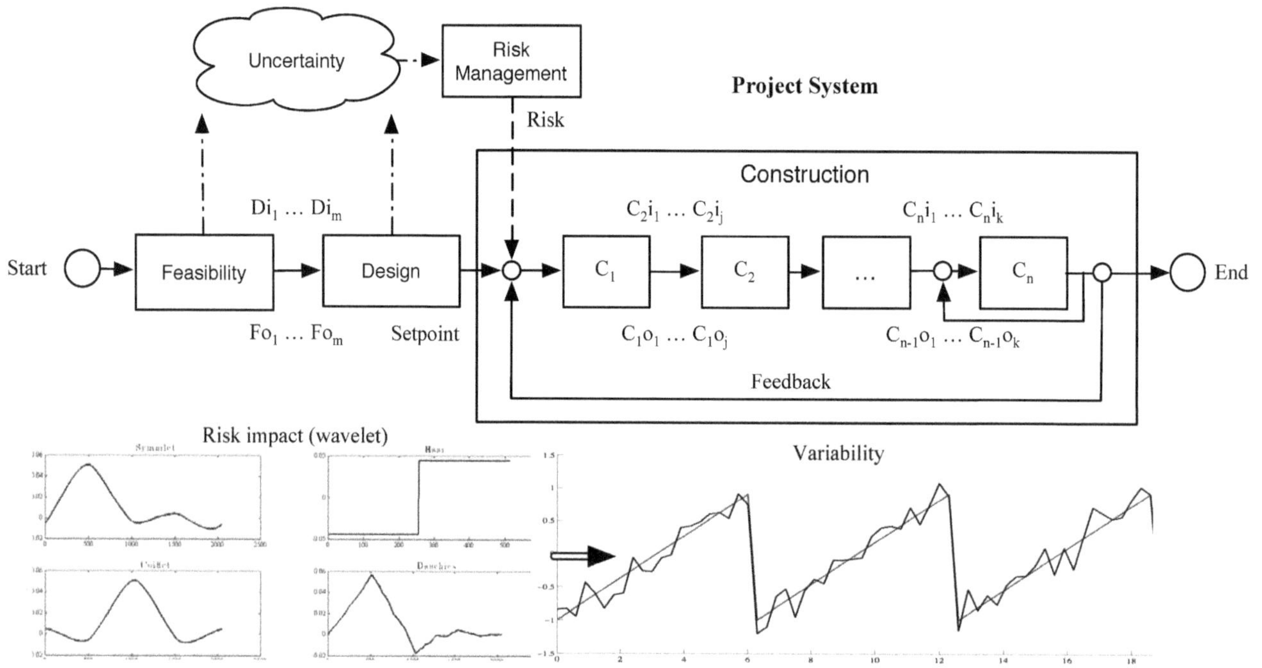

Figure 4. The project construction system from the theoretical framework.

The opportunity for improvement created by process repetitiveness reveals a similarity between the resulting theoretical framework and the concept of the control loop, or feedback loop, which controls the dynamic behaviour of the system. The block diagram shows the flow of signals across a system. The block diagram is common to control systems, where a system output may determine an input for the following system. Alternatively, a posterior system can affect the output of previous systems in the chain [100] as in the case of a closed-loop system. Possible categories of inputs on a construction system are tools, equipment, labour, management, time, conditions, product, e.g., a meter of the floor installed, or material, e.g., concrete mix, which may constitute the system output [13]. As an illustration, a manager in charge takes corrective action when an activity deviates from tolerances defined in the plan, *i.e.*, management by exception [34]. The control loop in the process control works in an identical manner. Once a set-point (equivalent to project plan) is defined, three tasks occur: measurement (how the activity is going), comparison (whether the activity is performing within the plan limits) and adjustment (corrective actions). If other feedback or an inner controller are added to the activity C_n (characterized by the interaction of the operational manager, for example), the system will be structured by two closely linked processes that encapsulate feedback, which constitutes a cascade-control system.

4. Conclusions

This literature review conducted an exploration of elements in the building construction project cycle, which possibly influence the project performance, mostly affecting production behaviour. Furthermore, connections and comparisons were made between projects and manufacturing, aimed at identifying key components that could serve as pivotal to the application of successfully proven theories and methods from manufacturing in construction projects. The outcome is a theoretical framework structured as a system. The system connects project stages—feasibility, design and construction—and activities (C_m), as well as proposes a flow of uncertainty to risk and the risk impact causing variability. This system structure representing the production provides a means to analyse the effects of interconnected elements in a micro and macro view. Although this literature review was comprehensive, the list of unmentioned factors that affect production in construction is still endless. Further research needs to be conducted to identify and describe the effect of construction project particularities on the construction production system. Additionally, further research needs to be conducted to develop a mathematical explanation of the production system that will supply the building construction industry with ways to measure and predict the performance of a production system. System identification approaches hereafter can be used to understand, model and predict tasks, processes and project performance within the proposed framework. Furthermore, the quantification of variability across the building construction system will enable a comprehensive understanding of the behaviour of repetitive processes in projects at the unsteady and steady state, applicable to any current production model that considers a system view. Moreover, in the future, the development of management practices to handle processes at both state conditions will take place.

Author Contributions

The study was part of a thesis undertaken by the author for correspondence, Ricardo Antunes. It was supervised by Vicente Gonzalez (principal supervisor).

Conflicts of Interest

The authors declare no conflict of interest.

References

1. Levy, S.M. *Project Management in Construction*; McGraw-Hill: New York, NY, USA, 2012; p 478.
2. United States Department of Labor. Bureau of Labor Statistics. Available online: http://www.bls.gov/ (accessed on 24 September 2014).
3. McCray, G.E.; Purvis, R.L. Project management under uncertainty: The impact of heuristics and biases. *Proj. Manag. J.* **2002**, *33*, 49–57.
4. Abdelhamid, T.S. The Self-Destruction and Renewal of Lean Construction Theory: A Prediction From Boyd-S Theory. In Proceedings of the 12th Annual Conference of the International Group for Lean Construction, Helsingør, Denmark, 3–5 August 2004.
5. Mills, A. A systematic approach to risk management for construction. *Struct. Surv.* **2001**, *19*, 245–252.
6. Lutz, J.D.; Hijazi, A. Planning repetitive construction: Current practice. *Constr. Manag. Econ.* **1993**, *11*, 99–110.
7. Koskela, L. *Application of the New Production Philosophy to Construction*; Stanford University: Stanford, CA, USA, 1992; pp. 1–81.

8. Bertelsen, S. Construction as a Complex System. In Proceedings of the 11th Annual Conference of the International Group for Lean Construction, Blacksburg, VA, USA, 21–24 July 2003.
9. Laufer, A. *Simultaneous Management: Managing Projects in a Dynamic Environment*; American Management Association: New York, NY, USA, 1997; p. 313.
10. Duenyas, I.; Hopp, W.J. Estimating variance of output from cyclic exponential queueing systems. *Queueing Syst.* **1990**, *7*, 337–353.
11. Hopp, W.J.; Spearman, M.L. *Factory Physics: Foundations of Manufacturing Management*, 2nd ed.; Irwin McGraw-Hill: New York, NY, USA, 1996; p. 726.
12. Kumar, S.A.; Suresh, N. *Operation Management*; New Age International: New Delhi, India, 2009.
13. Bernold, L.E. Simulation of nonsteady construction processes. *J. Constr. Eng. Manag.* **1989**, *115*, 163–178.
14. Koskela, L. *An Exploration towards a Production Theory and Its Application to Construction*; VTT Technical Research Centre of Finland: Espoo, Finland, 2000.
15. Walsh, K.D.; Sawhney, A.; Bashford, H.H. Production equations for unsteady-state construction processes. *J. Constr. Eng. Manag.* **2007**, *133*, 245–261.
16. Ohno, T. *O sistema toyota de produção além da produção em larga escala*; Bookman: Sao Paolo, Brazil, 1988; p. 139.
17. Womack, J.P.; Jones, D.T.; Roos, D. *The Machine that Changed the World*; Harper Perennial: New York, NY, USA, 1991; p. 323.
18. Poppendieck, M.; Poppendieck, T. *Lean Software Development—An Agile Toolkit*; Addison-Wesley: New York, NY, USA, 2003.
19. Ries, E. *The Lean Startup: How Today's Entrepreneurs Use Continuous Innovation to Create Radically Successful Businesses*; Crown Business: New York, NY, USA, 2011.
20. Hopp, W.J.; Spearman, M.L.; Woodruff, D.L. Practical strategies for lead time reduction. *Manuf. Rev.* **1990**, *3*, 1–15.
21. Hopp, W.J.; Wu, S.-C. Machine maintenance with multiple maintenance actions. *IIE Trans.* **1990**, *22*, 226–233.
22. Spearman, M.L.; Woodruff, D.L.; Hopp, W.J. CONWIP—A pull alternative to Kanban. *Int. J. Prod. Res.* **1990**, *28*, 879–894.
23. Simon, J.T.; Hopp, W.J. Availability and average inventory of balanced assembly-like flow systems. *IIE Trans.* **1991**, *23*, 161–168.
24. Koskela, L. In Proceedings of the Lean Construction, National Construction and Management Conference, Sydney, Australia, 17–18 February 1994; Wakefield, R.R., Carmichael, D.G., Eds.; Institution of Engineers: Sydney, Australia, 1994; pp. 205–217.
25. Project Management Institute. *Construction Extension to the Pmbok Guide Third Edition*, 2nd ed.; Project Management Institute: Newtown Square, PA, USA, 2007; p. 208.
26. Bossink, B.; Brouwers, H. Construction waste: Quantification and source evaluation. *J. Constr. Eng. Manag.* **1996**, *122*, 55–60.
27. Hoffman, T.L. Construction management: Added value or added cost? *J. Prop. Manag.* **1998**, *63*, 6531300.

28. Baily, M.N.; Gordon, R.J.; Nordhaus, W.D.; Romer, D. The productivity slowdown, measurement issues, and the explosion of computer power. *Brook. Papers Econ. Act.* **1988**, *19*, 347–431.
29. Hossain, M.A.; Chua, D.K.H. Overlapping design and construction activities and an optimization approach to minimize rework. *Int. J. Proj. Manag.* **2014**, *32*, 983–994.
30. Li, R.Y.M.; Poon, S.W. *Construction Safety*; Springer: Berlin/Heidelberg, Germany, 2013; p. 161.
31. Project Management Institute. *A Guide to the Project Management Body of Knowledge (Pmbok Guide)*, 5th ed.; Project Management Institute: Newtown Square, PA, USA, 2013; p. 616.
32. Project Management Institute. *A Guide to the Project Management Body of Knowledge (Pmbok Guide)*, 1st ed.; Project Management Institute: Newtown Square, PA, USA, 1996; p. 182.
33. International Project Management Association. *IPMA Competence Baseline Version 3.0*, 3rd ed.; International Project Management Association: Nijkerk, The Netherlands, 2006; p. 212.
34. Office of Government Commerce. *Managing Successful Projects with Prince2*, 5th ed.; The Stationery Office: London, UK, 2009; p. 346.
35. Cho, K.; Hong, T.; Hyun, C. Scheduling model for repetitive construction processes for high-rise buildings. *Can. J. Civil Eng.* **2010**, *38*, 36–48.
36. Bhoyar, S.; Parbat, D.K. Optimal scheduling for repetitive construction projects with multiple resource crews. *Int. J. Emerg. Technol. Adv. Eng.* **2014**, *4*, 302–307.
37. Tommelein, I.; Riley, D.; Howell, G. Parade game: Impact of work flow variability on trade performance. *J. Constr. Eng. Manag.* **1999**, *125*, 304–310.
38. Alarcon, L.F.; Ashley, D.B. Playing Games: Evaluating the Impact of Lean Production Strategies on Project Cost and Schedule. In Proceedings of the IGLC-7, Berkeley, CA, USA, 26–28 July 1999; pp. 263–274.
39. Halpin, D.W.; Kueckmann, M. Lean Construction and Simulation. In Proceedings of the 2002 Winter Simulation Conference, San Diego, CA, USA, 8–11 December 2002; pp. 1697–1703.
40. Farag, M.A.M. A bridge between increasing reliability and reducing variability in construction work flow: A fuzzy-based sizing buffer model. *J. Adv. Manag. Sci.* **2014**, *2*, 287–294.
41. Benjaoran, V.; Dawood, N. Intelligence approach to production planning system for bespoke precast concrete products. *Autom. Constr.* **2006**, *15*, 737–745.
42. Thomas, H.R.; Maloney, W.F.; Horner, R.M.W.; Smith, G.R.; Handa, V.K.; Sanders, S.R. Modeling construction labor productivity. *J. Constr. Eng. Manag.* **1990**, *116*, 705–726.
43. Handa, M.V.K.; Barcia, R.M. Linear scheduling using optimal control theory. *J. Constr. Eng. Manag.* **1986**, *112*, 387–393.
44. Bashford, H.H.; Walsh, K.D.; Sawhney, A. Production system loading-cycle time relationship in residential construction. *J. Constr. Eng. Manag.* **2005**, *131*, 15–22.
45. Gonzalez, V.; Alarcon, L.F.; Yiu, T.W. Integrated methodology to design and manage work-in-process buffers in repetitive building projects. *J. Oper. Res. Soc.* **2013**, *64*, 1182–1193.
46. Gonzalez, V.; Alarcon, L.F.; Molenaar, K. Multiobjective design of work-in-process buffer for scheduling repetitive building projects. *Autom. Constr.* **2009**, *18*, 95–108.
47. American Society for Testing and Materials. Building Economics. Available online: http://www.astm.org (accessed on 27 October 2014).

48. Smith, N.J.; Merna, T.; Jobling, P. *Managing Risk in Construction Projects*, 2nd ed.; Blackwell Publishing: Malden, MA, USA, 2006; p. 257.
49. Association for Project Management. *APM Body of Knowledge*, 5th ed.; Association for Project Management: Buckinghamshire, UK, 2006; p. 197.
50. Anderson, D.K.; Merna, T. Project management strategy—Project management represented as a process based set of management domains and the consequences for project management strategy. *Int. J. Proj. Manag.* **2003**, *21*, 387–393.
51. Project Management Institute. *A Guide to the Project Management Body of Knowledge (Pmbok Guide)*, 3rd ed.; Project Management Institute: Newtown Square, PA, USA, 2004; p. 405.
52. Gibb, A.; Isack, F. Re-engineering through pre-assembly: Client expectations and drivers. *Build. Res. Inf.* **2003**, *31*, 146–160.
53. Blismas, N.; Pasquire, C.; Gibb, A. Benefit evaluation for off-site production in construction. *Constr. Manag. Econ.* **2006**, *24*, 121–130.
54. Borg, R.; Gambatese, J.; Haines, K.; Hendrickson, C.; Hinze, J.; Horvath, A.; Koehn, E.; Moritz, S.; Mass, M.; Haughney, R. Rebuilding the world trade center. *Pract. Period. Struct. Des. Constr.* **2003**, *8*, 137–145.
55. Wyrik, D.A.; Eschenbach, T.G. Remote oil and gas facility construction. *J. Constr. Eng. Manag.* **1989**, *115*, 228–236.
56. Briscoe, G.; Dainty, A.R.J.; Millett, S. Construction supply chain partnerships: Skills, knowledge and attitudinal requirements. *Eur. J. Purch. Supply Manag.* **2001**, *7*, 243–255.
57. Said, H.; El-Rayes, K. Optimal utilization of interior building spaces for material procurement and storage in congested construction sites. *Autom. Constr.* **2013**, *31*, 292–306.
58. Baye, M.R. *Magerial Economics and Business Strategy*, 7th ed.; McGraw-Hill: New York, NY, USA, 2010; p. 658.
59. McGuigan, J.R.; Moyer, R.C.; Harris, F.H.D. *Managerial Economics: Applications, Strategy, and Tactics*, 12th ed.; South-Western: Mason, OH, USA, 2011; p. 792.
60. AACE International. Risk Analysis and Contingency Determination Using Expected Value. In *TCM Framework: 7.6—Risk Management*; AACE International: Morgantown, WV, USA, 2012; Volume 44R-08, pp. 1–14.
61. Project Management Institute. *Managing Change in Organization: A Practice Guide*; Project Management Institute: Newtown Square, PA, USA, 2013; p. 141.
62. Sutherland, J. Scrum Log Jeff Sutherland: Why Gantt Charts Were Banned in the First Scrum. Available online: http://scrum.jeffsutherland.com/2006/02/why-gantt-charts-were-banned-in-first.html (accessed on 14 April 2014).
63. Deemer, P.; Benefield, G.; Larman, C.; Vodde, B. The Scrum Primer. In *A Lightweight Guide to the Theory and Practice of Scrum*; Available online: http://assets.scrumfoundation.com/downloads/1/scrumprimer20.pdf?1352449266 (accessed on 2 July 2014).
64. Mandal, A.K. *Introduction to Control Engineering Modeling, Analysis and Design*; New Age International Publishers: New Delhi, India, 2006; p. 633.
65. Weiss, S.I. *Product and Systems Development a Value Approach*; John Wiley & Sons: Hoboken, NJ, USA, 2013; p. 275.

66. Vidal, L.-A.; Marle, F.; Bocquet, J.-C. Modelling project Complexity. In Proceedings of the International Conference on Engineering Design, ICED'07, Paris, France, 28–31 July 2007; pp. 1–10.
67. Baccarini, D. The concept of project complexity a review. *Int. J. Proj. Manag.* **1996**, *14*, 201–204.
68. Mitchell, M. *Complexity a Guided Tour*; Oxford University Press: Oxford, UK, 2009; p. 366.
69. Curlee, W.; Gordon, R.L. *Complexity Theory and Project Management*; John Wiley & Sons: Hoboken, NJ, USA, 2011; p. 419.
70. Project Management Institute. *Practice Standard for Work Breakdown Structures*, 2nd ed.; Project Management Institute: Newtown Square, PA, USA, 2006; p. 123.
71. Project Management Institute. *Practice Standard for Work Breakdown Structures*, 2nd ed.; Project Management Institute: Newtown Square, PA, USA, 2006; p 156.
72. Kezner, H. *Strategic Planning for Project Management Using a Project Management Maturity Model*; John Wiley & Sons: New York, NY, USA, 2001; p. 271.
73. Kaimann, R.A. Coefficient of network complexity. *Manag. Sci.* **1974**, *21*, 172–177.
74. Vidal, L.-A.; Marle, F.; Bocquet, J.-C. Measuring project complexity using the analytic hierarchy process. *Int. J. Proj. Manag.* **2010**, *29*, 718–727.
75. Project Management Institute. *Navigating Complexity: A Practice Guide*; Project Management Institute: Newtown Square, PA, USA, 2014; p. 114.
76. Temperley, H.M.V. *Graph Theory and Applications*; Halsted Press: New York, NY, USA, 1981; p. 130.
77. Nassar, K.M.; Hegab, M.Y. Developing a complexity measure for project schedules. *J. Constr. Eng. Manag.* **2006**, *132*, 554–561.
78. Saaty, T.L. *Multicriteria Decision Making: The Analytic Hierarchy Process: Planning, Priority Setting, Resource Allocation*, 2nd ed.; RWS Pubns: Pittsburgh, PA, USA, 1990.
79. Mello, P.C.D. *John Maynard Keynes, Presidente de Companhia de Seguros*; Pedro Carvalho de Mello: São Paulo, Brazil, 2011; pp. 1–152.
80. Keynes, J.M. *The General Theory of Employment, Interest, and Money*; Harcourt Brace & World: New York, NY, USA, 1936; p. 403
81. European Commission. *Value Management*; European Commission: Brussels, Belgium, 1995; p. 190.
82. Perminova, O.; Gustafsson, M.; Wikstrom, K. Defining uncertainty in projects—A new perspective. *Int. J. Proj. Manag.* **2008**, *26*, 73–79.
83. Knight, F.H. *Risk, Uncertainty and Profit*; Library of Economics and Liberty: Boston, MA, USA, 1921; p. 381.
84. Crouhy, M.; Galai, D.; Mark, R. *The Essentials of Risk Management*; McGraw-Hill: New York, NY, USA, 2005; p. 416.
85. Mun, J. *Modeling Risk: Applying Monte Carlo Simulation, Real Options Analysis, Forecasting, and Optimization Techniques*; John Wiley & Sons: Hoboken, NJ, USA, 2006; p 623.
86. Project Management Institute. *A Guide to the Project Management Body of Knowledge (Pmbok Guide)*, 4th ed.; Project Management Institute: Newtown Square, PA, USA, 2008; p. 507.
87. Jorion, P. *Value at Risk the New Bechmark for Managing Financial Risk*, 3rd ed.; McGraw-Hill: New York, NY, USA, 2007; p. 594.

88. Cleden, D. *Managing Project Uncertainty*; Gower: Surrey, UK, 2009; p. 146.
89. Wikipedia. Variability. Available online: https://en.wikipedia.org/wiki/Variability (accessed on 22 April 2014).
90. Radosavljević, M.; Horner, R.M.W. The evidence of complex variability in construction labour productivity. *Constr. Manag. Econ.* **2002**, *20*, 3–12.
91. Atkinson, R.; Crawford, L.; Ward, S. Fundamental uncertainties in projects and the scope of project management. *Int. J. Proj. Manag.* **2006**, *24*, 687–698.
92. McConnell, S. *Software Estimation: Demystifying the Black Art*; Microsoft Press: Redmond, WA, USA, 2006; p. 352.
93. Spencer, M.S.; Cox, J.F. An analysis of the product-process matrix and repetitive manufacturing. *Int. J. Prod. Res.* **1995**, *33*, 1275–1294.
94. Public Broadcasting Service (PBS). One World Trade Center. In *Super Skyscrapers*; PBS: Arlington, VA, USA, 2014.
95. Shaked, D. *Strength-Based Lean Six Sigma: Building Positive and Engaging Business Improvement*; Kogan Page Ltd.: London, GBR, 2013.
96. Taghizadegan, S. *Essentials of Lean Six Sigma*; Elsevier: Amsterdam, NL, USA, 2006; p. 275.
97. University of Alberta Dictionary. Universal Function Approximator. Available online: http://www.bcp.psych.ualberta.ca/~mike/Pearl_Street/Dictionary/contents/U/universalfn.html (accessed on 5 July 2014).
98. Xu, J.-X.; Tan, Y. *Linear and Nonlinear Iterative Learning Control*; Springer: Berlin, Germany, 2003; Volume 291, p. 177.
99. Dager, R.; Zuazua, E. *Wave Propagation, Observation and Control in 1-d Flexible Multi-Structures*; Springer: New York, NY, USA, 2006; Volume 50, p. 227.
100. Ogata, K. *Modern Control Engineering*, 5th ed.; Prentice Hall: Upper Saddle River, NJ, USA, 2010; p. 905.

7

Impacts of an Innovative Residential Construction Method on Internal Conditions

Roger Birchmore [1] *, Andy Pivac [2] and Robert Tait [2]

[1] Department of Construction, Unitec Institute of Technology, P.O. Box 92025 Auckland, New Zealand

[2] Department of Building Technology, Unitec Institute of Technology, P.O. Box 92025 Auckland, New Zealand; E-Mails: apivac@unitec.ac.nz (A.P.); rtait@unitec.ac.nz (R.T.)

* Author to whom correspondence should be addressed; E-Mail: rbirchmore@unitec.ac.nz

Academic Editor: George Baird

Abstract: New Zealand houses are known for producing sub-optimal internal thermal conditions and unacceptably high internal moisture levels. These contribute to poor levels of health, mould and can coincide with the decay of structural timber frames. A proposed solution is to provide an alternative structure utilising plywood, a vapour check on the internal face of the timber frame and an additional air gap, followed by the internal lining. The internal vapour check is designed to prevent moisture vapour diffusion from inside into the frame and to permit moisture diffusion from outside through the structure to the internal environment. Two full scale houses had temperatures, dew points and humidity levels monitored in passive, unoccupied conditions. The test case house incorporated the innovative construction solution. The control house was of identical design and location, using standard construction practice. The calculated internal moisture content profile appeared to be unrelated to the external moisture content as expected, instead following the profile of the changing internal temperature. Whilst the innovative construction appeared to prevent moisture diffusion into the structure in winter and permit it inside in summer, this resulted in a generally higher internal relative humidity than the control house.

Keywords: housing; internal moisture; innovative construction, vapour check; airtightness

---

1. Introduction

Mackintosh [1] summarises New Zealand's climate as:

> *"Warm subtropical in the far north to cool temperate climates in the far south, with severe alpine conditions in the mountainous areas. Mean annual temperatures range from 10 °C in the south to 16 °C in the north of New Zealand. Most of New Zealand would have at least 2000 sunshine hours annually."*

Auckland, New Zealand's largest centre of population, located in the north island experiences an average of 632 degree days (base 15.5 °C). In the alpine region in the South Island, Queenstown experiences an average of 2137 degree days to the same base [2].

This data does not describe harsh external conditions for much of the population but instances where combinations of low temperatures and high moisture levels lead to poor internal environments, in many parts or the country are documented widely by a number of authors [3–6]. The World Health Organisation [7] links poor internal conditions to a range of health problems that are also reported in New Zealand research. Reflecting the largely temperate nature of the climate, guidance from the Building Research Association of New Zealand (BRANZ) [8] is that vapour barriers are only a requirement in alpine regions or those with significant internal moisture generation linked with spa pools or other similar sources. Moisture in open roof spaces is also not considered to be a significant problem and New Zealand Building codes currently have no requirements for the ventilation of roof spaces. Since the publication of the guidance obviating the need for vapour barriers other changes to the Building Code have resulted in minimum insulations levels being raised. There are no requirements for the specific airtightness of buildings. In response to the challenges described, there has been research on solutions that tackle the problems of poor conditions and health directly or indirectly through improving the sustainability of homes [9–12]. This work has tended to focus on the thermal solutions and energy consumption aspects. Su [13,14] extended this and researched the prevention of winter mould growth in occupied New Zealand houses employing primarily passive and active ventilation and thermal insulation prevention measures. Comparing the static and dynamic simulation methods de Groot [3] expanded further to explore in detail the impacts of moisture transfer through the envelope. He cautioned that the increase of thermal insulation without the consideration of interstitial moisture might move the visible mould problem to an invisible one. Simulating alternative retrofit solutions over a three year period in Auckland, he demonstrated that a vapour barrier was effective in preventing interstitial condensation occurring to levels that might encourage mould growth. Leardini and van Raamsdonk [15] added to the concerns of occupant health to include structural degradation. They also outline fears that increasing levels of thermal insulation increases chances of interstitial condensation. Their Wärmer und Feuchte instationär (WUFI) [16] software simulation of a timber framed house in Auckland, retrofitted with insulation, indicated a clear risk of interstitial conditions when examined over a three year period. This analysis also indicated the tendency for humid summertime conditions to drive moisture inside. The conventional vapour barrier approach risks trapping this moisture vapour into the structure as shown in Figure 1B. They propose that a solution is to provide a vapour check that prevents vapour transfer from moist inside conditions into the wall structure (Figure 1A,C) but also permits externally driven vapour to pass through the structure to the inside (Figure 1D). This vapour check additionally provides all the benefits of an airtight barrier, so whilst reducing the possibility of interstitial condensation, may exacerbate the challenge of increased internal moisture levels and its associated risks. De Groot and Leardini [17] identified a lack of information on the success of retrofit solutions and the general need to improve understanding of the impacts of combining insulation airtightness and humidity control. This paper outlines the early findings of a research project that moves research from desktop simulation to exploring the impact of a construction employing such a vapour check on unoccupied conditions in a real house.

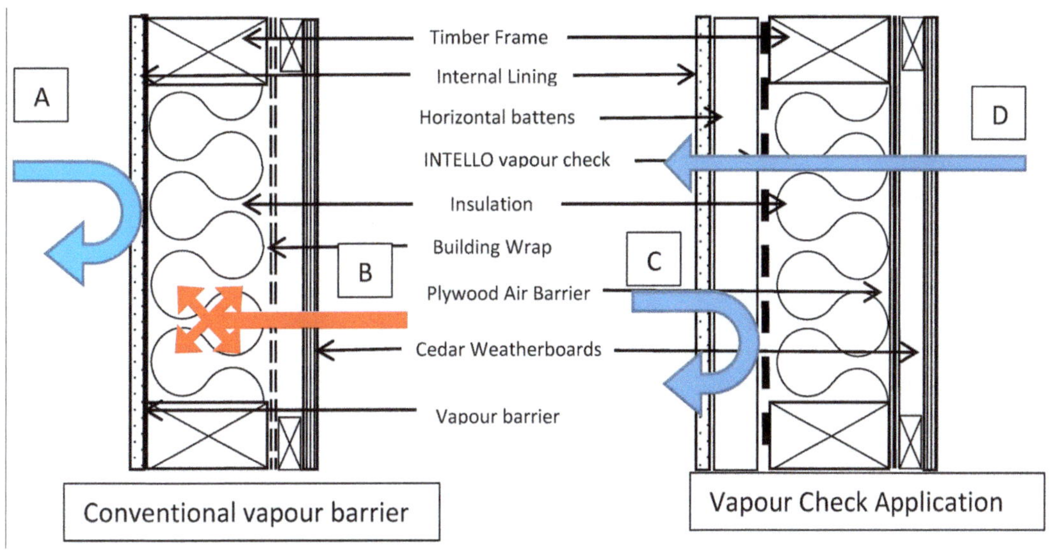

Figure 1. Moisture flows through the structure.

2. Research Design and Methods

The fundamental aim of this project is to allow comparison testing of individual or combinations of building materials and techniques that have the potential for improving the building performance of this standard New Zealand house type. In particular, this paper compares the use of a vapour check material not widely used in New Zealand, combined with a rigid air barrier, compared to conventional New Zealand construction methods. The particular focus is on impacts on the moisture levels and internal air temperatures. Initially, manual calculations employing the dew point method were conducted to investigate the likelihood of interstitial condensation occurring in the control house. Figure 2 shows the risk of condensation occurring in the structure 100 mm from the inside surface. Figure 3 shows how the vapour check removes this risk. This aligns with the more extensive analysis by Leardini and van Raamsdonk [15]. It was decided to employ the vapour check proposed in their work with a view to initially exploring its impact on the internal spatial conditions of the houses. [18,19] indicate the significant impact that occupants can have on the energy consumption and internal conditions of a space, quoting examples where variations were many times the base consumption figures. The two houses have therefore been tested in passive, unoccupied conditions to remove as many variables as possible. This data would provide the basis for predicting impacts of the construction methods on internal comfort conditions, risks of mould growth and also spatial heating and cooling energy use. This is the first stage of a project that will be followed with analysis of the construction method on the interstitial conditions.

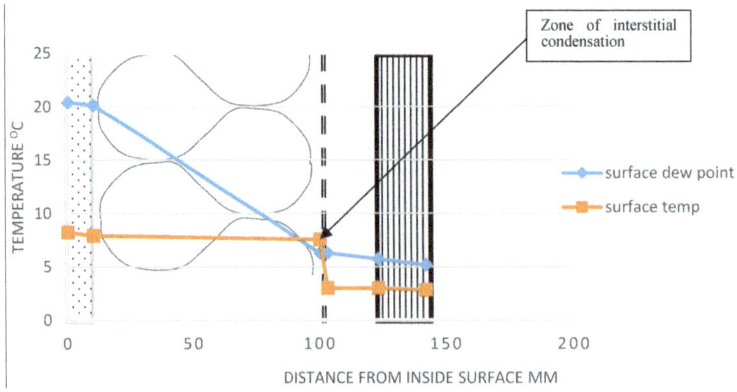

Figure 2. Interstitial condensation calculations in the control house.

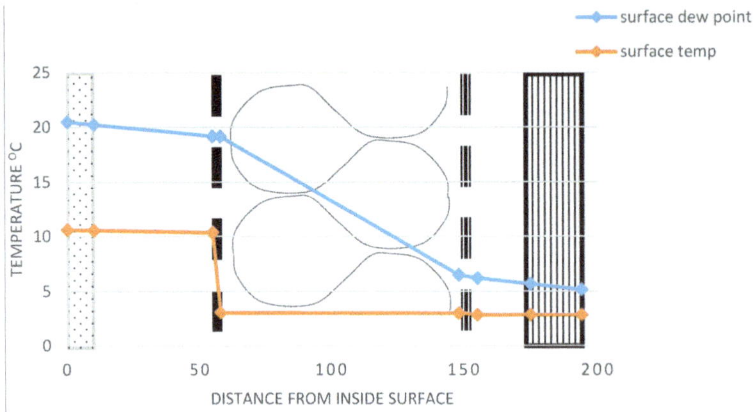

Figure 3. Interstitial condensation calculations in the test house.

2.1. Control House

The houses are single storied with three bedrooms and two bathrooms and are constructed as part of the Unitec carpentry programmes. The houses are completed by students to be relocated, and they are undecorated and without floor coverings or wall finishes. Electrical and plumbing fittings are installed but not connected. Figure 4 illustrates their overall nature.

Figure 4. Control house (foreground) and test house.

Table 1 summarises the materials used in the construction of these houses. Overhangs on the north side of the house provide complete shading from direct solar gain through double glazing during the hottest periods of the summer months. These houses are similar in design and construction to thousands of houses found in suburban areas and provide an ideal basis for examining the potential for improvements to a common housing type.

Table 1. Construction details for the control and test houses.

Common to Control House and Test House	
Timber Frame on pile foundation	
Sub floor cladding 150 mm × 25 mm radiata pine boards with 20 mm gaps	
Particle board, floor foil insulation draped between 190 mm × 45 mm joists	
140 mm polyester ceiling batts, 10 mm plasterboard	
Double glazed windows	
Control House	**Test House**
Trussroof (radiata pine treated) Coloursteel roofing on building paper	Trussroof (radiata pine treated) Coloursteel roofing on building paper, INTELLO wrap on bottom chord of trusses
cedar weatherboard cladding, natural finish	cedar weatherboard cladding, natural finish
20 mm cavity battens	20 mm cavity battens
Building wrap	7 mm Plywood
90 mm × 45 mm radiata pine framing	90 mm × 45 mm radiata pine framing
Polyester insulation	Polyester insulation
-	INTELLO Vapour check
-	45 mm × 45 mm battens
10 mm plasterboard	10 mm plasterboard

Table 2 provides dimensional thermal and hygroscopic properties of relevant materials in the external envelopes. These are values associated with readily available material sizes to ensure the overall construction *R*-values exceed minimums stated in verification method H/VM1 [20], ensuring compliance with the New Zealand Building Code. Minimum *R*-values for this context are:

- Roof 2.9 m^2K/W
- Wall 1.9 m^2K/W
- Windows 0.26 m^2K/W
- Floor 1.3 m^2K/W

Table 2. Thermal and hygroscopic properties of materials.

Material	Thickness t	Conductivity k	Resistance R	Vapour Resistivity r_v	Vapour Resistance G
-	mm	W/mK	m^2K/W	MNs/gm	MNs/g
External Walls	-	-	-	-	-
Plasterboard	10	0.22	0.05	60	0.6
Air gap (Test House)	45	-	0.18	-	5
Vapour Check (Test House)	3	-	-	-	see Table 3
Insulation	90	0.041	2.20	7	0.63
Building Wrap (Control House)	3	n/a	n/a	2300	6.9
Plywood Air Barrier	7	0.13	0.05	-	2.2
Ventilated Air Gap	20	-	0.09	-	-
Cedar Weatherboard	19	0.11	0.09	-	0.26

Table 2. *Cont.*

Material	Thickness t	Conductivity k	Resistance R	Vapour Resistivity r_v	Vapour Resistance G
-	mm	W/mK	m^2K/W	MNs/gm	MNs/g
Roof Structure	-	-	-	-	-
Plasterboard	10	0.22	0.05	60	0.6
Air gap (Test House)	45	-	0.18	-	5
Vapour Check (Test House)	3	-	-	-	see Table 3
Air Gap	300	-	0.16	-	5
Building Wrap	3	-	-	2300	6.9
Profiled Sheet Steel Roof	4	-	-	-	2000

2.2. Test House

The test house was constructed one year later than the control house and modifications incorporated the use of the INTELLO vapour check proposed by Leardini and van Raamsdonk [15]. In addition modifications were made were to replace the building paper with 7 mm thick plywood barrier treated to H3.2 CCA (Copper Chrome Arsenate) in accordance with AS/NZ 1604.3 [21] to meet AS/NZS 2269.0 [22]. Vertical sheet joints were sealed with flashing tape. This feature was felt to have significant potential as an alternative that provided the functions of bracing and rigid air barrier in a single element that offered enhanced seismic resistance. This had potential for use in the rebuild of houses since the Christchurch earthquakes. The vapour check was placed on the internal surfaces of external walls and ceilings. A 45 mm cavity batten was then added before fixing of the plasterboard. Details are shown in Figures 5 and 6. Ceiling insulation was placed directly on top of the ceiling material in the control house and on top of the INTELLO vapour check in the control house. In both houses the roof space that resulted was an uninsulated cavity. The INTELLO vapour check is intended to have two functions. The first is to control the passage of vapour through the timber to the structure. This minimises the chances of charging the frame with moisture that could then condense during cold spells, eventually resulting in the rotting of the frame. Its high level of resistance shown in Table 3 prevents vapour generated internally from passing into the structure in winter. The low level of resistance in the opposite direction permits the vapour to pass from high humidity conditions present in summer time though the structure to the inside. This again prevents the possible accumulation of moisture in the frame for future condensation. The second function is for the INTELLO vapour check to act as an additional barrier to infiltration through the fabric of the building. The minimisation of infiltration is proposed to be an important contributor to reducing unplanned heat loss and moisture ingress. The additional air gap between the INTELLO and the internal plasterboard lining provides a route for internal services to circulate with minimal penetrations through the INTELLO.

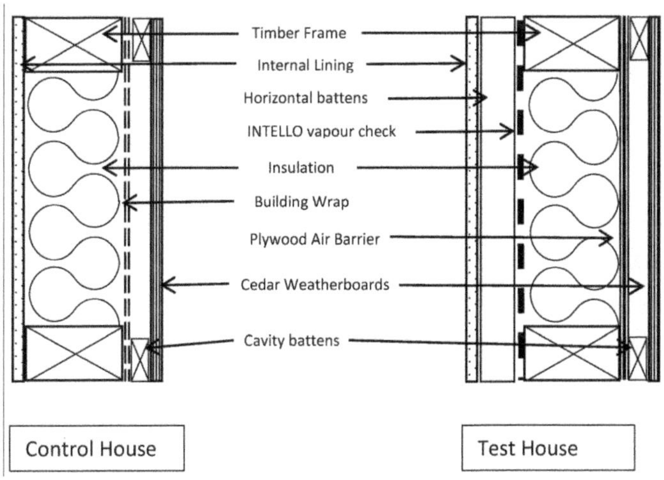

Figure 5. Construction detail—external walls.

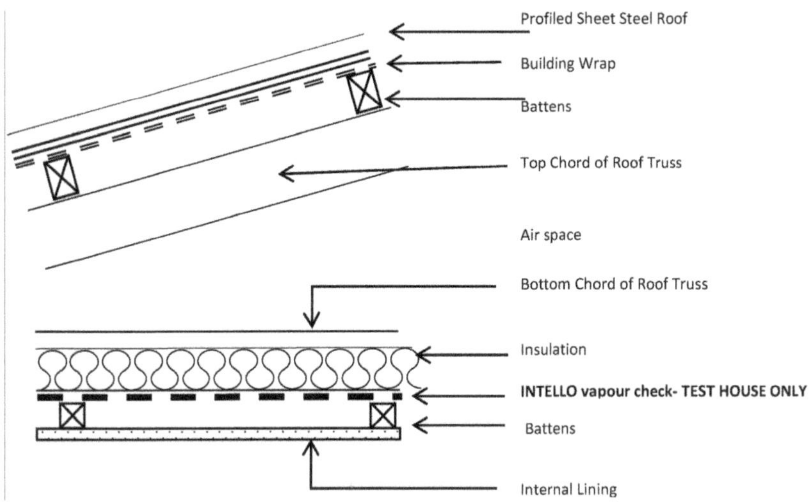

Figure 6. Construction detail—roof.

Table 3. Vapour diffusion resistance of the INTELLO membrane [23].

Testing Conditions	Vapour Diffusion Resistance (MNs/g)	
Average Ambient Humidity	20%	85%
Direction of Diffusion Flow	Out towards the air barrier	Inwards towards the air barrier
INTELLO	125	1.25

2.3. Site

The site is on the Unitec Institute of Technology campus in Mt Albert Auckland. The site is relatively exposed with an open grassed area to the northwest. Surrounding buildings are reasonably distant to the south, north and east. Behind the houses to the southeast is a hilly incline and the student building yard. The houses are located with identical orientations but separated to avoid mutual shading.

2.4. Monitoring Process

Temperature sensors have been set up to sample the internal air temperature at hourly intervals. Sensors used are Lascar EL-USB-2 Humidity and Temperature USB data loggers. These measure and store relative humidity (RH), dew point (DP) and dry bulb air temperature readings (DB) over 0% RH to 100% RH and −35 °C to +80 °C measurement ranges. Sensors were located identically in the two houses to align with practice outlined by Barley *et al.* [24] at a height of 1500 mm above ground level suspended from the ceiling by builders twine. Sensor layout is given in Figure 7. In order to check the appropriate test location for the sensor, a second sensor was located at the edge of the room to check initial operation and determine the degree of variability experienced across each space. It was found that the average variation between measurements from the centre of the room and from the edge of the room vary by an average of 0.2 °C over the 168 hourly measurements, with the maximum variation less than 0.5 °C. This is well within the accuracy stated for the sensors which is stated to be ±0.5 °C [25] and indicates that a single measurement in the chosen position is representative of the overall room conditions. Dew point measurements have been used as this single figure provides an indicator of absolute moisture content. Localised weather data is measured at a weather station indicated in Figure 8. They are monitored in a passive, unoccupied condition.

Figure 7. House plan with sensor locations.

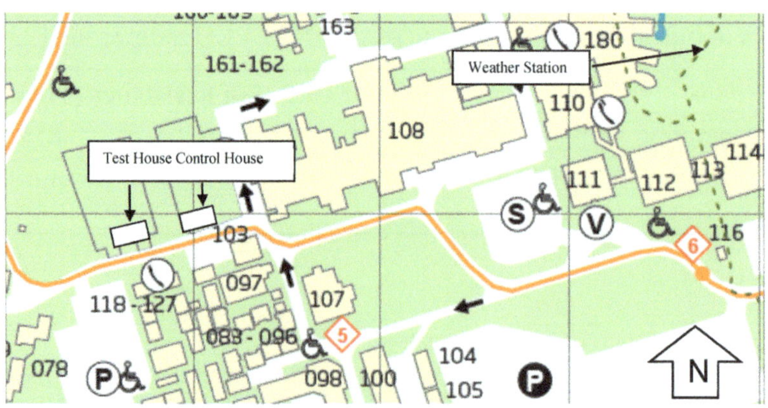

Figure 8. House and weather station location details.

2.5. Air Tightness

Both houses were tested for air tightness using the standard blower door test following European standard EN 13829:2000 [26]. Openings associated with extract ventilation and unconnected waste pipes were sealed for air pressure testing and for subsequent monitoring of air temperature, dew point and relative humidity. Table 4 shows the air changes per hour of the whole house volume under the standard test conditions of 50 Pa pressurisation and 50 Pa depressurisation. It indicates that the test house has an air leakage rate of less than a third of the control house. The control house sits just outside the airtight classification for New Zealand houses which peaks at an airtightness of 5 ac/h [27]. The test house is comfortably in the airtight classification but is still well above the requirements of the Passive House Institute [28].

Table 4. Results of airtightness testing.

Testing Conditions	Control House (ac/h)	Test House (ac/h)
Depressurisation	6.58	1.92
Pressurisation	6.93	2.10
Average	6.75	2.01

2.6. Rooms Being Analysed

The rooms chosen for analysis in this paper were the lounge kitchen dining room, bedroom 3, hall and the roofspace. Floor and wall areas are shown in Table 5. The lounge kitchen dining room is the largest space in the house and is has external walls on the South, East and North Face with glazing in each with a window to wall ratio of 29%. Its inclusion of the kitchen also examines a space where occupancy may generate significant additional internal moisture. Bedroom 3 is on the opposite, coldest corner of the house receiving minimal solar gain. With a window to wall ratio of 41% the hall is internal with the exception of its floor and ceiling. The roofspace provides data on the external side of the vapour check in the test house and the external side of the insulation in both houses.

Table 5. Details of rooms being analysed.

Room	Floor Area (m^2)	Net External Wall Area (m^2)
Lounge Kitchen Dining room	46.4	33.6
Bedroom 3	9.8	11.07
Hall	5.6	Nil
Roofspace	120.5	Nil

3. Results and Discussion

3.1. Seasonal Data

The means of dry bulb (DB) and dew point temperatures, (DP) measured at hourly intervals for the winter and summer seasons are summarised for each building in Figure 9. The summer season is defined as 1 December–28 February, spring as 1 March–31 May, winter as 1 June–31 August and spring as 1 September–30 November. The data indicates that as the mean temperatures drop towards winter and as expected rise up again in spring. Also as expected, the range of temperatures in the roofspace exceed those in occupied spaces indicating the smoothing effect of in wall and under floor insulation. The trend and interrelationship holds for all three internal spaces suggesting that the airspace in each room within each house is reasonably homogenous and that orientation and wall to window ratios have a small effect on mean temperatures. Absolute moisture content and dry bulb temperature are not parameters that are inherently connected. As the buildings have no internal moisture generation it was expected that absolute moisture contents, represented by the dew point temperature levels would follow external moisture readings. Figure 9 indicates that in fact they seem to follow the internal dry bulb readings. The detail of a few sample days in Figures 10 and 11 re-enforces this and indicates little or no connection between internal and external dew point temperatures.

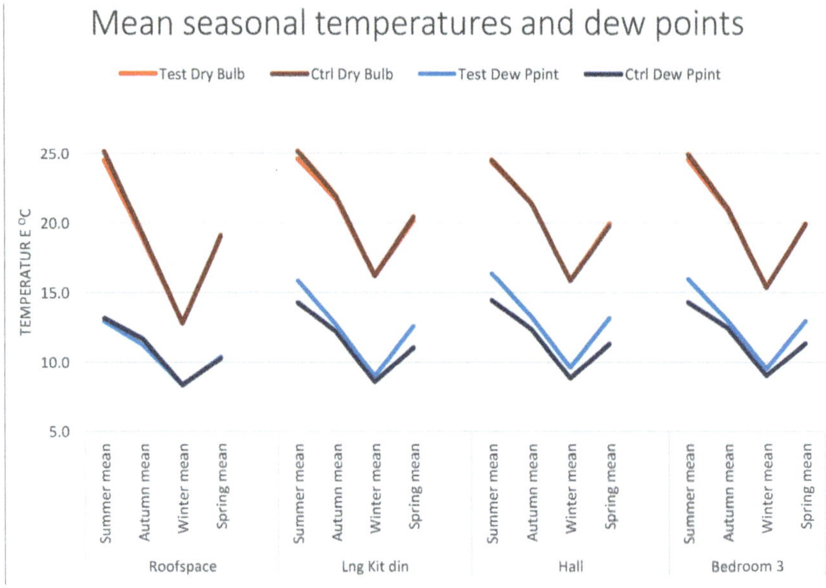

Figure 9. Comparison of space temperature and dew point seasonal means.

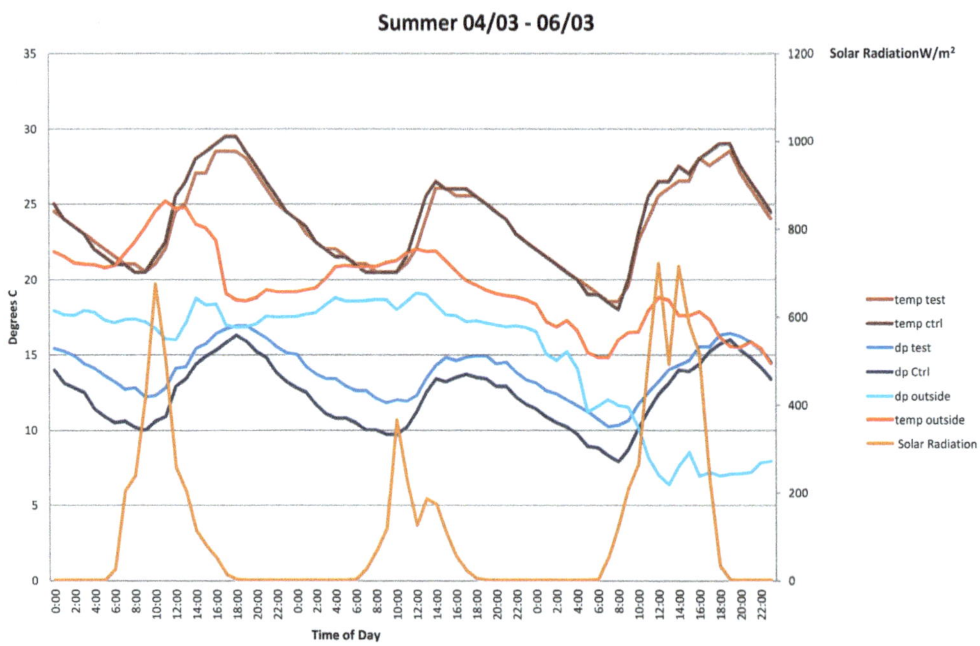

Figure 10. Dry bulb and dew point comparisons for the lounge kitchen dining room with external summer conditions.

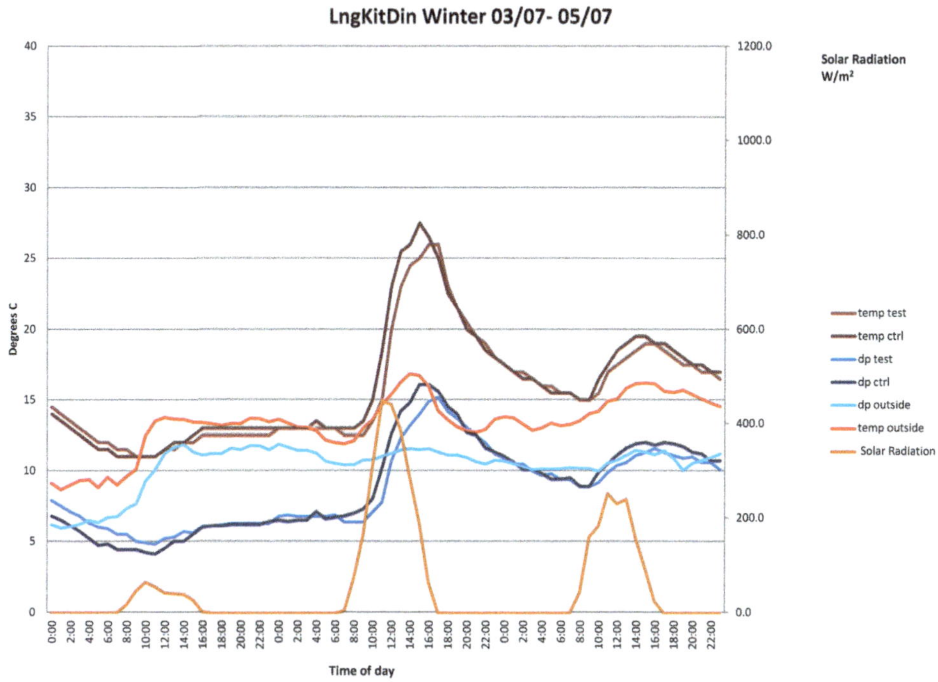

Figure 11. Temperatures and dew point comparisons for the lounge kitchen dining room with external winter conditions.

The significance of the differences between seasonal measures of control and test dry bulb temperatures and between control and test dew point temperatures were calculated using a two tailed *T*-test. Calculations were completed on two spaces, the roofspace and the lounge kitchen dining room being the most different spaces. Calculations were completed on seasonal means using an alpha of 0.05 with a hypothesis that the difference between two population means would be zero. *P* scores are quoted to four significant figures.

Scores where $p \geq 0.05$ indicates that the hypothesis is correct and that there is no significant difference between the populations. Scores where $p \leq 0.05$ indicates the hypothesis is incorrect and that there is a significant difference between the populations. Table 6 shows that in all spaces with the exception of the roofspace during winter and for dry bulb temperatures in spring the difference between the test and control house was statistically significant with a confidence of 95%.

Table 6. Summary of statistical significance of differences between control and test houses.

Season	Space	Dry Bulb	Dew Point
-	-	*p*	*p*
Summer	LoungKitDin	0	0
	Roofspace	0	0
Autumn	LoungKitDin	0	0
	Roofspace	0	0
Winter	LoungKitDin	0	0
	Roofspace	0.1007	0.3745
Spring	LoungKitDin	0	0
	Roofspace	0	0.0263

3.2. Detailed Results of Selected Days

3.2.1. Summer

The detailed results of a few days in each season are shown in Figures 10–13. They have been chosen to illustrate the apparent strong influence of solar gain on internal conditions. This indicates that the internal dry bulb temperatures follow the cyclical pattern of the external solar gain. The periods of the two cycles appear identical but with the dry bulb lagging between four and 6 h. The internal dew point also follows with the same period but with slightly reduced amplitudes. There is a very weak connection if any, with the dew point of the external air. As there were no occupant generated sources of internal moisture, the changing dew points over a daily cycle could be resulting from residual construction moisture. As the temperature rises, moisture still present in the construction evaporates into the air. As it cools it is re-absorbed into the structure. Whilst the test house vapour check is designed to reduce this, the floors of both houses are exposed particle board and the unpainted gypsum board will have some absorption capacity.

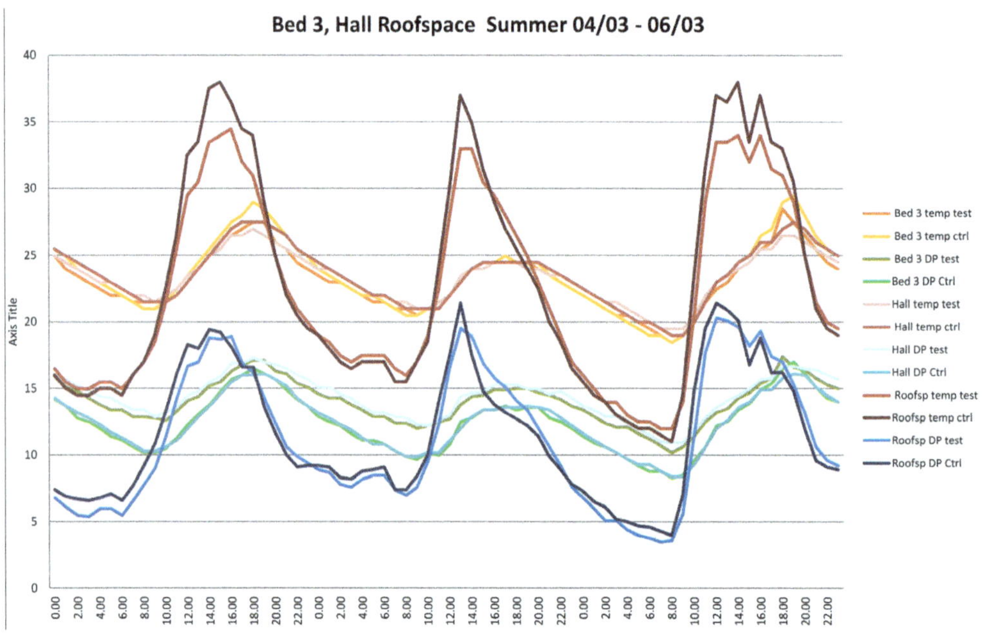

Figure 12. Dry bulb temperatures and dew point comparisons for the bedroom 3 hall and roofspace with external summer conditions.

Figure 12 shows the performance for three other spaces during the same period. The external conditions are the same as in Figure 10 but have been removed for clarity. Reflecting the seasonal means in Figure 9, the roofspace exhibits a wider range of variations than other internal spaces. The dry bulb and dew point temperatures in the bedroom 3 and hall cluster close together within similar ranges and cycle to those of the lounge kitchen dining room. This demonstrates that the seasonal homogeneity is reflected on a daily cycle. The reduction of extreme internal temperatures by thermal insulation in the roofspace is also very clear on this daily cycle. The four to 6 h lag between peak external temperature and internal temperature observed in the lounge kitchen dining room remains for other internal spaces. This is reduced to two hours for the roofspace. The reduction in lag is expected as the temperature is being measured on the external side of the insulation. Differences between dry bulb temperatures in the control house and the test house are small, being largest at extremes of conditions and very small between. Differences between the dew point readings in the houses appear more significant and more evenly spread over the daily cycle.

3.2.2. Winter

The data in Figure 11, Illustrates the strong link between solar gain, internal dry bulb and internal dewpoint temperatures where the solar gain on the second day leads and influences the plots of other parameters. Notably the thermal lag in both houses has reduced to approximately 2 h.

Figure 13 reflects parallels with Figure 12 but with lower temperature levels associated with the winter season. The clustering of the graphs continues to support the homogeneity of the internal spaces. The roofspace extremes are reduced. Time lags have also reduced to a maximum of two h and minimums of zero. Differences between control and test house dry bulb temperatures and dew points are very small especially on the first day with low solar radiation.

Moisture Contents, in kg of moisture per kg of dry air, can be read from a Psychrometric chart for given dew points. Using the mean density of the air and the space volume, the volume of moisture being evaporated and re-absorbed on a daily cycle can be estimated. 1 December was chosen as the day with the maximum daily change in dew point temperature and Table 7 indicates the result as 0.52 L.

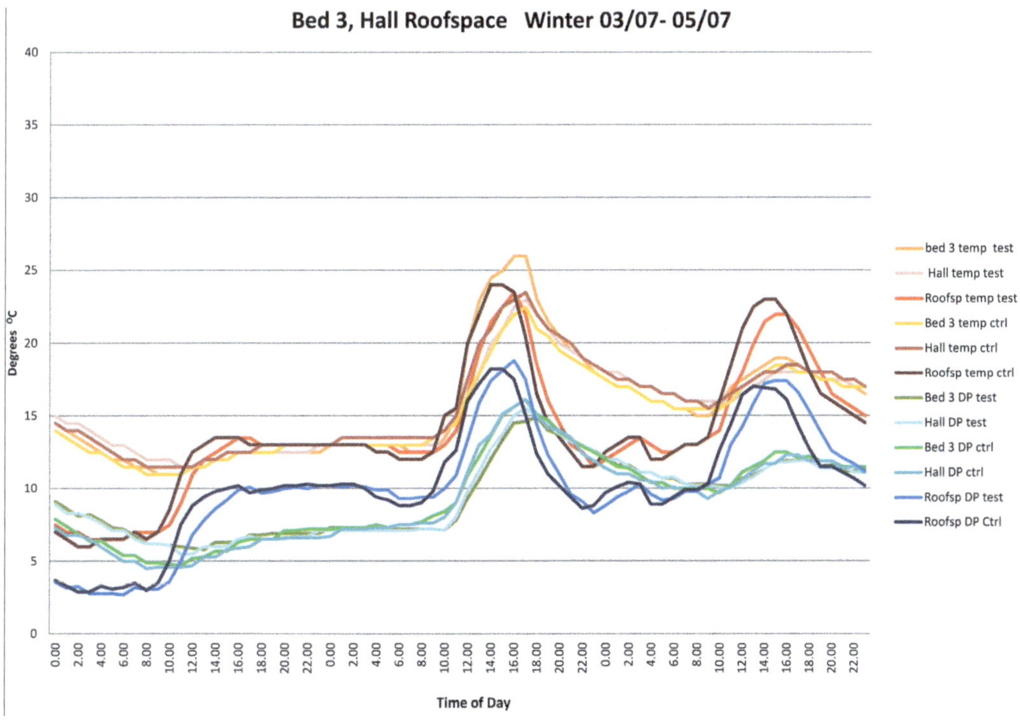

Figure 13. Dry bulb temperatures and dew point comparisons for the bedroom 3 hall and roofspace with external winter conditions.

Table 7. Data for estimation of the moisture volume being evaporated and re-absorbed.

Summer 1st Dec			
Dry Bulb (°C)	**Dew Point (°C)**	**Density (kg/m³)**	**Moisture Content (kg/kg)**
Max 25.5	15.4	1.163	0.0109
Min 15.5	8.6	1.211	0.0069
Difference	6.8	-	0.004
Mean density (kg/m³)	-	1.187	-
Space volume (m³)	108.54	-	-
Space mass (kg)	128.809	-	-
Moisture mass (kg)	0.52	-	-
Moisture volume (L)	0.52	-	-

4. Conclusions

The internal seasonal spatial data in Figure 9 along with the statistical analysis supports the expected performance of the test house structure regarding internal moisture levels. The summer season trends are reinforced by the daily analysis in Figure 10. In winter, higher internal dew points in the test house compared to the control house suggest that internal vapour is held within the occupied space and is not being permitted to enter the structure. The risk of interstitial condensation should therefore be reduced. In summer, higher internal dew points in the test house compared to the control house suggest that vapour is being allowed to pass through the structure to the inside. This prevents the moisture getting trapped within the timber frame, reducing the potential to cause interstitial condensation when the climate permits. The roofspace data supports this further. As the readings are on the external side of the vapour check, vapour passing through the check in summer should result in lower dew point temperatures within the control house roofspace compared to the test house. Figure 10 indicates this. They should also be lower in winter but this is not as evident when looking at seasonal means or the test of significance, but the daily analysis in Figure 12 does support this. Overall this differentiates the performance of the vapour check from a conventional vapour barrier which would trap the moisture within the timber structure, preventing it from entering the occupied space.

Compared to the control house, the test house offered improved thermal resistance from the additional sealed air gap and significantly reduced infiltration due to the vapour check properties. It was felt that this would result in warmer test house conditions in winter and cooler conditions in summer. Overall differences are not large however. Small differences between the average dry bulb temperatures over the winter and spring seasons between the control and test Houses suggest that the different properties are having a small effect on unoccupied thermal conditions. Figure 11 shows that on cloudy winter days the difference between internal and external temperatures was minimal and so differences between control and test houses could also be very small. However on sunny winter days, internal temperatures rose nearly ten degrees higher than outside. Under these winter conditions the test house was actually slightly cooler than the control house. As the test house infiltration rate was one third of the control house rate, this suggests that the heat losses due to infiltration through the closed structure in an unoccupied condition might be much smaller than anticipated or are not being prevented by increased airtightness of the envelope.

The calculated daily variation of 0.52 L of water is a small proportion of the possible moisture generation in a house. TenWolde and Pilon [29] suggest that a family of four would produce up to 15 L per day. However, 0.52 L represents the amount of moisture left to be absorbed by the air. The moisture absorptive properties on internal linings, quantified by Mitamura, Rode and Schultz [30] and termed moisture buffering, could account for a larger volume of daily moisture change in the unoccupied condition.

The tendency of the vapour check to produce slightly higher internal dew points emphasises the need to combine this type of structure with minimal ventilation rates to ensure the moisture is ventilated to outside. Internal surface temperatures are likely to be even lower than space temperatures, especially in winter. The combination of this fact and that moisture generated by occupation may aggravate conditions, further underlines the importance of ventilating moisture sources to outside.

Whilst the positive differences in conditions provided by the innovative structure are statistically significant they are still small and might be unlikely to be financially beneficial in the short term. Work is underway with the detailed monitoring of temperature and humidity in each layer of the wall construction of each building. This will enable tracking in detail of the passage of vapour throughthe envelope and the identification of interstitial condensation risk and help allocate the differences to the performance of the internal vapour check and the plywood rigid air barrier. Further work will include monitoring conditions under controlled, active heating, cooling and moisture generation which should provide a stiffer test and opportunity for the structure to show a financially beneficial difference.

Acknowledgments

This research has been carried out with the support of Thomas van Raamsdonk of ProClima and the Unitec Internal Research Fund.

Author Contributions

Roger Birchmore contributed to the research design and carried out the collection and analysis of the data. Robert Tait and Andy Pivac contributed to the research design experimentation set up and conducting measurements.

Conflicts of Interest

The authors declare no conflict of interest.

References

1. Mackintosh, L. NIWA Overview of New Zealand Climate. 2001. Available online: https://www.niwa.co.nz/education-and-training/schools/resources/climate/overview (accessed on 11 June 2014).
2. BizEE Degree Days. Weather Data for Energy Professionals. Available online: http://www.degreedays.net/# (accessed on 22 December 2014).
3. De Groot, H. Indoor Air Quality and Health. An Analysis of the Indoor Air Quality and Health in New Zealand's Homes. 2009. Available online: http://www.archigraphic.de/_originals/pdfs/indoor_air_quality_and_health.pdf (accessed on 11 June 2014).

4. Howden-Chapman, P.; Saville-Smith, K.; Crane, J.; Wilson, N. Risk factors for mould in housing: A national survey. *Indoor Air* **2005**, *15*, 469–476.
5. Howden-Chapman, P.; Crane, J.; Chapman, R.; Fougere, G. Improving health and energy efficiency through community-based housing interventions. *Int. J. Public Health* **2011**, *56*, 583–588.
6. New Zealand Business Council for Sustainable Development (NZBCSD). Media Release: $20 Billion Cost of Fixing Country's Homes Less than 4% of Their Value. 2008. Available online: http://www.scoop.co.nz/stories/BU0812/S00020.htm (accessed on 11 June 2014).
7. World Health Organisation. Guidelines for Indoor Air Quality: Dampness and Mould. 2009. Available online: http://site.ebrary.com/lib/unitech/Doc?id=10367463&ppg=55 (accessed on 11 June 2014).
8. Sargent, S. Getting Clear on Vapour Barriers and Underlays. 2007, Build 99. Available online: http://www.buildmagazine.org.nz/articles/show/getting-clear-on-vapour-barriers-and-underlays/ (accessed on 4 February 2015).
9. Howden-Chapman, P.; Matheson, A.; Crane, J.; Viggers, H.; Cunningham, M.; Blakely, T.; Cunningham, C.; Woodward, A.; Saville-Smith, K.; O'Dea, D.; *et al*. Effects of insulating existing houses on health inequality: Cluster randomised study in the community. *Br. Med. J.* **2007**, *334*, 460.
10. Easton, L.; Saville Smith, K. Homesmart Renovations—Testing Tools to Promote Sustainable Renovation. In Proceedings of the New Zealand Sustainable Building Conference, Wellington, New Zealand, 26–28 May 2010.
11. Callau, M. Upgrading Housing in NZ for Thermal Efficiency. In Proceedings of the New Zealand Sustainable Building Conference, Wellington, New Zealand, 26–28 May 2010.
12. Burgess, J.C.; Buckett, N.R.; Camilleri, M.J.T.; Burrough, L.J.; Pollard, A.R. Papakowhai Retrofit Project—Improving the Thermal Envelope and Space Heating. In Proceedings of the New Zealand Sustainable Building Conference, Wellington, New Zealand, 26–28 May 2010.
13. Su, B. Prevention of inter mould growth in housing. *Archit. Sci. Rev*. **2006**. *49*, 385–390.
14. Su, B. Indoor moisture control of Auckland houses with different ventilation systems. *Int. J. Civ. Archit. Sci. Eng.* **2013**, *7*, 411–415.
15. Leardini, P.M.; van Raamsdonk, T. Design for Airtightness and Moisture Control in New Zealand Housing. In Proceedings of the New Zealand Sustainable Building Conference, Wellington, New Zealand, 26–28 May 2010.
16. WUFI Pro, 2D, Plus Software 2013. Fraunhofer-Institute fur Bauphysik (IBP), Holzkirchen. Available online: http://www.wufi.de/index_e.html (accessed on 21 January 2015).
17. De Groot, H.; Leardini, P.M. Indoor Air Quality and Health in New Zealand's Traditional Homes. In Proceedings of the 44th Annual Conference of the Architectural Science Association, Unitec Institute of Technology, Auckland, New Zealand, 24–26 November 2010.
18. Love, J. Mapping the Impact of Changes in Occupant Heating Behaviour on Space Heating Energy Use as a Result of UK Domestic Retrofit. Retrofit Conference, University of Salford, UK, 2012. Available online: http://www.salford.ac.uk/__data/assets/pdf_file/0005/142385/023-Love.pdf (accessed on 21 January 2015).

19. Clevenger, C.M.; Haymaker, J. The Impact of Building Occupation on Energy Modelling Simulations, Joint International Conference on Computing and Decision Making in Civil and Building Engineering, Montreal, Canada, 2006. Available online: http://web.stanford.edu/group/peec/cgi-bin/docs/people/profiles/The%20Impact%20of%20the%20Building%20Occupant%20On%20Energy%20Modeling%20Simulations.pdf (accessed on 21 January 2015).
20. Department of Building and Housing Compliance document for New Zealand Building Code Clause H1 Energy Efficiency—Third Edition. Available online: http://www.dbh.govt.nz/UserFiles/File/Publications/Building/Compliance-documents/H1-energy-efficiency-3rd-edition-amendment-2.pdf (accessed on 21 January 2015).
21. *Specification for Preservative Treatment Part 3: Plywood*; SNZ 2012 AS/NZ 1604.3; Standards New Zealand: Wellington, New Zealand, 2012.
22. *Plywood—Structural Part 0: Specifications;* SNZ, 2012b, AS/NZS 2269.0:2012; Standards New Zealand: Wellington, New Zealand, 2012.
23. Moll, L.; van Raamsdonk, T. *A New Zealand Based Study on Airtightness and Moisture Management*; ProClima, NZ Ltd: Wellington, New Zealand, 2009.
24. Barley, D.; Deru, M.; Pless, S.; Orcellini, P. *Procedure for Measuring and Reporting Commercial Building Energy Performance*; Technical Report NREL/TP-550–38601; National Renewable Energy Laboratory, Battelle: Golden, CO, USA, 2005.
25. LASCAR Certificate of Calibration. Available online: http://www.lascarelectronics.com/pdf-usb-datalogging/data-logger0237687001331303304.pdf (accessed on 21 January 2015).
26. *Thermal Performance of Buildings—Determination of Air Permeability of Buildings—Fan Pressurization Method*; European Committee for Standardization CEN/TC 89 2000. EN 13829:2000; British Standards Institution: London, UK, 2001.
27. Stoecklein, A.; Bassett, M. *ALF3—A Design Tool for Energy Efficient Houses*; Building Research Association of New Zealand: Judgeford, New Zealand, 1999.
28. Passive House Institute Passive House Requirements. Available online: http://www.passiv.de/en/02_informations/02_passive-house-requirements/02_passive-house-requirements.htm (accessed on 11 June 2014).
29. TenWolde, A.; Pilon, C. The Effect of Indoor Humidity on Water Vapor Release in Homes. Available online: http://web.ornl.gov/sci/buildings/2012/2007%20B10%20papers/071_TenWolde.pdf (accessed on 11 June 2014).
30. Mitamura, T.; Rode, C.; Schultz, J. Full-Scale Testing of Indoor Humidity and Moisture Buffering in Building Materials. In Proceedings of the Indoor Air Quality 2001, Moisture, Microbes, and Health Effects: Indoor Air Quality and Moisture in Buildings, ASHRAE: Atlanta, GA, USA, 2001.

8

The Dissipative Column: A New Hysteretic Damper

Bruno Palazzo, Paolo Castaldo and Ivana Marino *

Department of Civil Engineering, University of Salerno, Via Giovanni Paolo II, 84084 Fisciano, Italy; E-Mails: palazzo@unisa.it (B.P.); pcastaldo@unisa.it (P.C.)

* Author to whom correspondence should be addressed; E-Mail: imarino@unisa.it

Academic Editor: Ali Memari

Abstract: A new replaceable hysteretic damper to better control seismic building damage, consisting of two or more adjacent steel vertical elements connected to each other with continuous mild/low strength steel shear links, is proposed and investigated in this paper. New Dampers, called Dissipative Columns (DC), continuously linked with X-shaped steel plates, provide additional stiffness and damping to a lateral system by using a basic and minimally invasive construction element: the column. Working in a way similar to coupled shear walls, the proposed element behavior is theoretically analyzed at linear and non-linear ranges. In fact, considering different restrained cases, a parametric analysis is developed in order both to evaluate the effect of the main geometrical and structural parameters and to provide the design capacity curves of this new damper. The DC can be considered a new damping device, easy to install in new as well as existing buildings in order to protect them from seismic damage.

Keywords: replaceable hysteretic damper; multi-bay dissipative column; damage control; amplified structural damping

---

1. Introduction

Strong earthquakes have shown that a large percentage of buildings in the affected areas, even if properly built and designed according to the most advanced codes, suffer such severe damages that they need to be demolished after the quake, since they would be expensive to repair. As is known, the acceptance of such level of damage due to severe earthquakes is related to the ductility-based design criteria that assume design seismic actions decrease by reduction factors. This approach may lead to high social and economic costs to the affected communities, and to a long recovery time for essential services and production activities. Inspired by new performance criteria, there is a growing belief that code design criteria are not sustainable for the high level of accepted damage, are impossible to repair, and that common buildings should be designed with a higher performance level. At the beginning of this century, performance based engineering [1] introduced new principles with the scope to select more articulated targets better corresponding to different building roles and use, defining a variety and complex subdivision of performance objectives for seismic events with different intensities and frequencies of occurrence. The "Direct Displacement Based Design" philosophy [2] relates the specified performance level to the strain or drift limits for a specified seismic intensity.

With the scope of minimizing structural damage, several dampers [3,4] as well as new strategies and integrated design approaches [5,6] have been developed in order to dissipate seismic input energy outside of the primary structure and new replaceable hybrid composite steel devices [7,8] have been recently proposed. Dampers can absorb a significant portion of the input energy reducing the hysteretic energy demand to the primary structural elements.

The damage in the main structural members can be prevented or minimized by connecting beams and columns with hinges, and by employing lateral-load resisting members to withstand lateral seismic load. In this case, most of the energy dissipation and structural damages caused by an earthquake will be concentrated on the lateral-load resisting members, and the main members will remain elastic. After the earthquake, the damaged lateral-load resisting members can be replaced easily at reasonable cost. The energy dissipation or damage prevention capacity of a steel framed structure can be greatly enhanced by employing buckling-restrained braces (BRBs). The structure system is composed of a hinge-connected main structure, which is designed to remain elastic under seismic load, and BRBs resisting all lateral loads. The BRBs dissipate most of the vibration energy through inelastic deformation while the other structural members remain elastic and undamaged. During a seismic event, the BRBs dissipate dynamic energy through stable hysteretic behavior [9].

In [10] Low Yielding Strength (LYS) steel is investigated for improving the ductility capacity of box-shaped steel bridge piers and experimental work is carried out for four specimens having different thickness and sectional configurations for cyclic loading test. The test results reveal that the LYS steel portion with longitudinal stiffeners greatly improves the strength and ductility capacity of box columns and it is observed that LYS steel has a great cyclic strain-hardening characteristic. The advantage of use of LYS steel is that it can effectively use large plastic deformation in component plates and the failure of column is concentrated at the LYS steel segment and the energy dissipation occurs far beyond the yield point.

The aim of this paper is to propose and investigate a new replaceable hysteretic damper having a basic form of the art of building, minimally architecturally invasive, consisting of two or more dissipative steel columns directly connected to two consecutive floors linked to each other with X-shaped low/mild steel plates. It will be shown that the new element is able to add significant stiffness and damping to the structural system to reduce seismic response and damage in primary structural members under severe earthquakes. The Dissipative Columns (DC) element will be investigated through non-linear pushover analyses useful to design the yielding properties of the proposed damper depending both on the characteristics of the primary structure and the expected performance as discussed by [11].

2. The Dissipative Column Concept

The proposed Dissipative Column models are shown in Figure 1 for two different configurations (models A and B). The elements can be considered as a sort of framed bi-pendulum with height equal to H, connected in parallel to the primary structure, able to react to the story drift Δ with a lateral force Q_D adding stiffness, strength and damping. The design concept of the DC element aims to obtain a lever mechanism by which a small inter-story drift provides an amplified vertical drift in the X-plate ends reacting with shear forces as shown in Figure 2. The X-shaped steel plates made of mild or Low Yielding Strength (LYS) steel, having length a, thickness t, width at the ends b and vertical distance i, are also used as shear links between coupled elements.

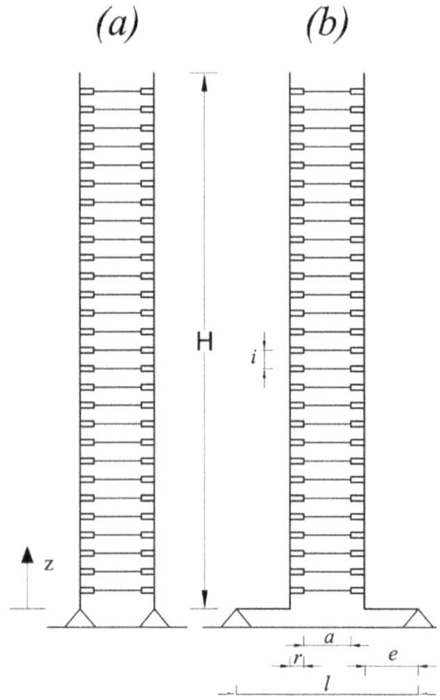

Figure 1. Dissipative Column Model: (**a**) Model A: Hinged at the ends; (**b**) Model B: Eccentrically Hinged at the base.

As for the model B, each lever arm is characterized by an eccentricity e. In both models, r is the rigid element representing half section of each column (Figures 1 and 2).

The top ends of the two models are linked to the upper floor through slotted bolted connections to allow large vertical displacements. As is known, well designed hysteretic dampers in framed structures should dissipate large rates of seismic input energy to control inter-story drift and damage in the primary structure.

By yielding a large volume of steel, the shear devices dissipate substantial input energy during earthquakes or strong winds, while also increasing damping in the entire system with the aim to reduce the damage in the primary system. With reference to reinforced concrete structures, the limit values of Inter-Story Drift Angle (ISDA) corresponding to different structural performance levels are suggested by Ghobarah [12].

The great advantages of DC elements, if compared with classical steel dissipative braces [9], are the reduced architectural invasiveness, so that they are able to be integrated in any building, the ease of installation everywhere, replacement after earthquakes and the stable behavior in cyclic reversal deformation. Axial forces should be designed in order to be transferred locally to a proper structural element.

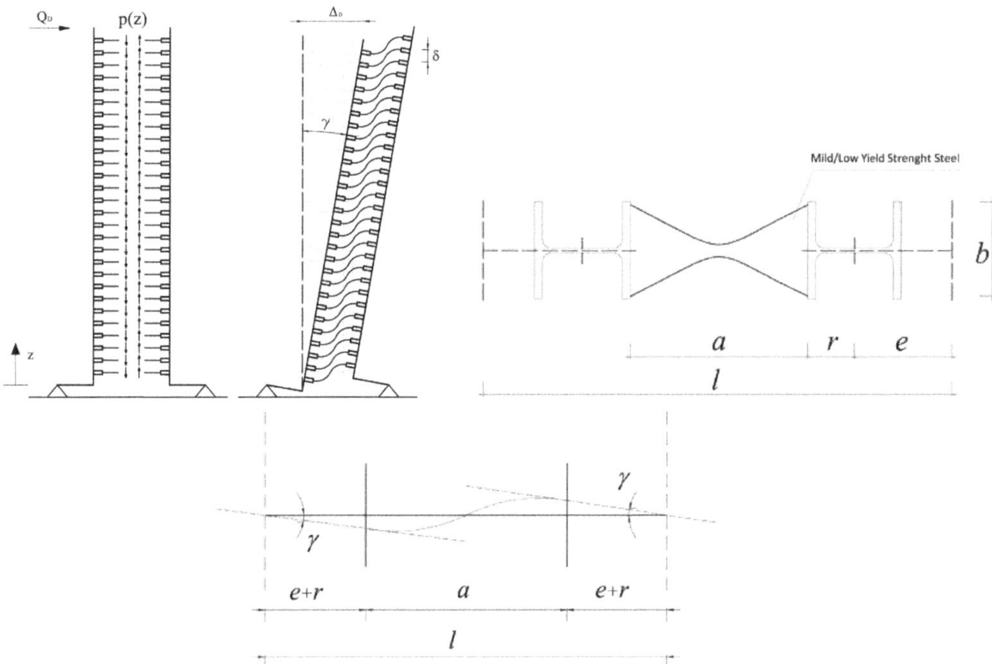

Figure 2. Eccentric Dissipative Column Model: Dissipative amplified mechanism.

3. Simplified Non-Linear Analysis of Doubly-Hinged DC

The vertical drift δ between the ends of generic shear link in the elastic range, being the curvature χ constant along each half plate, is related to shear force V_d developed by each X shaped plate, as:

$$\delta = 2\int_0^{a/2}\int_0^{a/2} \chi(x)dxdx = \frac{3}{2}\frac{a^3}{Ebt^3}V_d \tag{1}$$

where a, t, b, E are, respectively, the X-plate length, thickness, width at the ends and E is Young's modulus of steel plates. The axis x represents the barycentric axis of a generic steel plate. Hence, the single X-plate vertical stiffness is equal to:

$$K_d = \frac{V_d}{\delta} = \frac{2}{3}\frac{Ebt^3}{a^3} \tag{2}$$

With reference to the model A, a simplified analysis of the DC behavior subjected to relative displacement can be easily carried out assuming that the column flexural deformation is negligible in respect to the one of flexible inextensible links. An inter-story drift produces a shear drift angle γ and vertical drifts δ along the X-shaped steel plate. Under such simplified assumptions, the top-base relative displacement Δ of a DC element with height H is related to the drift angle γ as:

$$\Delta = \gamma \cdot H \tag{3}$$

Therefore, each X-plate undergoes a vertical drift equal to:

$$\delta = (l-a)\Delta/H \tag{4}$$

where l represents pin axes distance (Figure 2).

Therefore, the uniform distributed vertical load due to the shear drift angle γ applies:

$$p = \frac{2}{3}\frac{Ebt^3}{a^3 i}(l-a)\Delta/H \tag{5}$$

where i represents the plates vertical distance.

The term $(l - a)/2 = r$ represents a small lever arm that can be amplified using eccentricity e between the vertical axis and the supports as will be shown in the following. The axial force at the base of each column is equal to:

$$N_c = \frac{2}{3}\frac{Ebt^3}{a^3 i}(l-a)H\Delta/H \tag{6}$$

For the equilibrium, lateral force-displacement relationship is expressed as:

$$Q_D = \frac{2}{3}\frac{Ebt^3}{a^3 i}l\frac{\Delta}{H}(l-a) = K_D\Delta \tag{7}$$

where K_D represents the lateral stiffness of the Dissipative Column given by:

$$K_D = \frac{2}{3}\frac{Ebt^3}{a^3 i}\frac{l}{H}(l-a) \tag{8}$$

According to experimental tests [13,14], the load-deformation curve of the X-shaped mild steel plates can be idealized as a bilinear curve with a ratio of post yielding stiffness to the initial one equal to 0.03 and available displacement ductility ratio $\mu = \delta/\delta_y$ in the range between 3 and 5 [15]. Since yielding strength f_y is reached almost uniformly along the device, the yielding vertical load p_y can also be expressed as:

$$p_y = \frac{2M_{d,y}}{ia} = f_y\frac{bt^2}{2ia} \tag{9}$$

while the relative yielding displacement of each link can be written as:

$$\delta_y = \frac{V_{d,y}}{K_d} = f_y\frac{3a^2}{4Et} \tag{10}$$

Therefore, the lateral yield strength of the doubly hinged DC element can be evaluated as:

$$Q_{Dy} = f_y\frac{blt^2}{2ia} \tag{11}$$

and the yielding displacement applies:

$$\Delta_y = \frac{3}{4}\frac{f_y}{E}\frac{a^2 H}{t(l-a)} \tag{12}$$

The bending moment along each column under a lateral force Q_D results:

$$M_c = \frac{1}{2}(Q_D - pl)z \tag{13}$$

where z represents the vertical axis having origin in the hinge (Figure 1).

The dimensionless theoretical force-displacement relationship of a DC element consisting of two HE220 steel columns with $H = 3.5$ m linked to each other through X-shaped mild steel plates having $a = 200$ mm, $b = 200$ mm, $t = 10$ mm, $i = 50$ mm, $l = 480$ mm, is plotted in Figure 3.

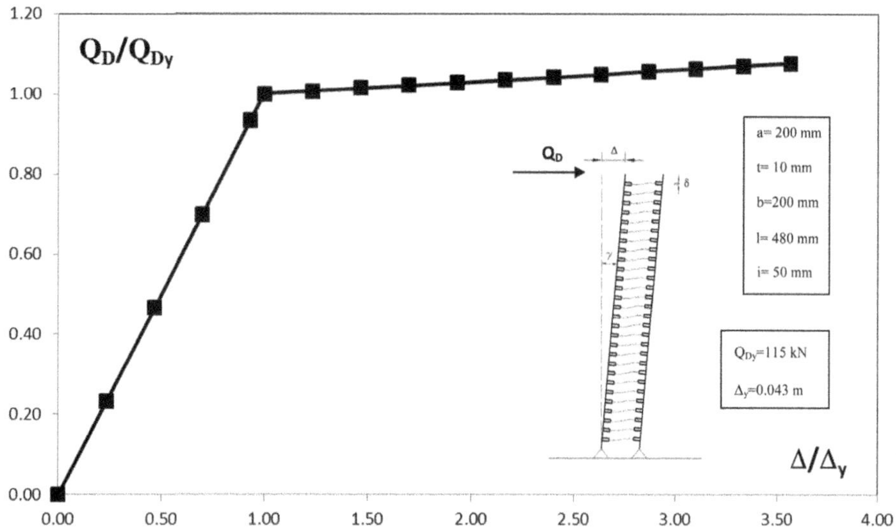

Figure 3. Model A: Theoretical force-displacement relationship.

Considering X-shaped LYS steel plates having yield strength 97.9 MPa at 0.2% offset strain with Young's modulus equal to 200 GPa, a parametric analysis of the DC element is developed in order to evaluate the effect of the flexible devices for different values of i and t. The results of the parametric analysis, illustrated in Figure 4, show that lateral stiffness and strength increase with greater values of t and with lower values of i.

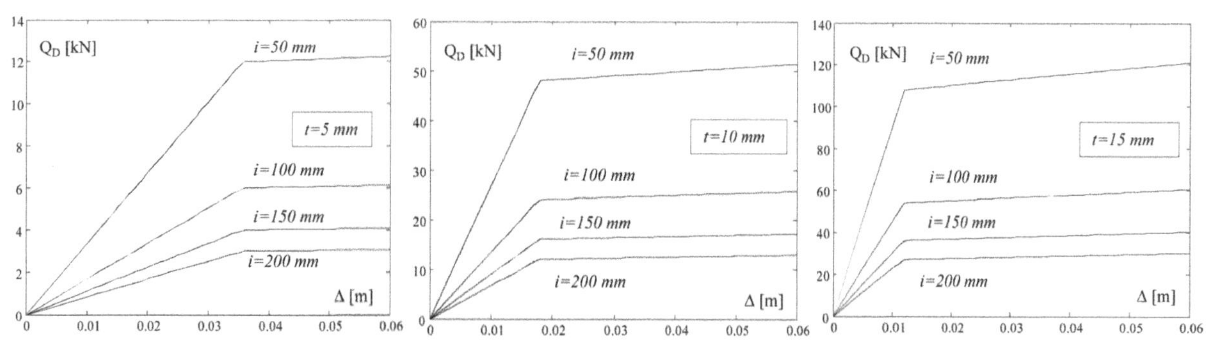

Figure 4. Model A: parametric analysis for different values of i and t.

Differently to the previous case, in presence of a significant eccentricity the column flexural deformation should not be neglected, therefore, the top-base relative displacement Δ of the DC element results being:

$$\Delta = \gamma \cdot H + Q_D\left(\frac{eH^3}{3EI_c l} + \frac{e^2H^2}{2EI_e l}\right) \tag{14}$$

and each X-plate undergoes a vertical drift δ (*z*):

$$\delta(z)=\gamma(l-a)+Q_D\left(2\frac{H}{l}\frac{e^3}{3EI_e}+\frac{H}{l}\frac{e^2}{EI_e}r+\frac{e}{l}\frac{z^2}{EI_c}r\right) \tag{15}$$

where E, I_c, I_e and $(l-a)=2(e+r)$ are, respectively, Young's modulus, inertia of each column, lever arm and the lever arm with the eccentricity e.

The uniform distributed vertical load applies:

$$p(z)=\frac{2}{3}\frac{Ebt^3}{a^3i}\left[\gamma(l-a)+Q_D\left(2\frac{H}{l}\frac{e^3}{3EI_e}+\frac{H}{l}\frac{e^2}{EI_e}r+\frac{e}{l}\frac{z^2}{EI_c}r\right)\right] \tag{16}$$

The axial force at the base of each column is equal to:

$$N_c=\int_0^H p(z)dz=\frac{2}{3}\frac{Ebt^3}{a^3i}\left[\gamma(l-a)+Q_D\left(2\frac{H^2}{l}\frac{e^3}{3EI_e}+\frac{H^2}{l}\frac{e^2}{EI_e}r+\frac{e}{l}\frac{H^3}{EI_c}r\right)\right] \tag{17}$$

For the equilibrium, lateral force can be expressed as:

$$Q_D=\frac{N_c l}{H}=\frac{\frac{2}{3}\frac{Ebt^3}{a^3i}\gamma(l-a)H\frac{l}{H}}{1-\left[\frac{2}{3}\frac{Ebt^3}{a^3i}\frac{l}{H}\left(2\frac{H^2}{l}\frac{e^3}{3EI_e}+\frac{H^2}{l}\frac{e^2}{EI_e}r+\frac{e}{l}\frac{H^3}{3EI_c}r\right)\right]} \tag{18}$$

4. Results of Non Linear Pushover Analyses

The responses of the DC elements have been analyzed by static pushover analyses using SAP2000 software [16]. The displacement-controlled pushover analyses have been performed until to reach 150 mm in terms of relative displacement modeling the non-linear response of the steel plates through the Ramberg-Osgood (RO) curve with a ratio of post yielding stiffness over the initial one equal to 0.03. The analyzed dampers are composed of two steel columns with different X-plates varying width, thickness and yield stress. Mechanical and geometrical properties of the tested models (models A and B) are shown in Table 1.

Table 1. Geometrical and structural properties of the Dissipative Column (DC) elements.

Model	Plate Thickness t (mm)	Height H (mm)	Plate Length a (mm)	Plate Distance i (mm)	Plate Width b (mm)	Plate Yield Stress f_y (N/mm²)	Column Distance l (mm)	Eccentricity e (mm)	Column Profile
A_1	5								
A_2	10	3500	200	50	200	240	480	0	HE220B S235
A_3	15								
A_4	20								
							360	0	
B_1	5						560	100	HE160B
B_2	10	3500	200	50	150	100	760	200	S460
B_3	15						960	300	

With reference to model A, the displacement-controlled pushover curves of the DC models until to reach 150 mm for different thickness values are illustrated in Figure 5.

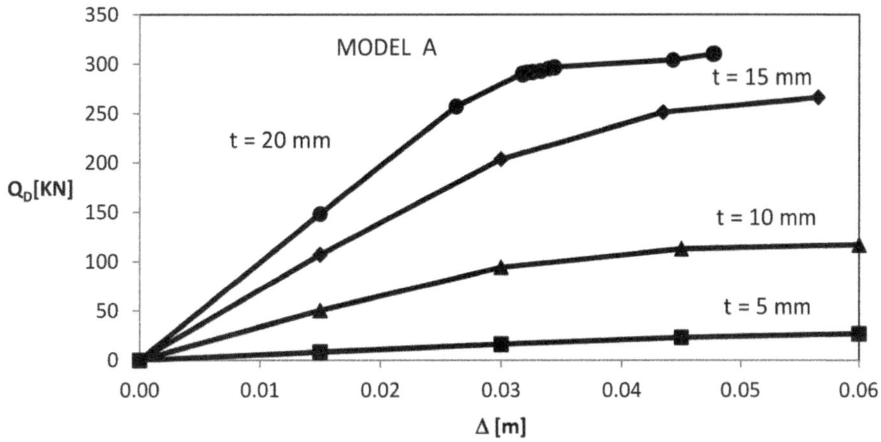

Figure 5. Model A: Pushover curves for different plate thickness.

The diagrams of the column bending moment, axial force, link bending moment related to model A_2 (t = 10 mm) are represented in Figure 6. The column bending moment is characterized by discontinuities in correspondence of each shear link due to the bending moments developed on each plate, as shown in Figure 6a. The axial force distribution along the columns is linear, in a manner consistent with the simplified assumptions previously examined, as shown in Figure 6b. Both columns remain quite far from their elastic limits (Figure 7) [17], while shear links immediately enter the plastic range as the lateral force increases. In particular, the steel plates located lower are able to dissipate more energy than the ones located higher, as shown in Figure 6c.

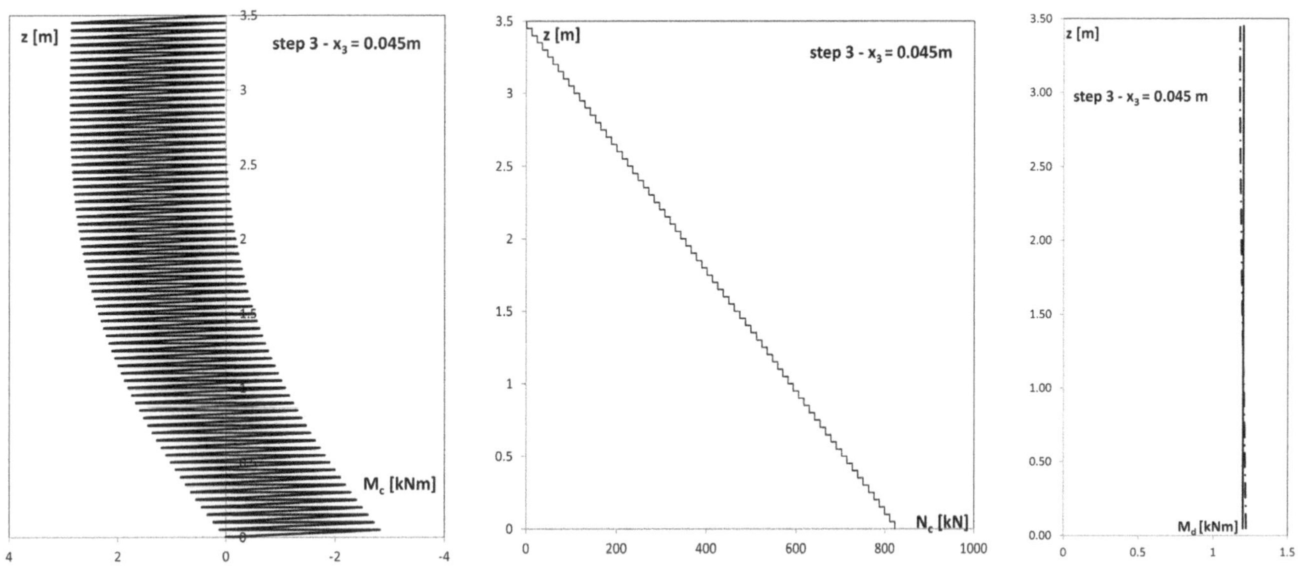

(a) (b) (c)

Figure 6. Model A_2: (**a**) Column Bending Moment; (**b**) Axial Force; (**c**) Link Bending Moment.

The model A_4 collapses before due to excess of non-linear deformations as demonstrated in Figure 7.

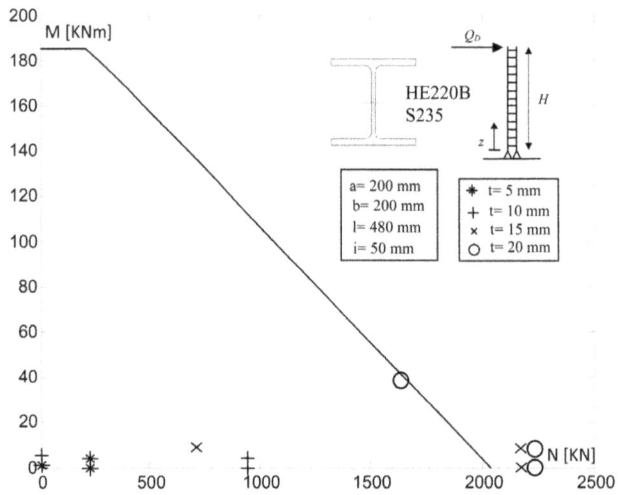

Figure 7. Model A_2: Stress Points on MN Domain at last pushover step.

Comparisons between theoretical and numerical DC behavior of model A are plotted in Figures 8 and 9. Force-displacement relationships result comparable and consistent with the assumption that the column deformation can be considered as a rigid neglecting the small curvature respect to the shear links' one. The maximum ductility factor $\mu = \delta/\delta_y$ demanded to model A X-plates, at the last considered step (150 mm), lower than the values suggested by [15], is shown in Figure 10.

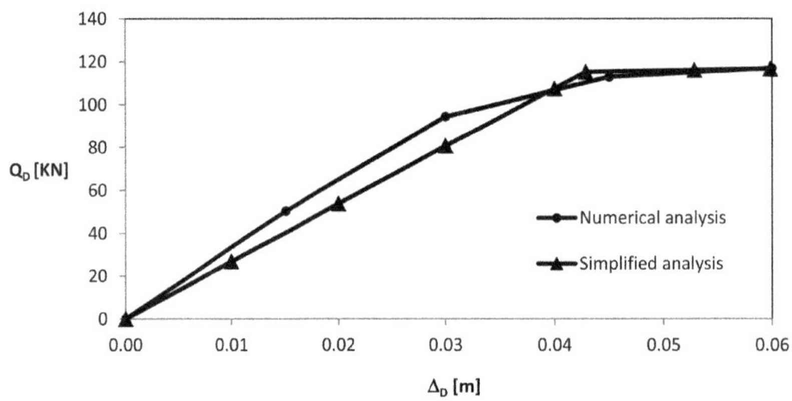

Figure 8. Model A_2: Theoretical *vs.* pushover analysis force displacement relationship.

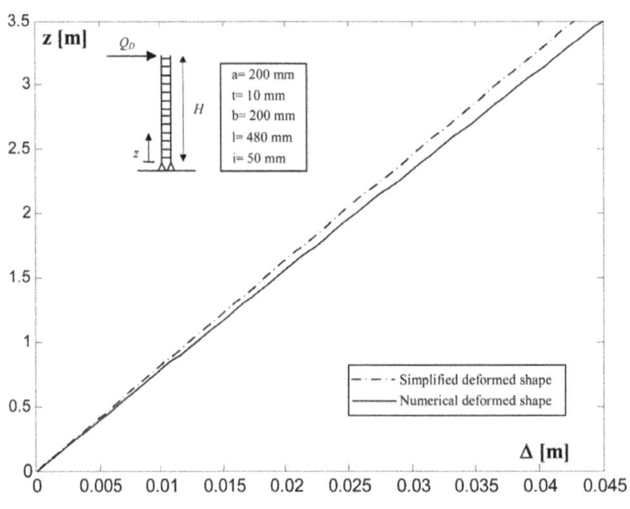

Figure 9. Model A_2: Simplified *vs.* numerical deformed shapes.

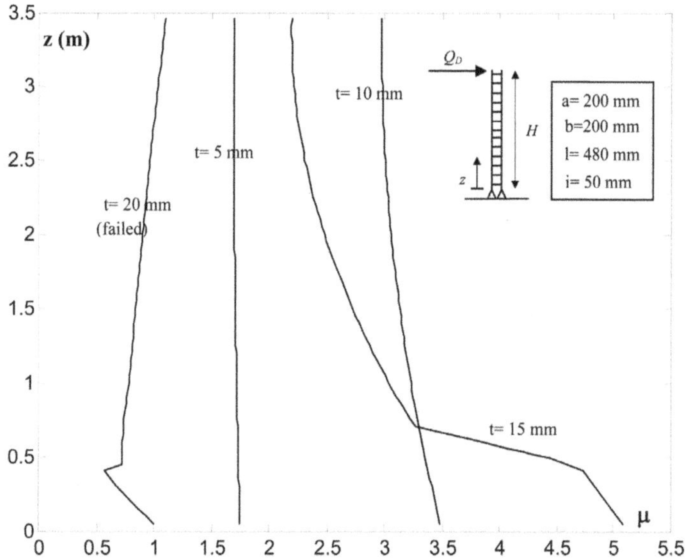

Figure 10. Model A: Displacement ductility requirements for steel plates at the last step.

With reference to model B, with eccentricity, the pushover curves for different X-plate thickness and eccentricity values, are shown in Figure 11a–c.

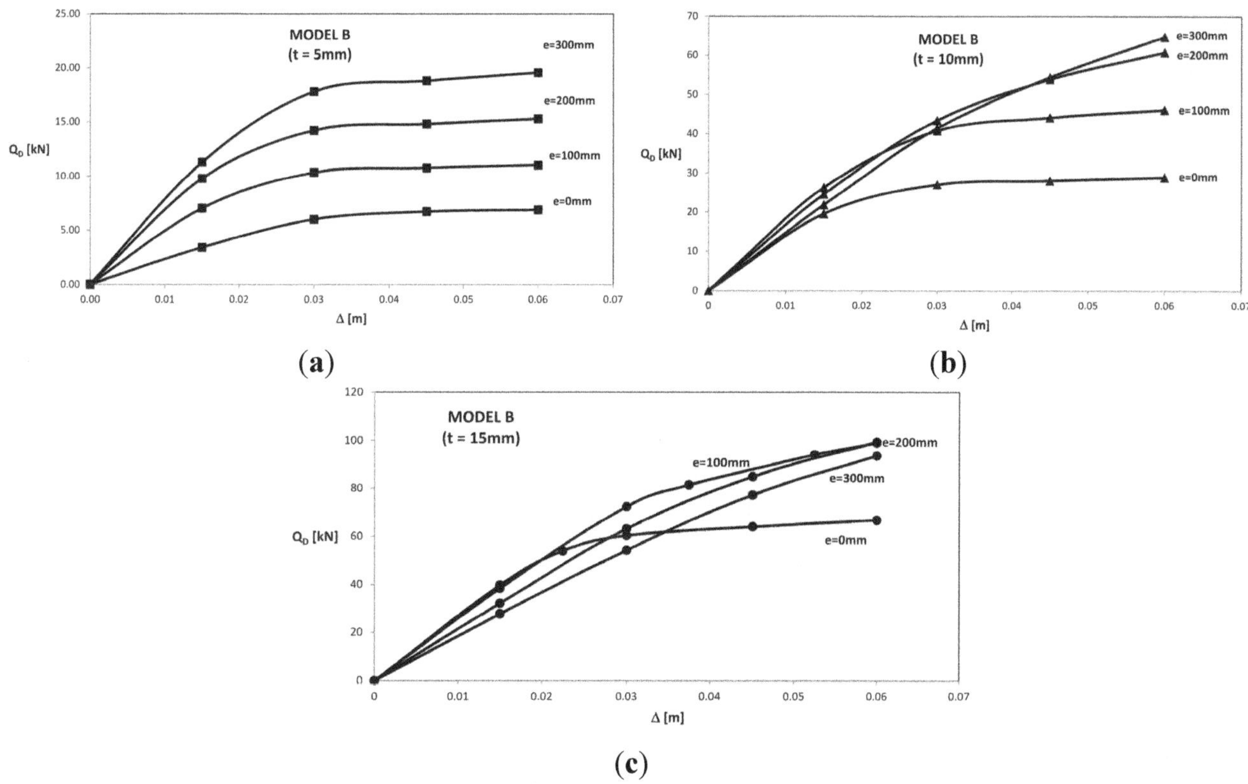

Figure 11. Model B: Pushover curves for different eccentricity *e* and plate thickness *t*: (**a**) *t* = 5 mm; (**b**) *t* = 10 mm; (**c**) *t* = 15mm.

Results show that the higher value of eccentricity produces an increase in both stiffness and lateral strength, while X steel plates yield lower displacement Δ. It follows that the DC element with significant eccentricity is able to add more stiffness and lateral strength to the primary structure and dissipate more energy than the small eccentricity case under the same displacement Δ. Figure 12 shows the displacement ductility factor along the height of the model B_2, required at displacement Δ equal to 0.045 m for different values of eccentricity. Figure 12 also shows that the steel plates located higher are able to dissipate more energy than the previous configuration presenting higher values of the demanded ductility factor and confirming the effectiveness of the eccentric DC element.

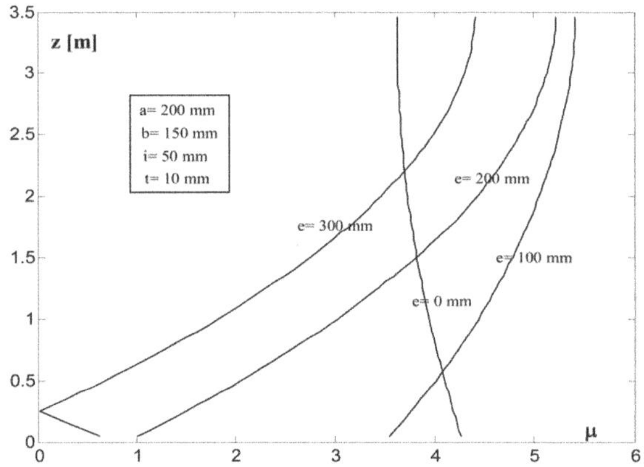

Figure 12. Model B_2: Displacement ductility requirements for steel plates at 4th step.

5. Multi-Bay Dissipative Columns

With reference to double-hinged model, in several cases it is possible to adopt multi-bay dissipative columns organized as represented in Figure 13. Let us consider the case of an assembled n DC elements, assuming the same simplified hypothesis used for the two adjacent columns case. The assembled system story drift is related to the shear drift angle γ as:

$$\Delta = \gamma \cdot H \tag{19}$$

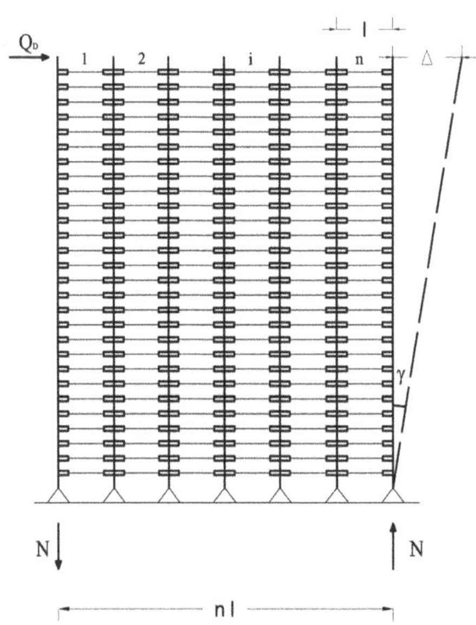

Figure 13. Multi-bay dissipative columns.

The i-th bay transmits to the column a uniformly distributed load p_i given by:

$$p_i = \frac{2}{3} \frac{Ebt^3}{a^3 i} \gamma (l - a) \tag{20}$$

Therefore, the internal columns are not subjected to axial force while only the two external ones reach, at the base, the axial force expressed by:

$$N = \frac{2}{3}\frac{Ebt^3}{a^3 i}(l-a)\Delta \tag{21}$$

The n-bays assembled dissipative columns react through a shear force given by:

$$Q_D = \frac{2}{3}\frac{Ebt^3}{a^3 i} nl \frac{\Delta}{H}(l-a) = K_D \Delta \tag{22}$$

where K_D is the sum the lateral stiffness developed by the links of each bay. Similarly, the post yield stiffness of the multi-bay element is equal to the sum of the single-bay one.

The responses of some multi-bay DC elements have been investigated through static pushover analyses. Adopting X-shape plates made of LYS steel, the yielding displacement and lateral strength obviously decrease as well as the axial force in both columns.

Multi-bay dissipative columns equipped with X-shape plates made of LYS steel represent a design solution to obtain higher lateral stiffness and low strength. Figure 14 comparatively shows the response curves for the case of 6-bays and 1-bay DC elements equipped with LYS steel X-shape plates between S235 HE220 steel columns having properties reported in Table 2.

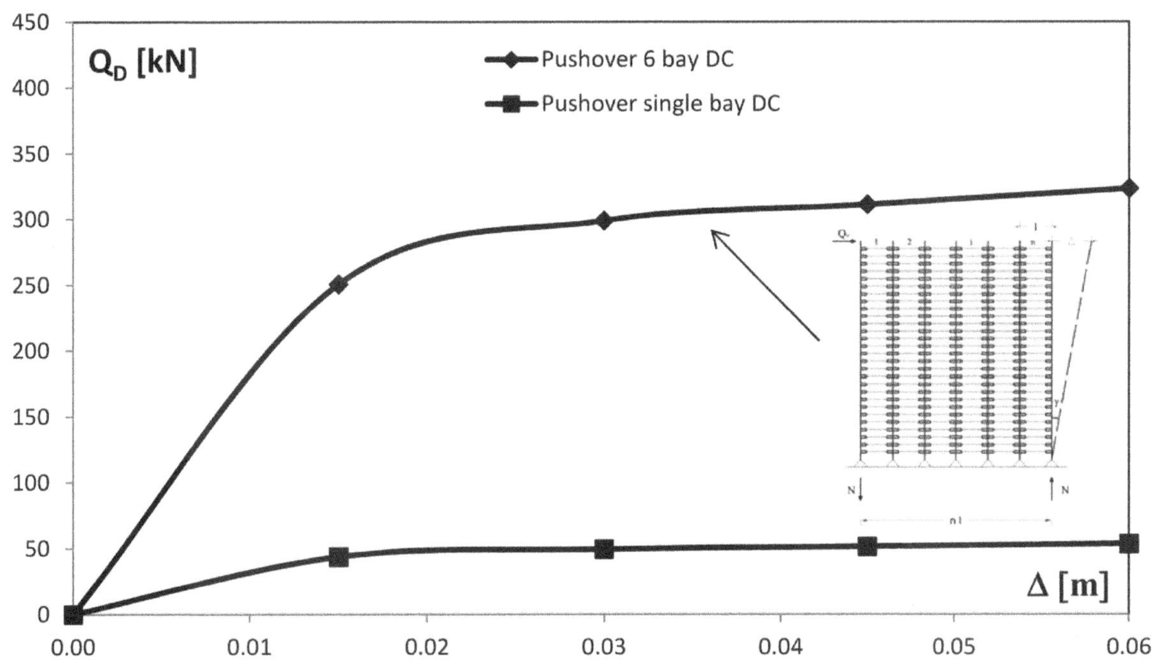

Figure 14. Six bays *vs.* single bay DC pushover curves.

Table 2. Geometrical and structural properties of the n-bay DC elements with Low Yielding Strength (LYS) steel plate.

Model	Height H (mm)	Column Distance l (mm)	Plate Length a (mm)	Plate Thickness t (mm)	Plate Distance i (mm)	Plate Width b (mm)	Plate Yield Stress f_y (N/mm²)	Restrains
1-bay DC	3500	480	200	10	50	20	100	model A
6-bays DC	3500	480	200	10	50	20	100	model A

Then a parametric analysis has been carried out in order to define the force-displacement relationship of the model for different values of the thickness *t*, number of bays and plates distance *i* (Figure 15).

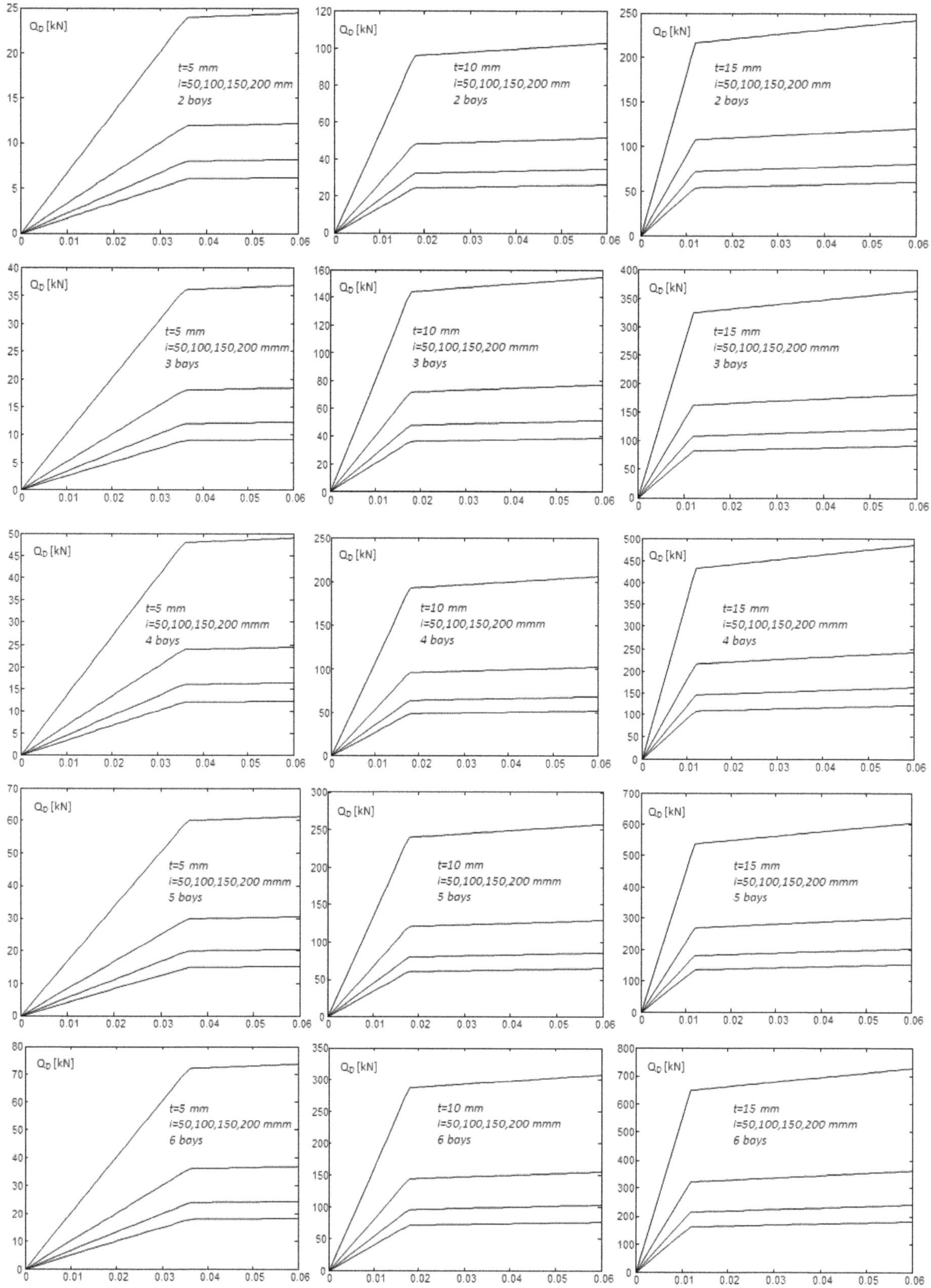

Figure 15. Parametric analysis of the force-displacement relationships of the model for different values of the thickness *t*, number of bays and plates distance *i*.

As shown in Figure 15, adopting X-shape plate made of LYS steel, the lateral stiffness and strength of the Multi-bay dissipative columns increase for lower values of *i* and for more bays. The yielding displacement decreases as device thickness increases.

6. Conclusions

A new replaceable hysteretic steel damper to control seismic building damage consisting of two or more adjacent steel Dissipative Columns (DC) connected to each other with continuous mild/low strength steel X-shaped plates, has been proposed and investigated.

The behavior of the proposed DC element with two different configurations has been investigated at linear and non-linear ranges, developing several parametric analyses in order to evaluate the hysteretic performances.

Numerical tests showed that doubly-hinged DC elements with low eccentricity can fully yield along their length while columns are subjected to axial strain and so, small bending moments can be neglected flexural deformation.

In the presence of significant eccentricity, large bending moments have been observed that affect the section size. Eccentricity acts as a mechanical lever arm in order to amplify energy dissipation increasing plate vertical drift producing easier yielding conditions in hysteretic dampers.

In any case, lateral stiffness increases for greater values of the X-shaped plate thickness and for lower values of their distance, and for greater eccentricities between bearings and column axes. For greater values of the device thickness, the yielding displacement decreases as well as for lower values of link steel strength. Strongly coupling dissipative affects can be obtained by using a mechanical lever to amplify the X-plate drift. The greater is the eccentricity, the higher the bending moment.

Multi-bay dissipative columns with X-shape plate made of LYS steel represent a design solution to obtain higher lateral stiffness and lower strength, while also strongly reducing the axial force in the internal columns. The number of DC elements should be designed for the buildings where they are to be installed.

The DC elements can be considered as new low-yield, ductile replaceable and minimally invasive dampers easy to install to new as well as to existing buildings, providing significant additional stiffness, strength and damping to a structural system potentially capable to reduce building seismic response and damage in primary structural members under severe earthquakes.

Author Contributions

Bruno Palazzo, Paolo Castaldo and Ivana Marino conceived the idea and designed the numerical analyses; Paolo Castaldo and Ivana Marino performed the numerical analyses; Bruno Palazzo, Paolo Castaldo and Ivana Marino analyzed the data; Bruno Palazzo, Paolo Castaldo and Ivana Marino wrote the paper.

Conflicts of Interest

The authors declare no conflict of interest.

References

1. California Office of Emergency Services (OES). *Vision 2000: Performance Based Seismic Engineering of Buildings*; Structural Engineers Association of California: Sacramento, CA, USA, 1995.
2. Priestley, M.J.N. Myths and fallacies in earthquake engineering-conflicts between design and reality. *Bull. NZNSEE* **1993**, *26*, 329–341.
3. Symans, M.D.; Charney, F.A.; Whittaker, A.S.; Constantinou, M.C.; Kicher, C.A.; Johnson, M.W.; McNamara, R.J. Energy dissipation system for seismic applications: Current practise and recent developments. *J. Struct. Eng.* **2008**, *134*, 3–21.
4. Wada, A.; Watanabe, A.; Connor, J.; Kawai, H.; Iwata, M. Damage Tolerant Structures. In Proceeding of the Fifth U.S.-Japan Workshop on the Improvement of Building Structural Design and Construction Procedures, ATC-15, San Diego, CA, USA, 8–10 September 1992.
5. Castaldo, P. *Integrated Seismic Design of Structure and Control Systems*; Springer International Publishing: New York, NY, USA, 2014.
6. Castaldo, P.; de Iuliis M. Optimal integrated seismic design of structural and viscoelastic bracing-damper systems. *Earthq. Engng Struct. Dyn.* **2014**, *43*, 1809–1827.
7. Pampanin, S. Emerging solutions for high seismic performance of precast-prestressed concrete buildings. *J. Adv. Concr. Technol.* **2005**, *3*, 202–222.
8. Pampanin, S. Reality-Check and Renewed Challenges in Earthquake Engineering: Implementing Low-Damage Structural Systems-From Theory to Practice. In Proceeding of the 15WCEE, Lisbon, Portugal, 24–28 September 2013.
9. Kim, J.; Seo, Y. Seismic design of low-rise steel frames with buckling-restrained Braces. *Eng. Struct.* **2004**, *26*, 543–551.
10. Susantha, K.A.S.; Aoki, T.; Kumano, T.; Yamamoto, K. Application of Low-Yield Strength Steel in Steel Bridge Piers. In Proceeding of the 13th WCEE, Vancouver, Canada, 1–6 August 2004.
11. Oviedo, A.J.A.; Midorikawa, M.; Asari, T. Earthquake response of ten-story story-drift-controlled reinforced concrete frames with hysteretic dampers. *Eng. Struct.* **2010**, *32*, 1735–1746.
12. Ghobarah, A. On Drift Limits Associated with Different Damage Levels. In Proceedings of the International Workshop on Performance-Based Seismic Design, Department of Civil Engineering, McMaster University, Bled, Slovenia, 28 June–1 July 2004.
13. Aiken, I.; Nims, D.; Whittaker, A.; Kelly, J. Testing of passive energy dissipation systems. *Earthq. Spectra* **1993**, *9*, 335–370.
14. Whittaker, A.; Bertero, V.; Alonso, J.; Thompson, C. *Earthquake Simulator Testing of Steel Plate Added Damping and Stiffness Elements*; Report No. UCB/EERC-89/02; Earthquake Engineering Research Center, University of California, Berkeley, CA, USA, 1989.
15. Whittaker, A.S.; Bertero, V.V.; Thompson, C.L.; Alonso, L.J. Seismic testing of steel plate energy dissipation devices. *Earthq. Spectra* **1991**, *7*, 563–604.
16. *SAP2000 vers*; Computers Structures Inc.: Berkeley, CA, USA, 2010.

17. European Committee for Standardization (CEN). *Eurocode 8 Part 1: General Rules, Seismic Actions and Rules for Buildings*; CEN: Brussels, Belgium, 2005.

9

Modeling a Decision Support Tool for Buildable and Sustainable Building Envelope Designs

Natee Singhaputtangkul [1,*] and Sui Pheng Low [2]

[1] School of Engineering and Technology, Central Queensland University, Sydney, NSW 2000, Australia
[2] Department of Building, School of Design and Environment, National University of Singapore, 4 Architecture Drive, Singapore 117566, Singapore; E-Mail: bdglowsp@nus.edu.sg

* Author to whom correspondence should be addressed; E-Mail: n.singhaputtangkul@cqu.edu.au

Academic Editor: Fitsum Tariku

Abstract: Sustainability and buildability requirements in building envelope design have significantly gained more importance nowadays, yet there is a lack of an appropriate decision support system (DSS) that can help a building design team to incorporate these requirements and manage their tradeoffs at once. The main objective of this study is to build such a tool to facilitate a building design team to take into account sustainability and buildability criteria for assessment of building envelopes of high-rise residential buildings in Singapore. Literature reviews were conducted to investigate a comprehensive set of the sustainability and buildability criteria. This also included development of the tool using a Quality Functional Deployment (QFD) approach combined with fuzzy set theory. A building design team was engaged to test the tool with the aim to evaluate usefulness of the tool in managing the tradeoffs among the sustainability and buildability criteria. The results from a qualitative data analysis suggested that the tool allowed the design team to effectively find a balance between the tradeoffs among the criteria when assessing multiple building envelope design alternatives. Main contributions of using this tool are achievement of a more efficient assessment of the building envelopes and more sustainable and buildable building envelope design.

Keywords: building envelope materials and designs; decision support system; quality function deployment; fuzzy set theory; high-rise residential buildings

---

1. Introduction

Building envelope systems, as the interface between interior space and exterior environment, generally serve the function of weather and pollution exclusion and thermal and sound insulation [1]. Their performance affects, for example, occupant comfort and productivity, energy use, strength, durability, fire resistance, and aesthetics appeal of a building [2,3]. Building envelopes have played an important role the success of private high-rise residential building projects, and this highlights the importance of assessment and selection of the building envelope materials and designs that can satisfy project requirements. The project requirements can be referred to as criteria for achieving sustainability and buildability in building envelope design. In brief, sustainability is a balance of social and economic activities and the environment [4], while buildability points to an ability to construct a building efficiently, economically and to agreed quality levels using its construction resources [5].

Each country may have different guidelines and legal responsibilities for sustainability and buildability in building design. In Singapore, sustainability and buildability regulations have been in place for several years [6,7]. However, it was found that these regulations do not cover all key project requirements [8]. Fundamentally, a design team can easily find the design solutions that meet minimum needs of the regulations if the team does not have to incorporate other key sustainability and buildability criteria that could affect their designs such as durability of materials, aesthetics, costs and so on [9]. Nevertheless, in practice, it is difficult to develop an optimal sustainable and buildable design that takes into account all key criteria without effective management of their tradeoffs [10]. This leads to a main research problem of this study, which is an absence of a decision support tool that can enable the designers in Singapore to assess building envelope designs based on all the sustainability and buildability criteria at once and to effectively manage the tradeoffs among such criteria.

This study therefore develops a new and robust decision support system (DSS) to fill this gap. The study applies a Quality Function Deployment (QFD) approach and fuzzy set theory to build the tool. This tool is known as a fuzzy QFD tool. The study begins by conducting literature reviews to identify a comprehensive set of the sustainability and buildability criteria for assessment of the building envelopes. This is followed by development of the fuzzy QFD tool embedded with a set of the identified criteria. The study then conducts a series of semi-structured interviews with 15 architects and engineers in Singapore in parallel with thorough literature reviews to acquire relevant knowledge for the tool and to fine-tune/validate the tool. The primary feedback from the interviews of these 15 architects and engineers is applied to adjust a decision-making process of the tool towards the achievement of more practical applications. This adjustment resulted in a more user-friendly tool being achieved.

A design team consisting of an architect, civil & structural (C&S) engineer, and mechanical and electrical (M&E) are then engaged to test this tool as part of a case study. The team is given a representative high-rise residential project with a task to use the tool to select a building envelope system for this project. After the exercise is completed, the study conducts semi-structured interviews to collect perspectives of the team members regarding effectiveness of using the tool for incorporating the criteria and managing their tradeoffs. The interview results from the case study are then analyzed qualitatively.

2. Sustainability and Buildablity

Sustainable development or sustainability is the fundamental principle underlying various efforts to ensure a decent quality of life for future generations. This concept aims to meet the needs of the present without compromising the ability of future generations to meet their needs. This implies that the environment and the quality of human life are as important as economic performance [1]. It was agreed that the mainstay of sustainability thinking is to strike a balance between three dimensions: environmental, social and economic impacts of the design. In the construction industry, awareness of sustainability among building professionals has increased [4]. This can be seen where the green market has been promoted to bring major improvements through developing green buildings, or where implementation of an energy rating guideline to assess environmental and energy performance of buildings has become more important. Benefits of adopting this concept in the construction industry include minimizing operating and maintenance costs, minimizing construction wastes, increasing occupant health and satisfaction, and so on [11,12].

Notwithstanding the concept of sustainability, buildability or constructability of a building also plays an important role in building design and construction. Buildability is widely accepted to be defined as the extent to which the design of a building facilitates ease of construction, subject to the overall requirements for the completed building [13]. Buildability takes into account all aspects of a building project, and enables the optimum utilization of construction resources. Several factors have been proposed over the years for achieving good buildability such as simplicity of design details, ease in material handling, ease in construction, *etc*. Benefits of buildability include lower costs of bidding, reduced site labor, increased cost effectiveness, and better resource utilization [14,15]. It was suggested that the design stage is critical for implementing buildability [16,17].

This study conducted extensive literature reviews to compile the comprehensive list of the sustainability and buildability criteria and this is shown in Table 1. These criteria are divided into sustainability and buildability categories. The sustainability category is also made up of three subcategories which are environmental impact, economic impact and social impact.

Table 1. Sustainability and buildability criteria for building envelope design.

Sustainability and buildability criteria			Selected references
Sustainability	Environmental impact	EN1: Energy consumption	Bryan [18], Scheuer *et al.* [19]
		EN2: Resource consumption	Chen *et al.* [20], Tsai *et al.* [21]
		EN3: Waste generation	Jaillon and Poon [22], Tsai *et al.* [21]
	Economic impact	EC1: Initial costs	Chua and Chou [3], Wang *et al.* [23]
		EC2: Long-term burdens	Das *et al.* [24], Lacasse *et al.* [25]
		EC3: Durability	Bryan [18], Kneifel [26]
	Social impact	SC1: Energy efficiency	Chua and Chou [3], Kibert [1]
		SC2: Appearance demands	Kaklauskas *et al.* [27], Yang *et al.* [10]
		SC3: Health, safety and security of occupants and society	Brock [28], Chew [2], Yu and Kim [29]
		SC4: Weather protection performance	Bryan [18], Yang *et al.* [10]
		SC5: Acoustic protection performance	Low *et al.* [15], Yang *et al.* [10]
		SC6: Visual performance	Low *et al.* [30], Nielsen [31]

Table 1. *Cont.*

Sustainability and buildability criteria		Selected references
Buildability	BC1: Health and safety of workers	Chen *et al.* [20], Hinze *et al.* [32]
	BC2: Community disturbance	Chew [2], Kibert [1]
	BC3: Simplicity of design details	Bryan [18], Yang *et al.* [10]
	BC4: Material handling	Chew [2], Gould [33]
	BC5: Ease in construction with respect to time	Chen *et al.* [20], Low *et al.* [15]
	BC6: Material deliveries from suppliers	Gould [33], Vrijhoef and Kisela [34]

3. Quality Functional Deployment (QFD)

QFD has been found useful in making decisions of organizations. QFD aims at satisfying customer needs and expectations, which are considered utmost importance for the organizations. Many companies have adopted a QFD approach to improve quality of their products to satisfy their customers [35,36]. This is also obvious in the building industry. As a building project is relatively unique in the sense that each building is tailor-made to meet the requirements and needs of the customers that refer to all stakeholders of a project, using a QFD approach seems to make good sense [37]. In brief, QFD links the customers' requirements to engineering characteristics. A conventional QFD tool controls quality of the project through its House of Quality (HOQ), which contains rooms to achieve targets of the project [35,36].

4. Development of a Fuzzy QFD Tool

Granted that QFD has been considered to be one of the most effective tools in facilitating group decision-making and suggesting optimal solutions, this study therefore taps into and, at the same time, expands this capability of a conventional QFD tool to build a more advanced QFD tool. In this regard, the convention HOQ was applied and modified to build an HOQ for Sustainability and Buildabilty (HOQSB) for the assessment of the building envelopes in this study. This HOQSB as shown in Figure 1 has five major rooms which are criteria for sustainability and buildability room (CR), building envelope materials and designs room (MR), relationships between the criteria and materials and designs room (RR), fuzzy inference engine for prioritizing design alternatives room (FR), and preference list room (PR). The CR storing the identified list of the sustainability and buildability criteria assists the Decision Makers (DMs) to identify the criteria for the assessment of the building envelopes, while the MR facilitates selection of the building envelope materials and designs. The RR contains the relationships between the criteria and design alternatives. These relationships include a matrix to indicate the parameters affecting each criterion and rules to guide the DMs when making the assessment decisions. The FR stores fuzzy calculation techniques operated by a fuzzy inference engine for making the tradeoffs between the criteria and fine-tuning the design alternatives. The PR records outputs of the FR in the form of a preference list of the design alternatives ranked by a Sustainability and Buildability Index (SBI). This index is a function of importance weights of the criteria and performance satisfactions of the design alternatives with respect to the criteria. This study then modeled the tool using Microsoft Visual Studio.

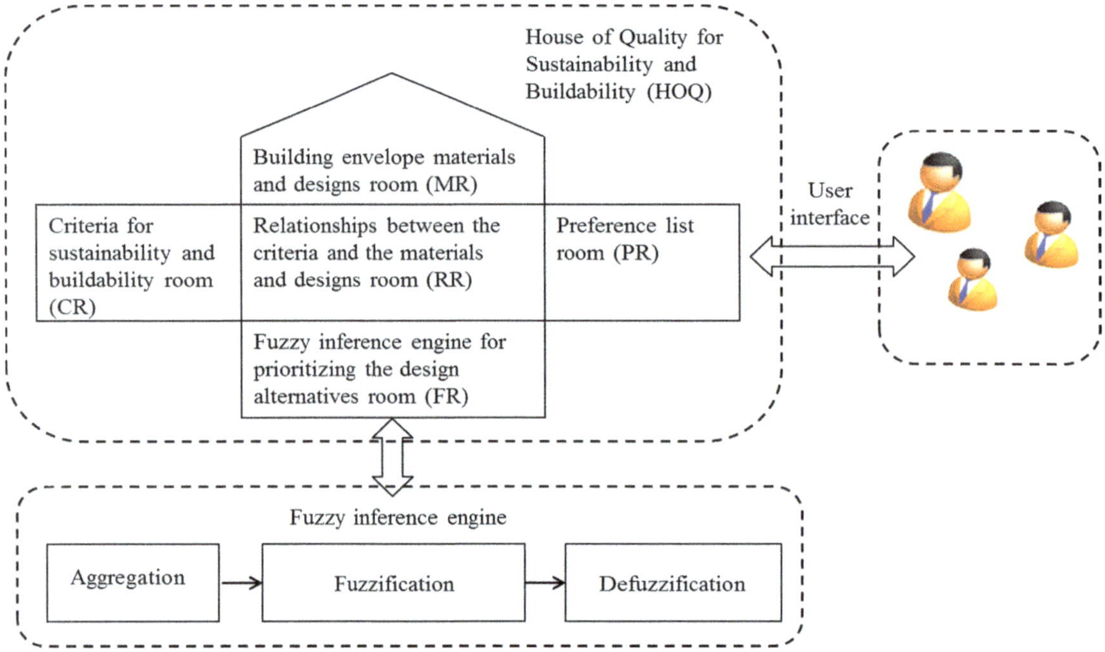

Figure 1. House of Quality for Sustainability and Buildabilty (HOQSB).

4.1. Fuzzy Inference Engine

In a real-world decision situation, making decisions has to process not only large amount of information but also subjective and uncertain requirements [38]. Particularly, DMs may encounter practical constraints from several criteria usually containing imprecision, subjective or vagueness inherent in the information [39]. It was suggested that the fuzzy set theory introduced by Zadeh [40] could mitigate this problem by translating subjective information, incomplete information and partially ignorant facts into the decision model. Hence, the study adopted the fuzzy set theory to build the fuzzy QFD tool, and this was embedded as part of the FR.

4.2. Fuzzy Linguistic Terms

The triangular fuzzy numbers and their fuzzy linguistic terms were used in this study as shown in the screenshot given in Figure 2 for assessment of importance weights of the sustainability criteria and performance satisfactions of the building envelope designs [39]. Specifically, the DMs determine the tradeoffs among the criteria through the use of these different fuzzy linguistic terms. Calculation examples of fuzzy operations can be found in Bayrak *et al.* [41] and Yang *et al.* [10].

Based on the HOQSB and fuzzy inference engine developed, this study formed a five-step decision making process for the DMs to provide their inputs through a user interface of the tool as follows:

Step 1: Input project relevant information, and set up the triangular fuzzy linguistic terms.

Step 2: Select the criteria for the assessment based on the CR.

Step 3: Assess the importance weights of all the criteria selected.

Step 4: Select the building envelope materials to form the design alternatives based on the MR.

Step 5: Assess the performance satisfactions of the design alternatives with respect to the selected criteria through the RR.

The FR embedded with the fuzzy inference engine then calculates the SBI and presents this through the PR.

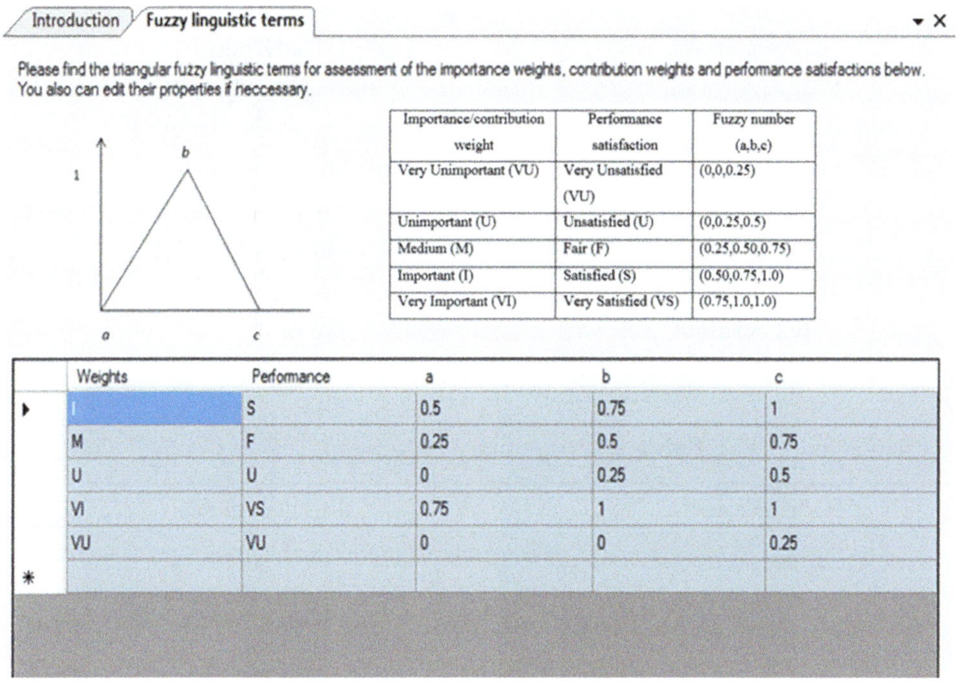

Figure 2. Fuzzy linguistic terms.

5. Case Study

This study employed a case study as a research design. This is because the case study approach meets the aim of this study, which is to investigate in-depth understanding of how the fuzzy QFD tool influences the decision-making process when the building design team assesses the building envelopes and the assessment outcomes. A case study tells a big story even if this is through the lens of a small case. This approach offers a close collaboration between researchers and participants, while enabling the participants to share their underlying perspectives [42]. Through this approach, the participants are able to express their views of reality, so much so that this allows the researcher to better understand the participants' actions and perspectives [42,43].

A hypothetical design team was engaged in the case study to determine a conceptual design of the building envelopes for a representative private high-rise residential building project. This design team consists of three DMs; namely architect (DM1), C&S engineer (DM2) and M&E engineer (DM3). The project general information and criteria preliminarily identified by the architect were given in Tables 2 and 3, respectively. For simplicity, the project criteria do not cover all the identified sustainability and buildability criteria. A main task of the team was to propose a building envelope system of this project to the developer by using the tool through its five-step decision making process to assess potential alternatives.

Table 2. General project information.

Developer	Condominium developer
Project title	High-rise residential building B
Contract type	Design-Bid-Build
Project location	Jurong East
Preferred external wall material	Precast/concrete block/claybrick
Orientation/Plan configuration	North-South/Square
WWR	0.3
Height	90 m
Floor-to-floor	3 m
Area per floor	400 m^2
Design and construction period	28 months

Table 3. Project key criteria.

Criteria category	Criteria name	Brief description
Environmental	EN1: Energy consumption	The building envelope material and design must minimize consumption of electricity and fuel during construction.
	EN2: Resource consumption	The building envelope material and design must minimize resources used during construction such as water, chemicals, sealants, *etc.*
	EN3: Waste generation	Waste generation especially air pollution and wastewater should be minimized to reduce the impacts on the surrounding environments.
Economic	EC1: Initial costs	The project budget must be minimized.
	EC2: Long-term burdens	The design must minimize long-term burdens particularly repairing and replacing costs.
Social	SC1: Energy efficiency	Energy efficiency of the design must be maximized to achieve high energy efficiency and occupant comfort.
	SC2: Appearance demands	Appearance demands of the design must be maximized and the design must be modern and represent positive image.
	SC3: Health, safety and security of occupants	Health, safety and security of the occupants and society must be maximized.
	SC4: Weather protection performance	The design should minimize negative influence from adverse weather during occupation phase.
	SC6: Visual performance	Visual performance of the design should be maximized to achieve high occupant comfort.
Buildability	BC1: Health and safety of workers	The building envelope material and design must maximize workers' health and safety during construction.
	BC4: Material handling	The building envelope material and design must maximize ease in off-site and on-site handling methods.
	BC5: Ease in construction with respect to time	The building envelope material and design must maximize ease in construction within a time given.

Step 1: Considering the information given in Table 2, the team entered relevant information of the project as shown in the screenshot in Figure 3 and set up the fuzzy linguistic terms.

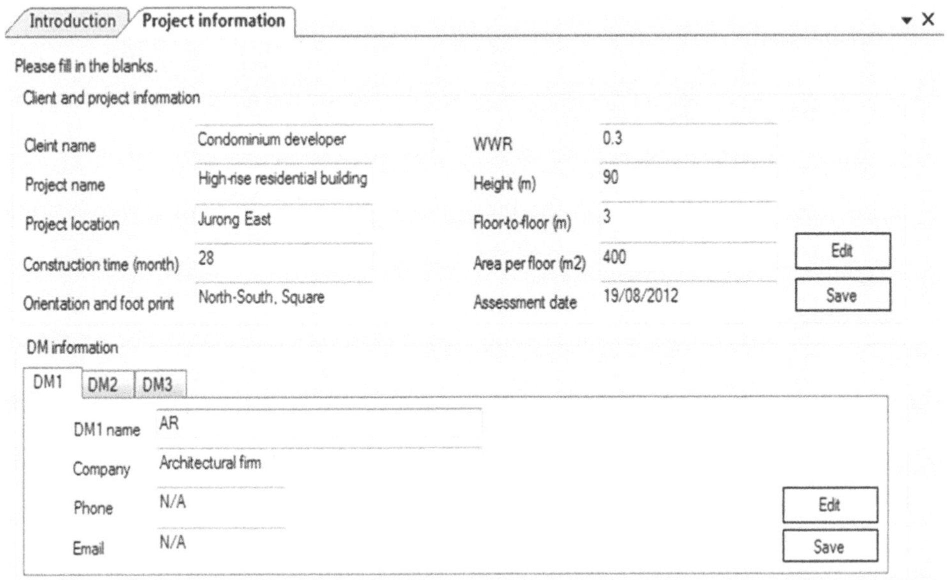

Figure 3. Project information.

Step 2: The design team inserted the sustainability and buildability criteria as given in Table 3 as the basic requirements of the project.

Step 3: The DMs determined the tradeoffs between the selected criteria by assigning their importance weights. Figure 4 shows the screenshot for rating the importance weights of the "EN1" Energy consumption, "EN2" Waste consumption, and "EN3" Resource consumption.

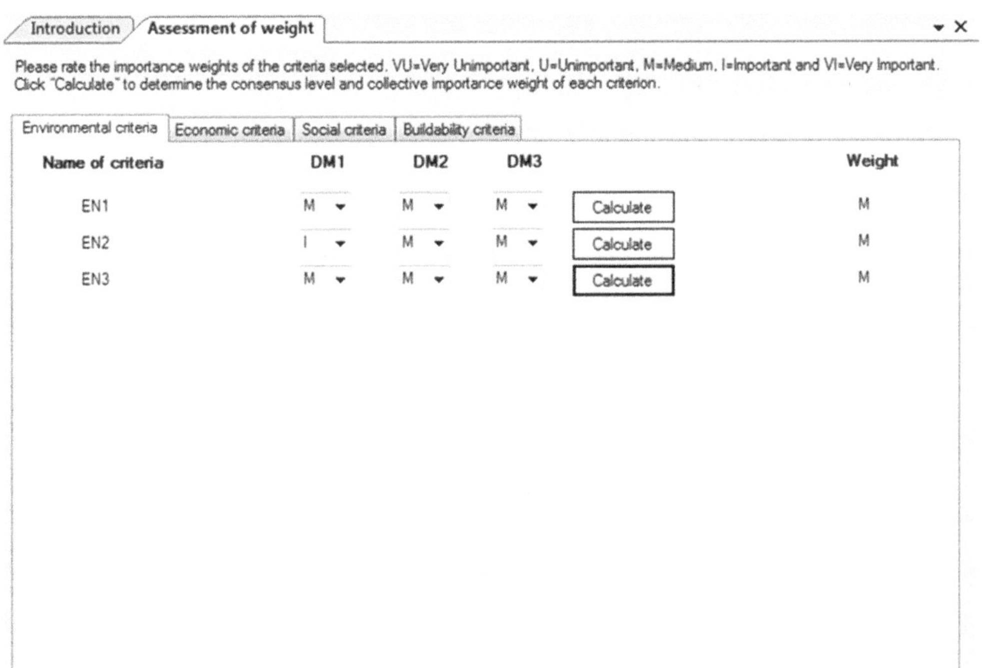

Figure 4. Assessment of the importance weights for all the criteria for the case study two.

Step 4: Based on the preferred external wall materials given in the Table 2, the team formed three alternatives for the assessment as shown in Figure 5. Alternative 1 consists of "BL1" Concrete block wall, "WG4" Double layer low-E glazing window, and the "SD1" Horizontal shading device. The team replaced the external wall of the Alternative 1 with "CB1" Claybrick to develop the Alternative 2. The Alternative 3 was then established. It comprises "PC1" Precast wall, "WG4" Double layer low-E glazing window, and "SD2" Horizontal shading device integrated with the precast panel.

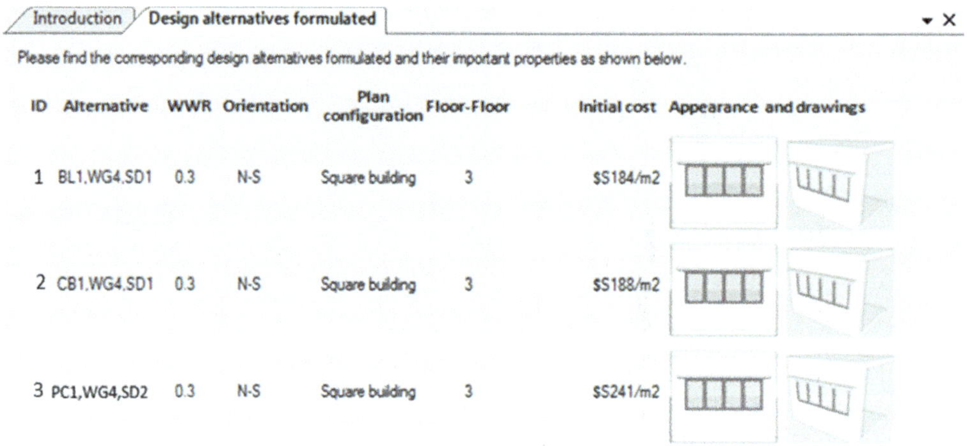
Introduction | Design alternatives formulated

Please find the corresponding design alternatives formulated and their important properties as shown below.

ID	Alternative	WWR	Orientation	Plan configuration	Floor-Floor	Initial cost	Appearance and drawings
1	BL1,WG4,SD1	0.3	N-S	Square building	3	$S184/m2	
2	CB1,WG4,SD1	0.3	N-S	Square building	3	$S188/m2	
3	PC1,WG4,SD2	0.3	N-S	Square building	3	$S241/m2	

Figure 5. Building envelope design alternatives.

Step 5: The DMs rated the performance satisfactions of the three design alternatives formulated with respect to the criteria. The screenshot given in Figure 6 reflects rating of the performance satisfactions of the design alternatives with respect to the "SC4" Weather protection performance.

The screenshot of the tool given in Figure 7 provides a summary of the importance weights of the criteria, performance satisfactions of the design alternatives, and their corresponding SBI. As can be seen in this figure, the ranking from the highest to lowest SBI of the design alternatives is the Alternative "3" PC1WG4SD2, "1" BL1WG4SD1 and "2" CB1WG4SD1. Comparing between the Alternative "2" CB1WG4SD1 and "1" BL1WG4SD1, the type of the external wall is the only difference between these two alternatives. However, the Alternative "2" BL1WG4SD1 received higher performance satisfaction with respect to a number of criteria particularly the "EN1" Energy consumption and "EN2" Resource consumption. This could be because the DMs viewed that the concrete blockwall requires less energy and resource consumption during construction as compared to the clay brickwall.

Furthermore, when it comes to comparison between the Alternative "3" PC1WG4SD2 and "2" BL1WG4SD1, there are two main differences which are the type of the external wall and type of the shading device. In brief, Alternative 3 using the precast concrete wall received higher performance satisfactions than Alternative 2 incorporated with the concrete blockwall with respect to the "EN1" Energy consumption, "EN2" Resource consumption, "SC4" Weather protection, and "BC6" Community disturbance. Similarly, based on its shading device, Alternative 3 obtained higher performance satisfaction with respect to a number of criteria such as the "EN1" Energy consumption, "EN2" Resource consumption, and "EN3" Waste generation. This was because the shading device of Alternative 1 would be integrated with the precast concrete panel by a manufacturer, while that of Alternative 2 would be installed on site. These collectively contributed to a higher SBI of the design Alternative 3 PC1WG4SD2. As such, the design team adopted this design alternative as a base case for further development of the detailed designs of this project. The design team took approximately two hours and a half to complete this case study exercise.

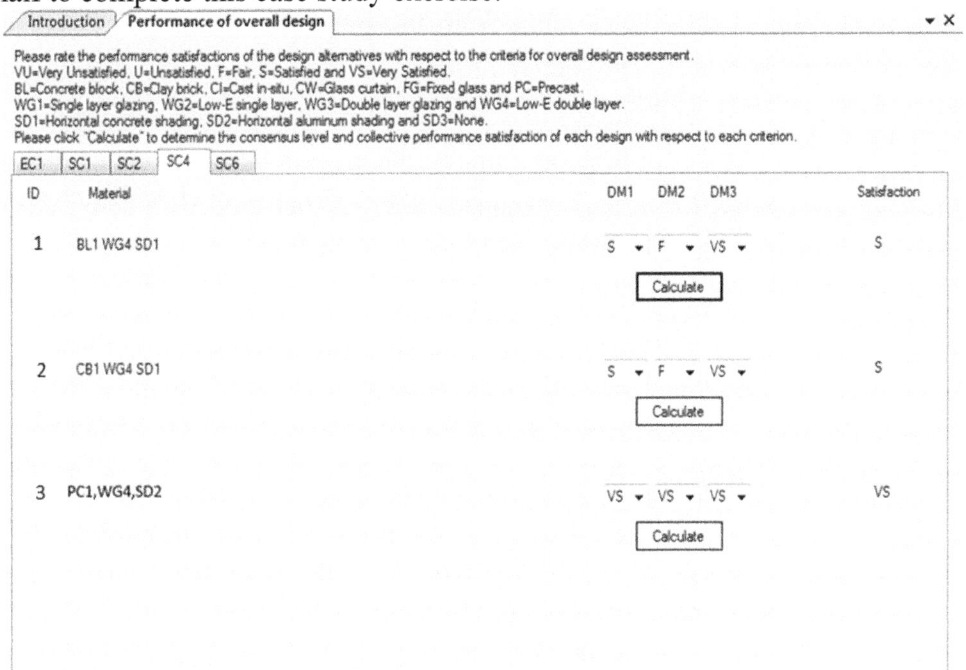

Figure 6. Assessment of the performance satisfaction of the design alternatives.

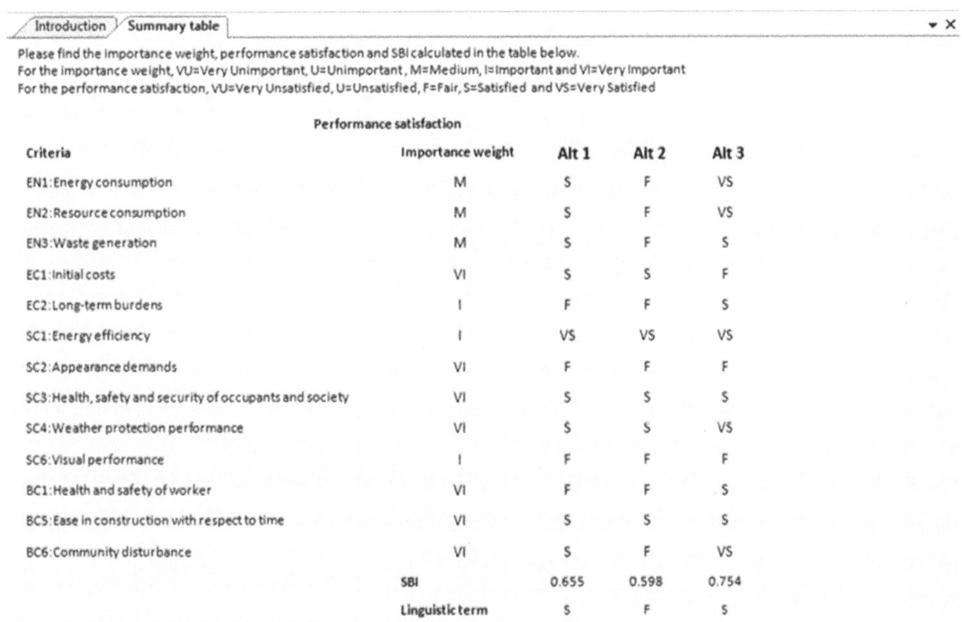

Introduction / Summary table

Please find the importance weight, performance satisfaction and SBI calculated in the table below.
For the importance weight, VU=Very Unimportant, U=Unimportant, M=Medium, I=Important and VI=Very Important
For the performance satisfaction, VU=Very Unsatisfied, U=Unsatisfied, F=Fair, S=Satisfied and VS=Very Satisfied

		Performance satisfaction		
Criteria	Importance weight	Alt 1	Alt 2	Alt 3
EN1:Energy consumption	M	S	F	VS
EN2:Resource consumption	M	S	F	VS
EN3:Waste generation	M	S	F	S
EC1:Initial costs	VI	S	S	F
EC2:Long-term burdens	I	F	F	S
SC1:Energy efficiency	I	VS	VS	VS
SC2:Appearance demands	VI	F	F	F
SC3:Health, safety and security of occupants and society	VI	S	S	S
SC4:Weather protection performance	VI	S	S	VS
SC6:Visual performance	I	F	F	F
BC1:Health and safety of worker	VI	F	F	S
BC5:Ease in construction with respect to time	VI	S	S	S
BC6:Community disturbance	VI	S	F	VS
	SBI	0.655	0.598	0.754
	Linguistic term	S	F	S

Figure 7. Summary of the design solutions.

6. Discussions

After the team completed the assignment, qualitative data analysis was applied to analyze the perspectives of the DMs collected through the interviews. From this analysis, the DMs presented the positive attitude that the fuzzy QFD tool provided a more structured decision making process based on both the decision making steps and fuzzy operations for assessing the building envelopes. This decision making process guided the DMs to deliberately go through each decision making step, and this increased efficiency and consistency in making the decisions. The tool also provided the DMs with more flexibility to instantly select different criteria from the provided list of the sustainability and buildability criteria to deal with specific project requirements, supporting the team to analyze the project criteria with respect to various alternatives in a limited timeframe. These as a whole significantly facilitated the DMs to discuss and sort out tradeoffs among the criteria in the forms of the importance weights and performance satisfactions. Success in management of such tradeoffs heavily lies in these discussions because the discussions gave the DMs a platform to fine-tune their perspectives including sharing and listening to each other's opinions.

The DMs further highlighted that the fuzzy linguistic terms well captured imprecise perspectives of the designers, and these were incorporated into the assessment in a systematic manner to determine the SBI of each alternative. This made the SBI representing a level of the sustainability and buildability of the design alternatives a good indicator for selection of a more sustainable and buildable building envelope design. Nevertheless, the DMs pointed out a few suggestions for future improvement of the tool. The first comment is that the efficiency of the assessment would be further improved if additional guidelines such as guided importance weights and guided performance satisfactions could be suggested by the tool. Such guidelines possibly stored in a Knowledge Management System (KMS) would help the DMs fasten their discussions and effectively arrive at agreed decisions [44]. Another point is that the DMs perceived that the tool seems to be a bit too complicated. It seems to be hard to update the knowledge of the tool, and the assessment appears to heavily rely (but understandably) on the team facilitator. Furthermore, in some circumstances, as there could be disagreements among DMs, a more systematic discussion procedure such as a consensus scheme should be embedded into the tool to mitigate this issue [45]. Future studies may consider these suggestions when developing their decision support tools. Although a limitation of this study is that the study engaged only one case study to proof the development of the tool, the study brings out key insights that could shed some lights on various possible scenarios. As such, the next phase is to expand the research scale to cover a larger number of the building design teams including use of actual building envelope design projects.

7. Conclusions

This study builds the fuzzy QFD tool to assist the building design to manage the tradeoffs among the sustainability and buildability criteria when assessing the building envelope materials and designs in the early design stage. This tool is modeled through the integration between the fuzzy multicriteria decision making method and QFD approach. The tool is tested in the case study of the building design team to determine the building envelope system for the representative project. The results from the interviews show that the tool has the potential to effectively strike a balance between the tradeoffs among the criteria. In brief, the fuzzy QFD tool helps the designers to simultaneously consider the sustainability and buildability criteria, and effectively assign the importance weights of the criteria and performance satisfactions of the designs. At the same time, the structured decision making process improves communication and integration among the designers. This helps generate effective and consistent decisions including more sustainable and buildable designs.

The tool developed in this study not only benefits the design team to manage the tradeoffs when assessing the building envelope materials and design in the early design stage but also expands the body of academic knowledge, as the study presents the advanced approach to combine the fuzzy multicriteria decision making method with QFD approach. However, the limitation of this study is that there is only one hypothetical case study used to prove the development of the concept for the decision-making tool. Future research is therefore recommended to apply the findings of this study to develop and further improve decision support tools to deal with other related research problems. This can include further investigations of the impacts of the fuzzy QFD tool on a larger research scale.

Acknowledgements

The authors gratefully acknowledge the suggestions and comments from three anonymous reviewers. Their recommendations have been most helpful.

Author Contributions

The first author was responsible for carrying out development of the tool and putting together the results/analysis. The second author provided guidance/direction on the research, as well as proof reading and editing of the paper.

Conflicts of Interest

The authors declare no conflict of interest.

References

1. Kibert, C.J. *Sustainable Construction: Green Building and Delivery*, 2nd ed.; John Wiley and Son, Inc.: New Jersey, NJ, USA, 2008.
2. Chew, Y.L. *Construction Technology for Tall Buildings*; World Scientific Publishing: Singapore, 2009.
3. Chua, K.J.; Chou, S.K. Evaluating the performance of shading devices and glazing types to promote energy efficiency of residential buildings. *Build. Simul.* **2010**, *3*, 181–194.
4. Bansal, P. Evolving sustainably: A longitudinal study of corporate sustainable development. *Strat. Manag. J.* **2005**, *26*, 197–218.
5. Low, S.P.; Liu, J.Y.; Oh, K.H. Influence of total building performance, spatial and acoustic concepts on buildability scores of facilities. *Facilities* **2008**, *26*, 85–104.
6. Building and Construction Authority (BCA). *Code for Environmental Sustainability of Building*, 3rd ed.; BCA: Singapore, 2012. Available online: http://www.bca.gov.sg/Envsuslegislation/Environmental_sustainability_legislation.html (accessed on 4 March 2015).

7. Building and Construction Authority (BCA). *Code of Practice on Buildability*; BCA: Singapore, 2013. Available online: http://www.bca.gov.sg/BuildableDesign/others/copbdsep2013.pdf (accessed on 2 March 2015).
8. Singhaputtangkul, N.; Low, S.P.; Teo, A.L. Integrating sustainability and buildability requirements in building envelopes. *Facilities* **2011**, *29*, 255–267.
9. Singhaputtangkul, N.; Low, S.P.; Teo, A.L.; Hwang, B.G. Criteria for architects and engineers to achieve sustainability and buildability in building envelope designs. *J. Manag. Eng.* **2014**, *30*, 236–245.
10. Yang, Y.Q.; Wang, S.Q.; Dulaimi, M.; Low, S.P. A fuzzy quality function deployment system for buildable design decision-makings. *Autom. Constr.* **2003**, *12*, 381–393.
11. Ahn, Y.; Pearce, A. Green construction: Contractor experiences, expectations, and perceptions. *J. Green Build.* **2007**, *2*, 106–122.
12. Lapinski, A.; Horman, M.; Riley, D. Lean processes for sustainable project delivery. *J. Constr. Eng. Manag.* **2006,** *132*, 1083–1091.
13. Wong, W.H.; Lam, T.L.; Chan, H.W.; Wong, K.W. Factors affecting buildability of building designs. *Can. J. Civil Eng.* **2006**, *33*, 795–806.
14. Lam, P.T.I.; Chan, A.P.C.; Wong, F.W.K.; Wong, F.W.H. Constructability rankings of construction systems based on the analytical hierarchy process. *J. Arch. Eng.* **2007**, *13*, 36–43.
15. Low, S.P.; Abeyegoonasekera, B. Buildability in design and construction through ISO 9000 quality management systems: Concepts and survey findings. *Arch. Sci. Rev.* **2001**, *44*, 355–366.
16. Fox, S.; Marsh, L.; Cockerham, G. Constructability runs les: Guidelines for successful application to bespoke buildings. *Constr. Manag. Econ.* **2002**, *20*, 689–696.
17. Nima, M.A.; Abdui-Kadir, M.R.; Jaafar, M.S.; Alghulami, R.G. Constructability implementation: A survey of the Malaysian construction industry. *Constr. Manag. Econ.* **2002**, *19*, 819–829.
18. Bryan, T. *Construction Technology: Analysis and Choice*; Blackwell Publishing Ltd.: West Sussex, UK, 2010.
19. Scheuer, C.; Keoleian, G.A.; Reppe, P. Life cycle energy and environmental performance of a new university building: Modeling challenges and design implications. *Energy Build.* **2003**, *35*, 1049–1064.
20. Chen, Y.; Okudan, G.E.; Riley, D.R. Sustainable performance criteria for construction method selection in concrete buildings. *Autom. Constr.* **2010**, *19*, 235–244.
21. Tsai, W.H.; Lin, S.J.; Liu, J.Y.; Lin, W.R.; Lee, K.C. Incorporating life cycle assessments into building project decision-making: An energy consumption and CO_2 emission perspective. *Energy* **2011**, *36*, 3022–3029.
22. Jaillon, L.; Poon, C.S. Sustainable construction aspects of using prefabrication in dense urban environment: A Hong Kong case study. *Constr. Manag. Econ.* **2008**, *26*, 953–966.
23. Wang, W.; Rivard, H.; Zmeureanu, R. Floor shape optimization for green building design. *Adv. Eng. Inf.* **2006**, *20*, 363–378.
24. Das, S.; Chew, M.Y.L.; Poh, K.L. Multi-criteria decision analysis in building maintainability using analytical hierarchy process. *Constr. Manag. Econ.* **2010**, *28*, 1043–1056.

25. Lacasse, M.A.; Vanier, D.J.; Kyle, B.R. Towards Integration of Service Life and Asset Management Tools for Building Envelope Systems. In Proceedings of the 7th Conference on Building Science and Technology: Durability of Buildings Design, Maintenance, Codes and Practice, Toronto, Canada, 20–21 March 1997; pp. 153–167.
26. Kneifel, J. Life-cycle carbon and cost analysis of energy efficiency measures in new commercial buildings. *Energy Build.* **2010**, *42*, 333–340.
27. Kaklauskas, A.; Zavadskas, E.K.; Raslanas, S.; Ginevicius, R.; Komka, A.; Malinauskas, P. Selection of low-e windows in retrofit of public buildings by applying multiple criteria method COPRAS: A Lithuanian case. *Energy Build.* **2006**, *38*, 454–462.
28. Brock, L. *Designing the Exterior Wall: An Architectural Guide to The Vertical Envelope*; John Wiley and Sons, Inc.: New Jersey, NJ, USA, 2005
29. Yu, C.W.F.; Kim, J.T. Building pathology, investigation of sick buildings: VOC emissions. *Indoor Built Environ.* **2010**, *19*, 30–39.
30. Low, S.P.; Liu, J.Y.; Wong, H.L. Relationship between buildability, indoor air quality and visual performance. *Struc. Surv.* **2008**, *26*, 38–54.
31. Nielsen, T.R. Optimization of Buildings with respect to Energy and Indoor Environment. Ph.D. Thesis, The Technical University of Denmark, Copenhagen, Denmark, 2002.
32. Hinze, J.; Devenport, J.; Giang, G. Analysis of construction worker injuries that do not result in lost time. *J. Constr. Eng. Manag.* **2006**, *132*, 321–326.
33. Gould, F.E. *Managing the Construction Process: Estimating, Scheduling, and Project Control*, 3rd ed.; Pearson Prentice Hall: New Jersey, NJ, USA, 2005.
34. Vrijhoef, R.; Koskela, L. The four roles of supply chain management in construction. *Eur. J. Purch. Supply Manag.* **2000**, *6*, 169–178.
35. Dikmen, I.; Birgonul, M.T.; Kiziltas, S. Strategic use of quality function deployment (QFD) in the construction industry. *Build. Environ.* **2005**, *40*, 245–255.
36. Xie, M.; Tan, K.C.; Goh, T.N. *Advanced QFD Applications*; ASQ Quality Press: Milwaukee, WI, USA, 2003.
37. Low, S.P.; Yeap, L. Quality function deployment in design/build projects. *J. Arch. Eng.* **2001**, *7*, 30–39.
38. Reilly, C. *Making Hard Decisions with Decision Tools*; Duxbury Thomson Learning: Pacific Grove, CA, USA, 2001.
39. Lam, K.C.; Ran, T.; Lam, M.C.K. A material supplier selection model for property developers using fuzzy principal component analysis. *Autom. Constr.* **2010**, *19*, 608–618.
40. Zadeh, L.A. Fuzzy sets. *Inf. Control* **1965**, *8*, 338–353.
41. Bayrak, M.Y.; Çelebi, N.; Takin, H. A fuzzy approach method for supplier selection. *Prod. Plan. Control* **2007**, *18*, 54–63.
42. Yin, R.K. *Case Study Research: Design and Methods*, 4th ed.; Sage: Thousand Oaks, CA, USA, 2009.
43. Creswell, J.W. *Research Design Qualitative, Quantitative and Mixed Methods Approaches*, 2nd ed.; Sage Publications: Thousand Oaks, CA, USA, 2003.
44. Turban, E.; Aronson, J.E.; Laing, T.P.; Sharda, R. *Decision Support and Business Intelligence Systems*, 8th ed.; Prentice Hall: New Jersey, NJ, USA, 2007.

45. Elbarkouky, M.; Fayek, A. Fuzzy similarity consensus model for early alignment of construction project teams on the extent of their roles and responsibilities. *J. Constr. Eng. Manag.* **2011**, *137*, 432–440.

10

Driving Innovative Thinking in the New Zealand Construction Industry

Mohammadali Noktehdan [1], Mehdi Shahbazpour [2,*] and Suzanne Wilkinson [1]

[1] Department of Civil and Environmental Engineering, The University of Auckland, Private Bag 92019,Auckland 1142, New Zealand; E-Mails: mnok946@aucklanduni.ac.nz (M.N.); s.wilkinson@auckland.ac.nz (S.W.)

[2] Department of Mechanical Engineering, The University of Auckland, Private Bag 92019, Auckland 1142, New Zealand

* Author to whom correspondence should be addressed; E-Mail: m.shahbazpour@auckland.ac.nz

Academic Editor: George Baird

Abstract: This paper examines the relationship between innovation and productivity improvement in the construction industry. It is argued that this relationship is not well understood due to lack of in-depth understanding of innovation in construction. To overcome this obstacle, the authors present a multi-dimensional innovation classification system which aims at better defining and classifying what is meant by innovation in construction. The use of this classification system is demonstrated by applying it to a database of 500 innovations reported by the construction alliance, the Stronger Christchurch Infrastructure Rebuild Team. The results clearly demonstrate the diversity of types, degree of novelty and performance improvement benefits among construction innovations. Such diversity means that the impact of the reported innovations on productivity and performance are of different levels of significance. The classification system developed in this study can be used by construction organisations and alliances in the future to develop more detailed methods of calculating innovation performance indicators, based on the innovation type, novelty and benefits factors. By using this system, they can also put in place mechanisms to influence the types of innovation developed in their projects with the aim of maximising their productivity performance.

Keywords: innovation classification; key performance indictor (KPI); productivity; alliance; construction

1. Introduction

The Construction industry is critical to the functioning of a domestic economy in many countries [1]. In New Zealand, the construction industry is one of the largest sectors of the economy accounting for 8% of total employment in the country [2]. However, in spite of its importance to the national economy in terms of size, the construction industry seems to be lagging other sectors in terms of productivity [3]. In 2010, the construction industry in New Zealand established the Building and Construction Sector Productivity Partnership to actively address the issue of low productivity in the sector. Although the early focus was on identifying and quantifying the problems that lead to low productivity, over the recent years the focus has shifted to problem solving and addressing the cultural and mechanistic change that is needed to resolve the well documented issue [3].

One of the areas that the partnership has identified as critical for achieving significant improvements in the sector's productivity is innovation. The ultimate goal of 20% productivity improvement by 2020 cannot be achieved by repeating the old ways of doing things. New innovative approaches are required in order to significantly improve performance at the same cost or maintain the same level of performance at much lower cost. Unfortunately the construction industry is one of the least innovative sectors compared to other industries such as manufacturing and traditional services [4]. The research and development report produced by Statistics New Zealand indicates that R&D expenditure in the construction industry accounts for a low 5% of the total expenditure in the sector [5]. Indeed, this problem is not limited to New Zealand as internationally the construction industry is seen as a traditional or low-technology sector with low levels of expenditure on activities associated with innovation [6].

Furthermore, it appears that the link between innovation and productivity in construction industry is not well understood. Loosemore [7] observes that in spite of many recent developments in the field of construction innovation, none of these studies feature in the mainstream construction productivity literature.

In contrast, the link between productivity and innovation has been well researched and explored in the manufacturing literature [8]. However, while certain parallels can be drawn from the manufacturing literature, it is generally agreed that there are significant differences between manufacturing and construction. Koskela and Vrijhoef [9] identified the one-of-a-kind nature of construction project, site production, temporary multi organization and regulatory intervention, as differentiating characteristics of the construction industry accounting for lower innovation. Reichstein *et al.* [4] also identified a number of "liabilities" which sets construction industry apart from other innovative industries such as manufacturing. These are Immobility, Project-based, Uncertainty of demand, Smallness, Separation and Assembly.

There is a need for more construction specific research to explore the relationship between innovation and productivity. The authors contend that while productivity is a well-defined term, innovation in contrast is poorly defined for the construction industry and is only understood in abstract terms. This is a major obstacle in the attempt to clarify the relationship between innovation and productivity in the construction industry. The study presented in this paper, aims to make a contribution in this area by demonstrating the diversity of various types of innovation present in a construction project and the varying degrees of impact these innovations have on productivity indicators.

The paper will first present this classification system from a theoretical perspective and illustrates how it can be used to better understand the relationship between innovation and productivity. The use of this classification system is then demonstrated by applying it to a database of 500 innovations reported by the construction alliance, the Stronger Christchurch Infrastructure Rebuild Team (SCIRT), responsible for rebuilding the horizontal infrastructure after the earthquakes in Christchurch. SCIRT is an organisation who put innovation as one of the main key performance indictors (KPIs) for its construction project delivery team. The results will highlight the diversity of innovation types that have been reported by the alliance group, and the classification system used in the paper makes sense of these innovations, and suggests ways of improving productivity through innovation use in construction.

2. Defining Innovation

Productivity is generally defined as a ratio between output produced and input used [10]. For instance, labour productivity can be calculated as a ratio of hours of labour worked (input) and units produced (output). More holistic definitions of productivity include a number of performance dimensions for units of input and output, such as quality, cost, time and safety. Innovation on the other hand is more ambiguous in its definition. Brown [11] provides a broad definition of innovation as "*doing things differently or better across products, processes or procedures for added value and/or performance*". West and Altink [12] provide a similar, but perhaps more complete, definition of innovation as the "intentional introduction and application within a role, group or organisation of ideas, processes, products or procedures, new to the relevant unit of adoption, designed to significantly benefit the individual, the group, the organisation or the wider society".

When comparing the above definitions and other popular definitions of innovation in the literature, three key defining elements of innovation can be identified. First is the type of innovative idea or invention. The Oslo Manual on collecting and interpreting innovation data [13] outlines four types of innovations: product, process, marketing and organisational. Second is novelty. The minimum entry level for an innovation is that it must be novel or "new to the firm" [13]. Degree of novelty or "newness" can be defined as the extent of uncertainty associated with the implementation of the innovative idea within the context of its application [14]. Slaughter [15] divided the Novelty of the construction innovation in five different levels of Incremental, Modular, Architectural, System and Critical. The third is benefit or improvements in performance. In construction, traditionally performance has been measured in terms of cost, time, quality and safety [16]. Recently, environmental measures have also become an important performance indicator in construction projects [17].

Each of these elements (type, novelty, and benefit) can alter the significance and type of the impact that a given innovation has on productivity. In terms of type, whether the innovative ideas are a product, process or an organisational method results in different type of impact on productivity. For instance, process and organisational innovations are mainly concerned with the construction phase and thus will have a direct impact on the project's productivity performance. On the other hand, product innovation may have much less significant impact on the production phase but much larger impact once the building or infrastructure is in operation. In terms of novelty, the degree of novelty determines the significance of the impact the given innovation may have on productivity. For instance, an incremental change results in much less significant impact on performance than a more radical and disruptive change. Finally, different innovations would improve different set of performance indicators, thus having very different impacts on productivity.

It is therefore proposed that in order to better determine the impact and nature of the relationship between given innovations and overall project productivity, types of innovation, novelty and the specific performance indicators being improved must first be determined. This requires an innovation classification system, which can be used to categorise the innovations implemented in a project.

3. Innovation Classification System

A multi-dimensional classification system is presented, based on review of relevant construction and innovation management literature regarding the elements discussed previously. After a review of the literature, it was found that the most commonly accepted innovation classifications have mainly been developed by analysing innovation within the manufacturing and services context. As mentioned previously, there are significant differences between construction and other industries. Consequently, in order to develop an appropriate innovation classification system for the construction industry, modifications were made to these classification systems and where appropriate new categories were developed. This section will briefly outline each dimension and provide a definition for each classification category.

Innovation Type: The authors found it necessary to distinguish between development or utilisation of innovative construction materials or componentry and the development of innovative designs and features for buildings or infrastructures. Therefore, the product innovation category was limited to cover new materials or products used in the construction phase, and a new category was added, called Design, to account for the innovative design features introduced at the design phase of the project. Furthermore, guided by the construction technology classification system developed by Tatum [18], the process innovation category was divided into two sub-categories of Tools and Function.

- **Product:** Product Innovation involves all new construction materials and products developed in the project or introduced to the project and used within the construction process.
- **Design:** Design Innovation is related to new and innovative plans, designs, sketches or concepts for the final building or infrastructure that is being developed in the project.
- **Tool:** The Tool Innovation involves the development or implementation of novel construction machinery equipment or tools in the construction project.
- **Function:** The Functional Innovation refers to new tasks developed or introduced in the construction project and associated management processes.

After further consultation with industry experts two other categories were added where a combination of the previous sub-categories could exist:

- **Technology (Design + Product):** The New Technology refers to the new design that is coupled with a new material or product.

- **Method (Tool + Function):** The method innovation is the combination of the Tool and Function Innovation that involve both a new tool or equipment and new tasks that are usually related to the new tool.

Innovation Novelty: Typically the innovation literature distinguishes between incremental and radical innovations. However, Slaughter [15] provides a more detailed categorisation of novelty within construction innovation context. These categories are Incremental, Modular, Architectural, System and Critical.

- **Incremental:** Incremental Innovation is a small change, based upon current knowledge and experience. It is often the result of continuous improvement initiatives and on-the-job problem solving.
- **Modular:** Modular Innovation entails a significant level of novelty in one area of a system, but without impacting the other components of the system. Modular innovations may be developed within an organization and implemented without much negotiation with parties involved in the development or selection of other components.
- **Architectural:** Architectural Innovation involves a small change within a component of a system, which results in major changes in the links to other components and systems. The distinction between modular and architectural innovations is made on the region of the change and, specifically, the degree of interaction with other components of the system.
- **System:** System Innovations are identified through their integration of multiple independent innovations that must work together to perform new functions or improve the facility performance as a whole.
- **Critical:** Critical Innovation is a breakthrough in science or technology that often changes the character and nature of an industry. While incremental innovations occur constantly, critical innovations are rare and unpredictable in their appearance and in their impacts.

Innovation Benefit: This dimension of the classification system deals with the type of performance improvement that is achieved through implementation of the new innovative ideas. As mentioned previously, the following performance indicators will be used to distinguish between the types of benefits that are delivered by the given innovations.

- **Cost:** Direct cost savings or better utilisation of resources.
- **Time:** Reduction in lead-times or increasing speed for the project or sub-tasks.
- **Quality:** Improvements in degree of conformance with specifications and/or satisfaction of stakeholders with the outputs of the construction project.
- **Safety:** Improving safety, health and wellbeing of the employees and public during and after the construction project.
- **Environment:** Reducing adverse impact of the construction processes as well as the final building or infrastructure on the natural environment.
- **Community:** Reducing adverse impact on communities affected by the construction project and improving communication with the stakeholders.

4. Case Study

SCIRT is an organisation established under an alliance agreement and is responsible for rebuilding horizontal infrastructure in Christchurch following the earthquakes of 2010 and 2011. Innovation was given a special consideration from the outset, when the SCIRT alliance was formed. In fact, members of the alliance were encouraged to innovate and report on their innovations on a monthly basis as one of their KPIs. These KPIs were linked directly to the pay/reward aspect of the contract. As a result, the alliance members had ample motivation to report their innovations. To date more than 500 innovations have been reported by SCIRT. This has provided a unique opportunity to analyse and better understand the relationship between construction innovation and productivity improvements. The authors were given full access to SCIRT's innovation database. For each reported innovation, the database contained a unique identification number, description of the innovative idea, its potential benefits and information regarding which member organisation had initiated the innovation. Some of the reported innovations were also accompanied by pictures or sketches to better describe the innovation. Each of the reported innovations were analysed and categorised based on the innovation classification system outlined in the previous section. The categorisation process was carried out mainly by one researcher to ensure consistency of interpretation across the data set. Once completed, 20 innovations were selected randomly and categorised by two other researchers using the classification system. This was done to make sure the classification system was being applied appropriately. Any uncertainty around the definition was clarified through this process.

5. Results

In order to better understand the spread of various categorisations of innovations developed by SCIRT, the data was analysed using pivot-table functionality of MS Excel. A pairwise analysis was also conducted to identify any potential trends which may lead to insights about the nature of innovation activity at SCIRT. The following section presents the summary of the results from this analysis.

5.1. Spread of Innovation

Figures 1–3 represent the spread of innovation categories in each dimension of the classification system: type, novelty and benefit. Most innovations in the SCIRT database seem to be made up of tools or functions in terms of innovation type, and modular or architectural in terms of novelty. They also appear to deliver a wide spread of performance benefits, mainly dominated by quality, time and cost.

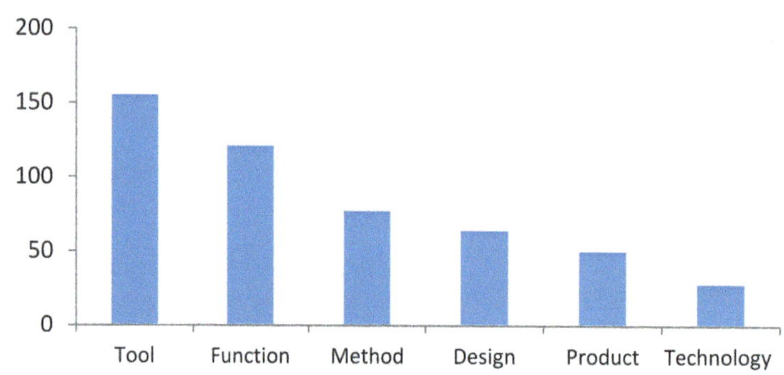

Figure 1. Innovation classification based on Type.

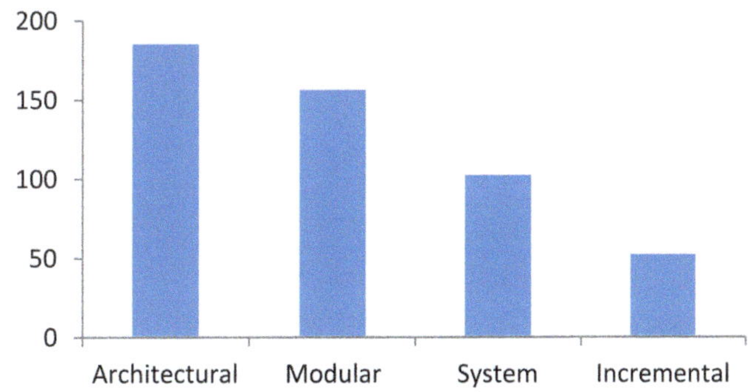

Figure 2. Innovation classification based on Novelty.

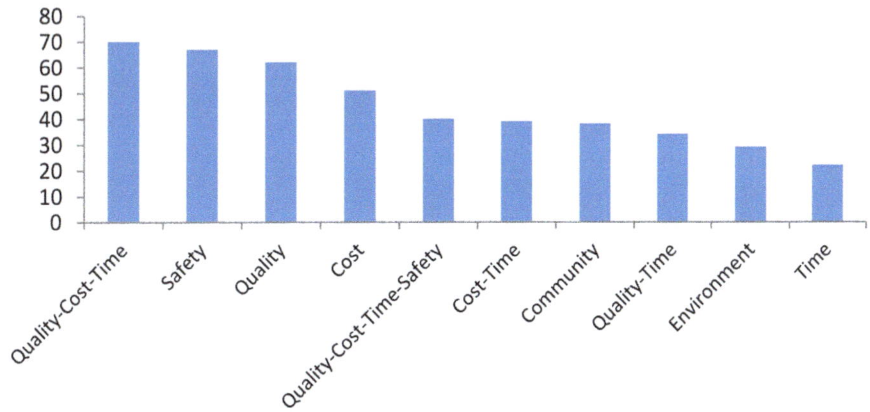

Figure 3. Innovation classification based on Benefit.

5.2. Pairwise Analysis

The data was also analysed to identify emerging trends that would provide more insight in to the relationship between the three dimensions of the innovation classification system. When looking at the spread of innovation based on a pair of two dimensions of benefit and novelty, an interesting trend emerged. As illustrated in Figure 4, it appears that architectural and modular categories of innovation are more focused on delivering a single benefit. On the other hand, system category of innovation seems to be the one that mostly delivers a combination of quality, time and cost benefits (see Figure 5). Modular innovation also appears to be the most prevalent category which focused on the indirect productivity improvements through safety, environment and community (see Figure 6).

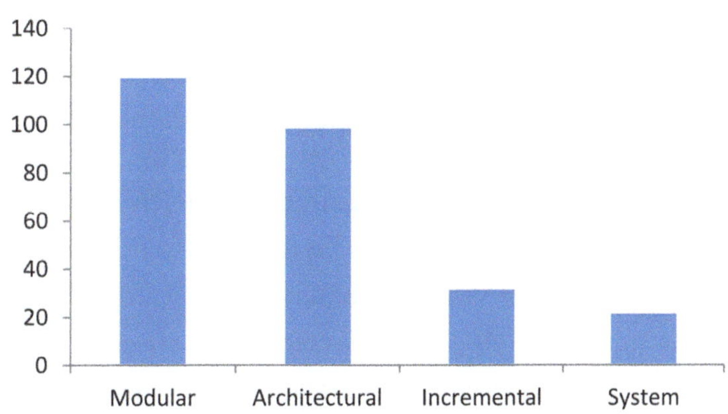

Figure 4. Innovation novelty categories that focused on delivering a single benefit.

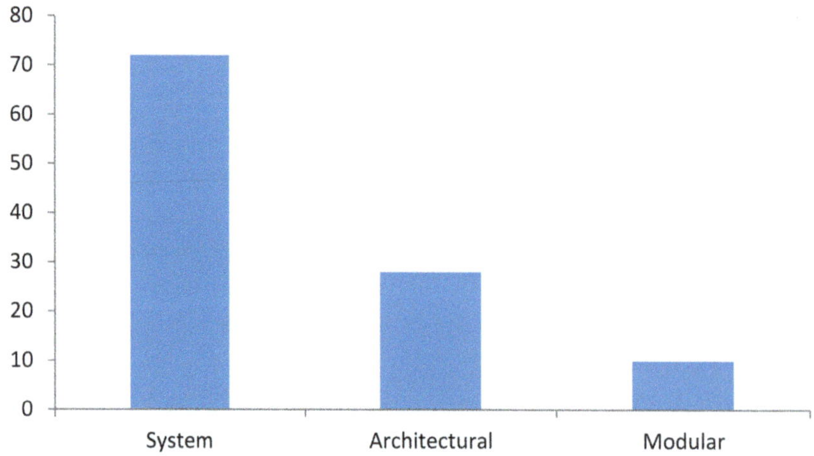

Figure 5. Innovation novelty categories that delivered a combination of quality-time-cost benefit.

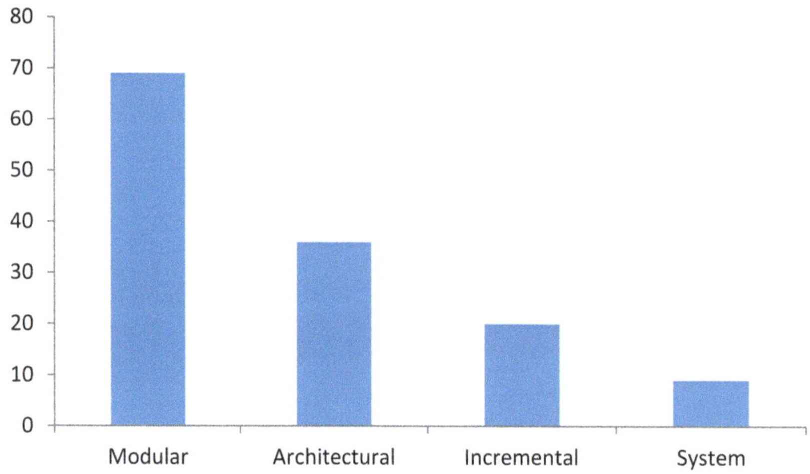

Figure 6. Innovation novelty categories that focused on either sustainability, safety or community.

When looking at the pair of innovation types and benefits similar trends appear. As illustrated in Figure 7, the majority of innovations are delivering a combination of direct productivity improvement benefits such as quality, time and cost, were from the two categories of design and method. On the other hand, function and tool categories seem to be more focused on delivering a single benefit (see Figure 8). Tools also appear to be the most prevalent type of innovation that deliver either safety, environment or community benefits (see Figure 9).

Analysis of the pair of type and novelty also reveals some interesting trends. As illustrated in Figure 10, design innovation was dominated by architectural level of novelty, while product innovation was split between modular and architectural. Technology innovation (design + product) on the other hand was mainly dominated by system level of novelty. Modular and architectural innovations made up the majority of the tools and functions, while those innovations classified as method were mainly system or architectural innovations. Figure 11 also illustrates that the system level of novelty was mainly prevalent in methods, while modular innovations were mostly found under the tool and function categories. Architectural innovations were spread amongst tools, designs, functions and methods. Also, it was found that most incremental innovations were under the tools category.

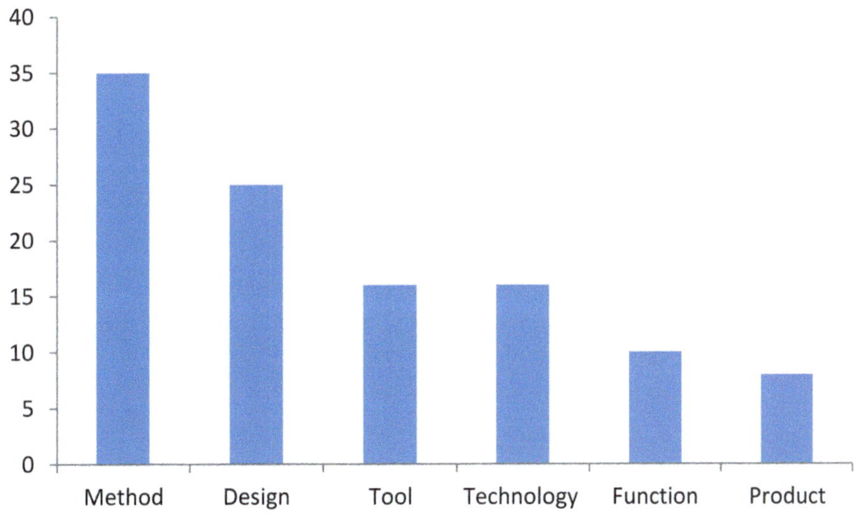

Figure 7. Innovation types that delivered a combination of quality-time-cost benefit.

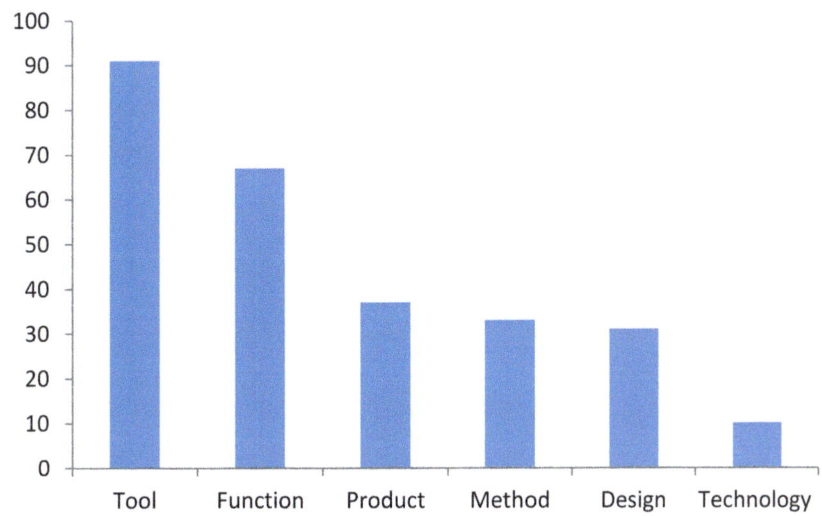

Figure 8. Innovation types that focused on delivering a single benefit.

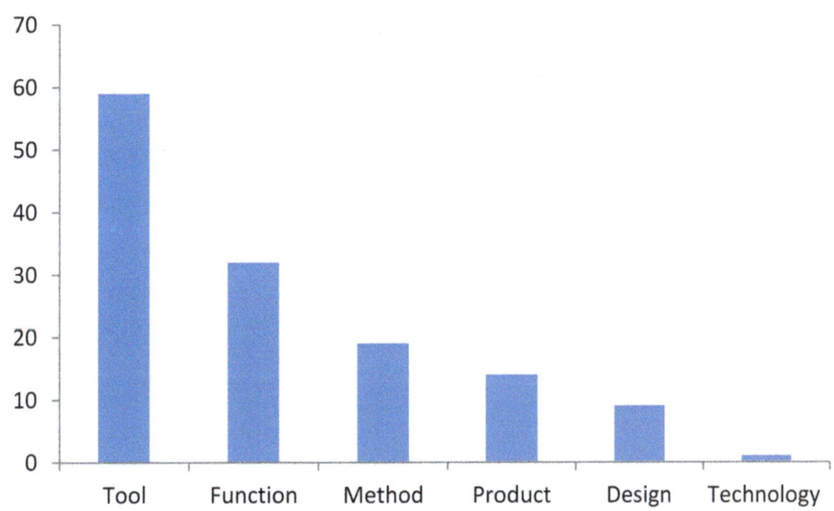

Figure 9. Innovation types that focused on either sustainability, safety or community.

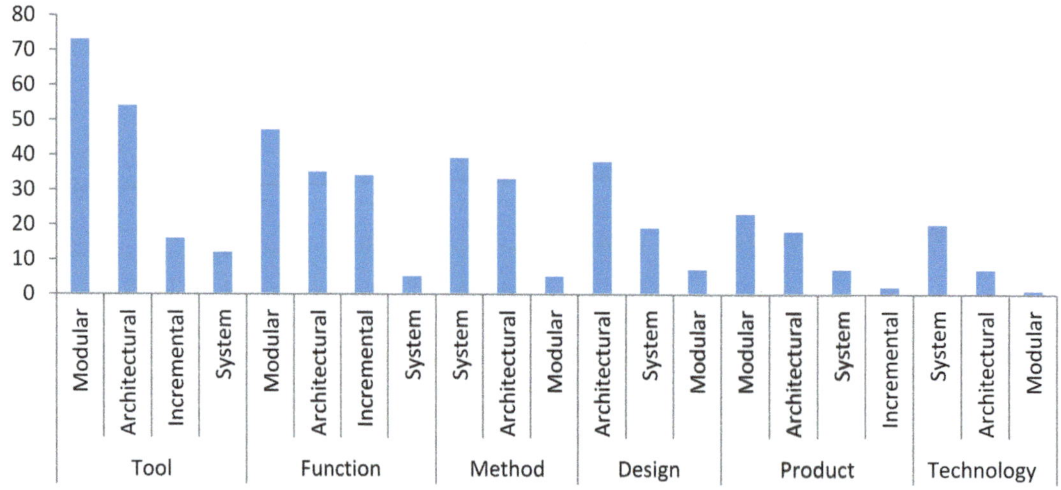

Figure 10. Degree of novelty of various types of innovation.

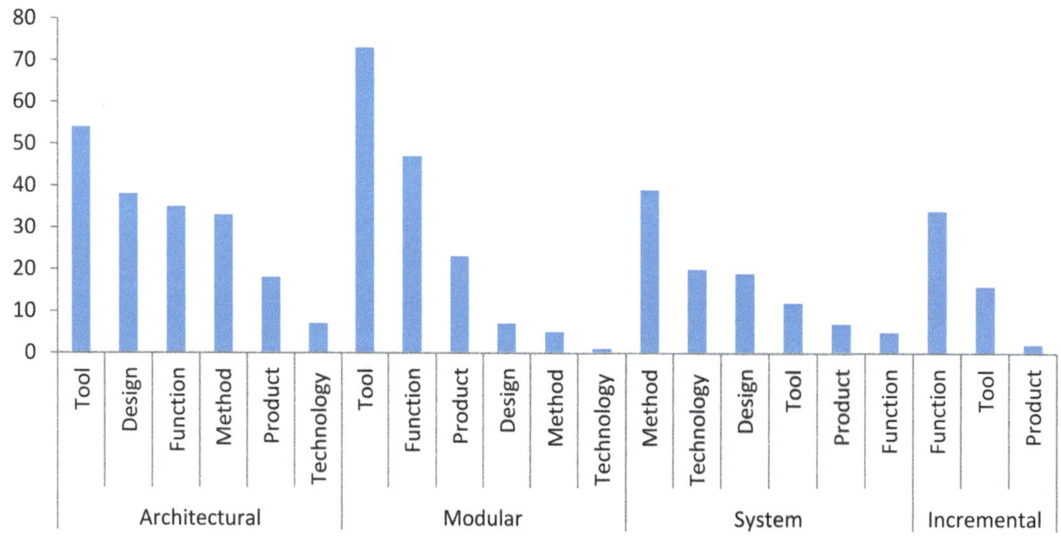

Figure 11. Spread of innovation types for each level of novelty.

6. Discussions

The results of the analysis of 500 reported innovations by the alliance group, SCIRT, clearly demonstrate the diversity of types, degree of novelty and performance improvement benefits among construction innovations. The trends presented in the previous section have all emerged naturally based on the organisational dynamics and culture present among the member organisations as well as within the virtual alliance organisation. Given that innovation KPI reporting was linked to pay/reward system for the member organisations, there was motivation for all parties to look for opportunities to innovate. However, the data shows that most of the reported innovations were tools or functions that were developed to overcome immediate problems facing the operational teams. As a result, most of these innovations were modular or architectural in terms of novelty. This indicates that most reported innovative solutions were developed to either solve localised problems or issues arising at the interface of operational sub-systems. The data also shows that when architectural and modular innovations were dominant, the reported innovations were mainly focused on a single aspect of performance improvement. In contrast, the results show that when more sophisticated types of innovation such as technology and methods were developed, the impact was more widespread and significant, delivering benefits along multiple dimensions of performance such quality, time and cost.

Given that the reported innovations in the SCIRT database were developed without consideration given to the types of innovation, degrees of novelty or performance improvement benefits, they were all treated the same. The KPI associated with innovation simply counted the number of innovations reported by the member organisations. The main reason for this equal treatment of the reported innovations is lack of in-depth understanding of various categorisations of innovation by the SCIRT management team. Consequently, they missed the opportunity to guide and direct the innovation efforts of the alliance member organisations towards specific types of innovation which could lead to specific aspects of productivity and performance improvement.

One important point with respect to future application of the classification system to other construction projects is the uncertainty associated with matching the definitions of the various categories of the classification system with the information provided about the particular innovation. Although this was not a major issue in the case of innovations in the SCIRT database, it is still a notable concern for future applications. While applying the classification system in the case study, it was clearly evident that this uncertainty was directly related to the amount and quality of information provided about the innovation. Given that in this case the classification system was applied retrospectively, the researchers did not have any influence over how data about the innovations was collected and reported. It is therefore suggested that if the proposed classification system is used for development of a KPI system, a data collection template is designed based on the definitions of the various categories. This could for instance take the form of a questionnaire with specific questions around novelty, benefits and type of innovation, which would ensure that adequate information is available for classifying the particular innovation.

7. Conclusions

In conclusion, the results of the analysis illustrate the need for a more in-depth understanding of innovation in construction. The innovation classification system presented in this study provides a practical tool for researchers and practitioners in the field of construction management, to not only better understand the relationship between innovation and productivity, but to also put in place mechanisms to influence the types of innovation developed in their projects with the aim of maximising their productivity performance. This classification system can be used by construction organisations and alliances in the future to develop more detailed methods of calculating innovation KPIs, based on the innovation type, novelty and benefits factors. In such methods, different weightings can be given to each factor as a mechanism to encourage certain types of innovation, with higher levels of novelty and focusing on specific combination of performance indicators that are tailored to the specific requirement of the construction project. For instance, in projects where the Green Construction approach is adopted, the system can be set up to encourage more environmental benefits, whereas if Prevention through Design (PtD) approach is utilised, it can be configured to emphasise safety related innovations in products and designs. Future research should look into expanding the application of the innovation classification system from productivity improvement to these emerging construction paradigms.

Acknowledgments

The authors would like to express their gratitude to Building Research Association of New Zealand (BRANZ) for providing the funding for this research. This work would not have been possible without access to the SCIRT innovation database. The authors thank SCIRT and University of Canterbury Quake Centre for providing access.

Author Contributions

Mohammadali Noktehdan: Paper writing, data analysis, development of classification system; Mehdi Shahbazpour: Paper writing, development of classification system; Suzanne Wilkinson: Paper writing.

Conflicts of Interest

The authors declare no conflict of interest.

References

1. Murray, M.; Murray, M.; Langford, D. Rethinking construction: The Egan report (1998). *Constr. Rep.* **2003**, *1944–1998*, 178–195.
2. Smith, B. Valuing the Role of Construction in the NZ Economy. Available online: http://www.constructionstrategygroup.org.nz/downloads/PwC%20Report%20-%20Construction%20Sector%20Analysis%20Final%204%20Oct.pdf (accessed on 30 March 2015).
3. Wilkinson, S.; Kempton, T.; Gleeson, A. Identifying Canterbury Rebuild Project KPI's—Baseline Report. Available online: http://buildingvalue.co.nz/sites/all/themes/productivity/images/Wrap/Reports/1211Canterbury_KPI_Project_Nov12.pdf (accessed on 30 March 2015).
4. Reichstein, T.; Salter, A.J.; Gann, D.M. Last among equals: A comparison of innovation in construction, services and manufacturing in the UK. *Constr. Manag. Econ.* **2005**, *23*, 631–644.
5. Statistics New Zealand. Research and Development Survey. Available online: http://www.stats.govt.nz/browse_for_stats/businesses/research_and_development/info-releases.aspx (accessed on 30 March 2015).
6. Seaden, G.; Guolla, M.; Doutriaux, J.; Nash, J. Strategic decisions and innovation in construction firms. *Constr. Manag. Econ.* **2003**, *21*, 603–612.
7. Loosemore, M. Innovate or perish? Exploring some of the myths of construction innovation. *Australas. J. Constr. Econ. Build. Conf. Ser.* **2014**, *2*, 44–55.
8. Huergo, E.; Jaumandreu, J. Firms' age, process innovation and productivity growth. *Int. J. Indust. Org.* **2004**, *22*, 541–559.
9. Koskela, L.; Vrijhoef, R. Is the current theory of construction a hindrance to innovation? *Build. Res. Inf.* **2001**, *29*, 197–207.
10. Park, H.-S.; Thomas, S.R.; Tucker, R.L. Benchmarking of construction productivity. *J. Constr. Eng. Manag.* **2005**, *131*, 772–778.
11. Brown, M. *Introduction to Innovation—Managing Ideas into Action*; Henley Management Center/Price Waterhouse: London, UK, 1994.

12. West, M.A.; Altink, W.M.M. Innovation at work: Individual, group, organizational, and socio-historical perspectives. *Eur. J. Work Org. Psychol.* **1996**, *5*, 3–11.
13. Organisation for Economic Development. *Oslo Manual: Guidelines for Collecting and Interpreting Innovation Data*; OECD Publishing: Pairs, France, 2005.
14. Shahbazpour, M. Strategic Manufacturing System and Process Innovation: A Framework for Small and Medium Sized Enterprises. Ph.D. Thesis, The University of Auckland, Auckland, New Zealand, February 2010.
15. Slaughter, E.S. Models of construction innovation. *J. Constr. Eng. Manag.* **1998**, *124*, 226–231.
16. Bassioni, H.A.; Price, A.D.F.; Hassan, T.M. Performance measurement in construction. *J. Manag. Eng.* **2004**, *20*, 42–50.
17. Environmental Protection Agency. The Lean and Environment Toolkit. Available online: http://epa.gov/lean/environment/toolkits/environment/resources/LeanEnviroToolkit.pdf (accessed on 30 March 2015).
18. Tatum, C.B. Classification system for construction technology. *J. Constr. Eng. Manag.* **1988**, *114*, 344–363.

11

Aiming for a Better Public Realm: Gauging the Effectiveness of Design Control Methods in Wellington, New Zealand

Morten Gjerde * and Brenda Vale

School of Architecture, Victoria University of Wellington, P.O. Box 600, Wellington 6140, New Zealand; E-Mail: brenda.vale@vuw.ac.nz

* Author to whom correspondence should be addressed; E-Mail: morten.gjerde@vuw.ac.nz

Academic Editor: George Baird

Abstract: Design review aims to improve the quality of urban settings, principally by seeking to influence the design of individual buildings positively during the planning approval stages of development. Design review systems were first set up in New Zealand in the mid-1990s in Wellington. The aims of design review are laudable; even if the process is not set up to secure the best possible design outcome it should be able to prevent the worst outrages, so as to ensure a minimal visual quality of streetscapes. However, does design review really achieve what it sets out to do? After a brief summary of design review practices, this article considers whether current design control practices in Wellington are helping to foster well-liked urban streetscapes. Using the results from a recent case study, comparing the aesthetic preferences of the public to those of professionals who participate in design control, the article notes where preferences overlap and where they are different. In the process, the characteristics of buildings and streetscapes that are considered positive and negative are identified. The article goes on to speculate how design review could be made to work more effectively in Wellington.

Keywords: design review; urban design guidance; environmental aesthetics; urban streetscapes

1. Introduction

It is never difficult to find opinions critical of the appearance of buildings and cites. When discussing Sydney a decade ago, John Punter made the observation that the city had succumbed to "spectacularly ordinary commercial development" in the period after World War II and that this poor aesthetic quality is only excused by an otherwise discerning public because of Sydney's spectacular setting [1] (p. 406). Making a more general observation, Bentley [2] suggested that the public are united in their dislike of the ways cities have been transformed in recent times. When discussing New Zealand architecture, Layla Dawson of the *Architectural Review* recently expressed surprise that Auckland has the appearance of an unplanned jumble of high-rise blocks. Why, she asked, would the public put up with such poor architecture [3]? Criticism of New Zealand urban design is even older than this. In 1958 when the architectural historian Nicholas Pevsner visited New Zealand he was hardly complimentary, suggesting there was no town planning in New Zealand, and certainly none in evidence in the central areas of the nation's cities. Even if there were, he wrote, it would be viewed by the non-designer public with suspicion [4]. In a 1958 radio broadcast, Pevsner directed his critique specifically toward Auckland, noting that the city's grid layout was run through the centre with scant regard for the slope toward the Waitamata Harbour or, more generally, for the undulating topography [5]. However, these are the opinions of design specialists not the general public. For the latter does the appearance of the built environment really matter?

The appearance of the built environment can have a tangible influence over property values, directly affecting the financial well-being of those who own property or conduct a business in an area. The choices people make every day usually include an aesthetic dimension. These may include deciding which city to live in, which house to buy or even which street to walk along [6] (p. 164). It seems that visually attractive places are more sought after by the public, which serves to increase demand and therefore economic values. Increasingly, the built environment is being used as a tool in battles fought between cities for business investment, tourism and high-calibre workers [7,8]. People regularly make choices around matters of aesthetics and it is clear that the visual attractiveness of cities affects the financial well-being of individuals as well as the wider community.

Like many other cities, Wellington, New Zealand has adopted an explicit design control regime as part of its regulatory planning system. Through its District Plan the Wellington City Council seeks to "*encourage positive growth that promotes the City's comparative advantages*" and lists eight principles to guide development [9]. One of these principles acknowledges that the quality of the public realm is influenced by the buildings and structures that define it. The primary tool used by Council to achieve a high quality public realm is the Central Area Urban Design Guide, the aim of which is to ensure that new developments are coherently designed, make a considered response to their setting and establish positive visual effects [10].

2. What Is Design Control and How Is It Managed?

Design control is generally administered as part of the wider regulatory planning system. Regulation of private development emerged in the wake of the Industrial Revolution as it became clear that the free market could not be relied on to deliver appropriate results in a consistent manner. Many urban environments had become unhealthy as well as unattractive through development that had been undertaken to support rapidly expanding industrial activities [11] (p. 25). The bases of contemporary planning regulation can be traced to efforts made in the wake of the Industrial Revolution to ensure that the interests of society as a whole are served when changes to the urban environment are proposed. Under the umbrella of planning regulation, design control is undertaken by local government on behalf of those who must live with the consequences but do not otherwise have a say in the design of new buildings [12,13]. Depending on the planning context, regulation is nothing more than a backstop to help ensure that individual projects are built in a manner that does not create untoward environmental effects. Coupled with reliance on the private sector to initiate changes, this means that planning regulation tends to be reactive, rather than proactive, when it comes to managing the built environment.

A number of methods for regulating and controlling the aesthetic outcomes of urban development have been developed and successfully operate in different parts of the world. Punter [14] advises that design review practices can be either regulatory or discretionary. Regulatory (also referred to as administrative) systems are an add-on to zoning controls, where design outcomes are controlled by rules and objective measures. Examples of this are maximum height limits, prescriptions for location of buildings in relation to boundaries, and floor area ratios. These systems provide high levels of certainty to all parties and regulation can be conducted largely through administrative functions, reducing demand on local authority resources. However, they are also criticised for being coarse in nature with a tendency toward monotonous environments as there is a tendency for new projects within an area to be built to the same prescribed limits [6,12].

Discretionary systems are more "pragmatic", as they allow regulation to refer to the immediate setting and other variable factors that cannot be accounted for in most rules based regimes. The more successful discretionary systems, in terms of meeting the needs of different stakeholders, are informed by design guidelines or briefs that provide both a target for designers and a reference for those assessing the proposal [15]. While discretionary systems often allow local conditions to be taken into account they are often criticised by the development industry and designers for being arbitrary [16].

Over the past 30 years regulatory design control has tended to become more discretionary, creating scope for development design quality to be scrutinised more closely. Likewise, design guidance informing discretionary systems has become more prescriptive in response to calls for more certainty around the process for applicants. In short, design control systems appear to be converging as they are modified to include both discretionary and regulatory characteristics [14].

The literature outlines a number of challenges for design review. Scheer [16] notes that design review is time consuming and expensive, that it is easy to manipulate through persuasion and pretty pictures, and that it is administered by overworked and inexperienced staff. However, she also suggests that these matters are easy to solve, mainly with additional financial or human resources, through education and by ensuring the process is adequately audited for political involvement. Presenting more substantial challenges for design review are issues relating to power, freedom, justice and aesthetics. These issues exist as tensions between competing social, political or legal forces. The *power* issue raises questions around who makes the key decisions. Arguing from the perspective of a design professional, Scheer makes the observation that design review is the only field in which lay people (or those not specifically trained in design) can be left to rule over professionals in their areas of expertise.

Freedom of expression in the built environment is another matter that troubles design review. Concerns about freedom are triggered by societal values aligned with property ownership, particularly in the "new world" countries. This concern is most pronounced in the United States although the courts there have consistently upheld the rights of government to control design outcomes [17]. Concerns about *justice* arise when the reviewer is seen to hold the power, which may limit the extent to which a "fair hearing" is possible. This issue is minimised where written guidelines underpin the planning process. Finally, the murky area of *aesthetic preference* questions whose values should be referenced. Scheer notes that in many cases guidance remains unhelpfully abstract and projects tend toward soft solutions and mimicry in order to navigate the review process.

Traversing the legal issues raised by design review, George and Campbell [18] identify the various interests that are active in it. Their work addresses the subject in an American context, so freedom of expression is a consistent theme. They highlight the need for variety in the built environment from a physiological perspective. As human wellbeing is affected by the environment, the case for regulation suggests that new designs should avoid inducing unnecessary stress, anxiety or fear in people. However, people also learn and develop their senses through environmental experience and so it can be useful for environments to be challenging. More complex settings, which arise through unfettered design decisions, can create such opportunities. It is on this basis that George and Campbell challenge control methods that require new developments to conform to the patterns and design expressions already present in the setting. In addition, pushing the current boundaries of design creates opportunities for development of new theories, new methods and general advancement of the field, which they argue would have wider social benefits. To acknowledge and balance the various interests properly, George and Campbell [18] (pp. 171–173) describe four criteria that a sound design review process must address:

1. A clearly articulated and demonstrable public interest;
2. Demonstrable links to the stated intent;
3. Application early in the design or decision process;
4. Encouraging a variety of acceptable decisions.

Punter [14] incorporates these criteria into a comprehensive framework for design review and development management (Table 1). The twelve principles are grouped under four headings; community vision, design planning and zoning, substantive design principles and due process. The principles under the first heading aim to capture the views of the community as a step in generating a comprehensive view to guide development and control. The community vision becomes the reference for all decisions. Secondly, three principles inform development of a suitable process that includes incentives as well as the requirements necessary to regulate for high quality design outcomes. The next three principles address the nature of the relationship between the comprehensive plan and the development industry charged with its implementation. Punter argues that the relationship should be pluralistic and not overbearing in order to allow creative solutions to emerge. Under the final heading four principles address issues of fairness in administering the design control regime.

Table 1. Principles for progressive design review. Source: Punter [14].

Heading	Principle
Community Vision	1. Committing to a comprehensive and coordinated vision of environmental beauty and design; 2. Developing and monitoring an urban design plan with community and development industry support and periodic review;
Design Planning and Zoning	3. Harnessing the broadest range of actors and instruments (tax subsidies, land acquisition) to promote better design; 4. Mitigating the exclusionary effects of control strategies and urban design regulation; 5. Integrating zoning into planning and addressing the limitations of zoning;
Broad, Substantive Design Principles	6. Maintaining a commitment to urban design that goes well beyond elevations and aesthetics to embrace amenity, accessibility, community, vitality and sustainability; 7. Basing guidelines on generic design principles and contextual analysis and articulating desired and mandatory outcomes; 8. Not attempting to control all aspects of community design but accommodating organic spontaneity, vitality, innovation, pluralism: not being over-prescriptive;
Due Process	9. Identifying clear a priori roles for urban design intervention; 10. Establishing proper administrative procedures with written opinions to manage administrative discretion, and with appropriate appeal mechanisms; 11. Implementing an efficient, constructive and effective permitting process; 12. Providing appropriate design skills and expertise to support the review process.

3. Does Design Control Help Achieve Well-Liked Streetscapes?

Concerns about the poor visual quality of many contemporary buildings, as noted in the popular media and by other researchers, must question the effectiveness of design control processes. In a 2003 study, McIndoe [19] audited the effectiveness of two residential character area design guides that had been introduced into the Wellington District Plan. The report reviewed the resource consent process and monitored the architectural merit of the completed project. Following the review of 16 projects implemented over a three year period, the report concluded that the planning rules and associated design guidelines had been effective in achieving desired character outcomes. These results could have been predicted however, as the study investigated the two character areas with the most prescriptive design guides in the District Plan. The report went on to note a positive correlation between the involvement of an architect and a successful design outcome. Similarly, there appeared to be a strong correlation between the absence of a skilled professional and a poor design outcome. The findings of this report thus question the chance of achieving high or even passable outcomes in those projects that do not involve skilled professionals and that are being undertaken in areas with no design guidance.

A year later, Rae [20] reviewed completed projects in the central area of Wellington including the Courtenay and Cuba Character Areas in a similar, subjective review of 20 completed projects. The study concluded that the guidelines were not effective when used in conjunction with controlled activity status under the District Plan. Controlled activity status is where the proposed building or activity essentially cannot be refused, regardless of the aesthetic qualities of the built form. While the earlier study found pre-application consultation to be widespread and beneficial for project outcomes, the second study found early consultations to be ineffective for a variety of reasons. The architects Rae consulted noted that the advice given by Council officers in the early stages was inconsistent and conservative. He also found that when it was undertaken, consultation did not take place at a critical, site-specific analysis stage. Instead, he found, it only took place once preliminary design proposals and economic budgets had been formed. As a consequence, Council urban design input was unable to prompt significant changes, should they be necessary, to improve outcomes. Finally, the research suggested that the discipline specific approach to evaluating projects could have detrimental effect on projects when compared to assessment of proposals in a more holistic manner. Rae cited changes in project particulars to satisfy singular and conflicting advice from heritage, traffic, wind, and urban design advisers as having diminished the quality of design outcomes.

While interesting, both of these earlier studies relied on the expertise of the authors to evaluate the design outcomes. The evaluations appear to have been done in the absence of a design framework, instead relying on the expert opinions of each author. Neither study sought input from others, including from members of the public. This would please Scheer [16], who argues that design review should be the exclusive domain of experts trained in the visual arts. In particular, she advocates for the process of design review, if it is undertaken at all, to be fully entrusted to architects, the experts in this field. However, there is little evidence to suggest that such experts speak on behalf of the wider population. Indeed, the little research that has been done to compare the aesthetic preferences of lay people and professionals suggests that there are differences in the way each group evaluates the built environment [21,22]. The paper will go on to discuss the findings of a study that was aimed at understanding people's perceptions of urban streetscapes and the individual buildings of which they consist. The study findings are then discussed in relation to the regulatory planning context to firstly understand the extent to which design control leads to well-liked urban streetscapes in Wellington and then to consider areas where improvement could be made.

4. College Street Case Study

Late in 2013 a survey of people's visual perceptions of the buildings lining the two sides of College Street in Wellington was conducted. The survey was designed to understand people's preferences for the designs of individual buildings, for the relationships formed between adjacent buildings and for the overall streetscape. The study also sought to identify areas where the aesthetic preferences of members of the different professional groups involved in design review might differ from those of the lay public. College Street was selected for the study for a number of reasons. Primarily, it was seen as a typical urban street, with buildings built to the street edge and from side boundary to side boundary. As it is not a main thoroughfare for motorcars there would be little traffic to distract the attention of people as they made their evaluations of the buildings. As there had been a moderate number of redevelopment projects undertaken over the past 20 years, College Street was also seen as a setting that had been influenced by the city's design control practices.

Survey participants were recruited through notices placed in cafes and other semi-public places throughout a small geographic area. Others were recruited through notices placed in the newsletters of local community groups, business organisations and city council departments. Recruitment sought to ensure participation from a range of demographic backgrounds, including age and occupation. All participants were offered a pre-paid hot drink of their choice when they returned their survey form to one of the local cafes that had agreed to collect these. These efforts led to 75 responses being received.

Initial analysis of the survey data brought forward several streetscape circumstances for further discussion with the participants. Discussion of these issues took place in two focus groups, one comprising trained professionals working in the fields of planning, architecture and urban design and the other only made up of un-trained lay members of the public. The rationale for convening separate groups was to allow the discussions to run more freely, unhampered by a need to consider the views of the other group. The College Street case study is part of a larger research project design to identify characteristics in the design of individual buildings and their relationships with others that are valued by the public as well as those that are not. If those that are highly valued can be encouraged through design review and those that diminish aesthetic experience can be avoided then the chance that design control can have a positive effect on the visual qualities of Wellington would be improved.

Preferences for two individual buildings stood out. The first (Figure 1) can be seen to typify the changing context of this part of the city in that it is a building converted from its earlier light industrial use to one that is primarily residential. However, contrary to several other recent developments, it is of a reasonably small scale. The second most liked building is one that accommodates a well-known food market (Figure 2). With the business having been established in the area for more than 20 years, the building has recently been renewed and expanded. The key reasons given by survey respondents for their preferences for these two buildings were the façade details and their overall shapes. Through the focus groups it came to light that these characteristics help both buildings create a positive relationship with the street space and that fenestration helps people relate to the activities taking place inside. The two focus groups were in agreement that these qualities enabled people to forge positive associational meanings for the two buildings.

Figure 1. Residential building in College Street. This building has the highest preference rating among the public.

Figure 2. Detail from the street facade of the building of a prominent food retailer, the second best liked building in the street.

Although several buildings, like the two discussed above, were perceived positively, nearly half the buildings along the street were not liked by the public. The two most disliked buildings were a small, dark coloured structure between the two found to have the greatest appeal (Figure 3) and a multi-storey commercial building at the eastern end of the street (Figure 4). Both were considered impersonal buildings. In the case of the smaller structure, the mirrored glass and dark colour together helped generate these feelings for people. The larger building is large in relation to the street width, has very flat surfaces and finishes and at street level does not forge a relationship with the street because of its blank walls and driveway. Participants in the focus groups identified all these as reasons for the negative evaluation. Both buildings predate the current design control processes and were realised in a planning context where regulation of floor area ratio was the principal method of design regulation.

Figure 3. The least liked building in College Street was also judged by survey participants to form a positive relationship with the building to the left.

Figure 4. This large commercial building dating to the 1980s is the second least liked building in College Street.

There appear to be several positive relationships between adjoining buildings along College Street. The relationship between two buildings at the western end of the street (Figure 5) was established decades ago, when warehousing and light industrial activities were dominant. Known in the survey as buildings A & B, this harmonious relationship is enhanced not only by similarities in height but also by similarities in colour and design expression. As can be seen, the fenestration patterns, façade detailing and alignment between floor levels help create links between the two sites. Not far behind in this preference category is the well-liked relationship between the food market and the least liked building in College Street. This is an interesting result as the two are at opposite ends of the spectrum of public preference. Factors that made this relationship more likable include similarity in building heights, colour and rhythm of façade articulations. These links help overcome differences created by the design approaches taken by the respective designers.

Figure 5. The relationship between these two buildings is seamless and was well-liked by the survey respondents.

The least liked inter-building relationship is considered to be representative of the tensions that arise as the built environment changes to accommodate the growing popularity of an area. One half of this relationship is a large apartment building constructed in 2005, figuratively "riding the wave" of popularity of living close to the city centre. The project amalgamated several titles and was built to 4 m above the allowable height limit for the zone, which was 27 m. This building was built alongside a single storey, vividly coloured metal workshop with the resultant relationship judged to be very poor by the survey respondents (Figures 6 and 7). Discussions within the professional focus group suggested this poor relationship could be more the perceived "intrusion" of the smaller building, despite it having been there first. In terms of all respondents, the poor relationship was mainly attributed to differences in height and, to a lesser degree, differences in cladding materials. The professional focus group considered that the workshop activity is no longer appropriate for College Street, whereas the lay group felt the opposite.

Figure 6. This building for a metal recycling business is typical of the older buildings in College Street. The new apartment building to its left creates a poor relationship between the two.

Figure 7. The Trinity Apartments building, in the middle, establishes a poorly regarded relationship with its neighbours. The public also dislike the large blank side wall and the private balconies projecting beyond the site and into the public street space.

The final scale at which perceptions of the College Street environment were examined was across all buildings, in other words the *streetscape*. Respondents were asked to evaluate the collections of buildings along each side according to relationships between building heights, the three dimensional alignment between facades and the surface designs of these. Definition of these three levels or scales is based on environmental psychology literature [23,24]. The research findings suggest that, while some variation in height along a streetscape can be attractive, people are generally dissatisfied with significant variations within a single street. The southern side of College Street exhibits greater variation and was rated more poorly than the north. Respondents to the survey and participants of both focus groups were united in this opinion. Variations in height were given as evidence of poor relationships between buildings and disruptive of aesthetic coherence. While the southern side of the street has undergone more extensive redevelopment than the northern side during the past 20 years, it is perceived negatively at each of the three spatial scales.

As noted, the study was designed to enable comparisons between the aesthetic preferences of those who are directly involved in design review and the lay public, whose views are rarely, if ever, represented. It appears that the opinions of building designers and planners align with the lay public in several areas examined by the survey, with agreement for all but one of the six questions on streetscape. That is, both groups liked or disliked a particular streetscape characteristic, based on a comparison of mean preference scores. Building designers and planners appeared to be marginally more extreme in their ratings, particularly where the mean score fell below neutral. For example, Table 2 shows this group rated the relationships of façade styles on both sides of College Street lower than lay people did, so in this sense they could be considered more critical. When considering the individual buildings and their façade designs similar patterns were observed. In addition to designers and planners being more critical about buildings that were disliked their scores were also generally higher when both groups liked a building or façade design. These patterns align with an earlier study [25] that found professionals were more extreme in their views about streetscapes than lay people. Reasons for this were thought to be training and familiarity, which meant professionals were more confident of their opinions and could express them more clearly.

Table 2. Comparison of mean scores for the streetscape characteristics, on 5-point scale.

Streetscape Characteristic	Change Professionals	Lay Public
South side	–	–
Building Heights	2.95	2.77
Facade Alignment	2.42	2.63
Façade Styles	2.47	2.79
North side	–	–
Building Heights	3.16	3.53
Facade Alignment	2.79	3.31
Facade Styles	2.32	2.84

There were however, two areas where there were differences of opinion between these groups. Firstly, the lay public, unlike designers and planners, appeared to like the older, light industrial buildings in College Street, and this was true for every building belonging to the light industrial typology lay. One interpretation of this, supported by the focus group discussion, is that the designer group sees the street character as emergent and considers the older, industrial buildings to be unwanted remnants and possible barriers to further development of the new street character. Secondly, the same group were both more positive about sharp differences in height along the southern side of the street and less positive about the settled pattern of building heights on the northern side. This finding was also supported by the focus group discussion with the possible reasons given for this being a higher wish for variety amongst those directly involved in urban transformation.

Discussions about the merits of the large apartment building seen in Figure 7 deserve special mention, as its design raised several concerns with the public. Those trained in design and planning were somewhat ambivalent about it but the survey responses suggest that the lay public liked the design. Despite this, members of the layperson's focus group gave several reasons for disliking the building. The key issues were its overall height, the poor relationships formed with the buildings on each side, the large blank side walls, the projecting private balconies over what was felt to be public space (the pavement), and the large areas of glass in the street façade. Interestingly, while people consider fenestration an important contributor to the design, it was generally agreed that the areas of glass are too large for the residential activity in this setting. People were both empathetic to the way glazing can restrict privacy for the buildings' residents as well as cause overheating. While they could appreciate the need for residents to address both concerns by drawing blinds during the day, people also thought the designers could have done more to avoid this.

5. Discussion

Building height came out as an important factor influencing people's perceptions of urban streetscapes. The surveys went some way toward identifying this and it was further established during the two focus groups. Where height had a negative influence it was either because this led to an overbearing presence in the street, essentially creating a sense of claustrophobia amongst the respondents, or because excessive variations in height meant large, blank side walls were easily seen. Pronounced differences in height, usually created by the more recent intervention, were considered to be visually awkward and gave an impression of overdevelopment.

More than any other characteristic, height leads to tensions in development and change processes. Height tends to be controlled by area-wide prescriptions. In the urban context, height is where the landowner stands to improve profitability of a redevelopment project. The Wellington design guides do not stress the importance of height relationships between buildings or height in relation to street dimensions. Accordingly new development is prone to push to the limit, and many projects push through it by putting forward an argument that the additional effects of the extra height are minimal. Nevertheless, height differences, such as those that emerge between the building seen in Figure 7 and its neighbours, appear to be recognised by people moving along the street.

The regulatory planning regime of Bergen, Norway is an example of how building height could be addressed in the design review process to limit, if not altogether avoid, potentially negative visual outcomes. The Bergen *Regulerings Plan* [26] allows for site-specific height controls. Allowable heights are set out in the plan and the current prevailing maximum is 27 m. However, there is also an overarching requirement for new buildings to fit closely with the context and this enables the regulators to limit the height of new buildings to no more than 3 storeys above the neighbouring structure, irrespective of the stated maximum. Another restriction limiting height is the requirement for façades to have a relationship with the street width of no more than 1.1 to 1. Clearly the emphasis is on the public space surrounding the site rather than giving maximum development potential. While the vision of the plan is a denser, more compact city, the approach seeks to maintain existing building scale relationships while encouraging incremental change.

Wellington has established blanket allowances for building height, varying from 27 m in Te Aro to 65 m in the central area. The only additional factor influencing height appears to be the effect the proposed structure would have on wind patterns at street level. In most circumstances these can be mitigated by the addition of design features, such as sidewalk canopies, that interrupt the flow of wind down to street level. The regulatory context in Wellington does not enable building heights to be controlled in relation to adjoining building heights if they are at or below the blanket height restrictions. And yet, differences in height between adjoining buildings was found to be a factor leading to streetscapes that people dislike. It seems that streetscape outcomes could be enhanced if blanket height controls were to be moderated through consideration of how the proposed heights of new buildings relate to those of adjoining buildings such as is currently done in Bergen.

Visual interest in façade composition appears to be a key success factor. Where the street façade was flat and visually uninteresting, the public tended to judge the building poorly. The survey responses revealed strong correlations between people's preferences for the façade design and their preference for the building overall. The design review process appears to be effective in fostering interesting façade treatments in new buildings. The food market building redevelopment was a good example of this. Uninfluenced by design guidance, most warehouse buildings incorporate flat, utilitarian façade treatment as the emphasis is on a cost efficient enclosure of space. While the food market building in College Street fits with the warehouse typology, the new parts of the building that face onto the street incorporate surface treatments, fenestration and detailing that help generate visual interest. The public responded positively to this, making the street level façade design the best liked of all and ranking it overall the second best liked building in the street.

When it comes to building height new developments that are markedly higher than their neighbours mean the side walls become clearly visible from the street. Side walls are often kept blank for practical, economic and fire protection reasons. While the design guides in the Wellington District Plan seek to avoid uninteresting boundary walls, the recommendations tend to focus on surface treatment and small scale surface modelling. Fundamentally, these remain blank because of a lack of fenestration. It seems that design review has been unable to limit creation of new, visually uninteresting side walls in College Street. This may in part be attributable to a general acceptance by designers and planners that visual interest can be generated in these through application of surface patterns or textures. The public were not willing to accept such treatments as appropriate and expressed their disdain for these additions to College Street through the focus group.

Encroachment into public space by private development was considered to be negative in spatial terms as well as generate negative meanings about the building. The layperson's focus group was particularly concerned about the way the balconies of the large apartment building had been allowed to encroach into public space over the street (Figure 7). When it was pointed out that other buildings, such as the one judged to be favourite amongst all of those in College Street, also had balcony projections into public space, their views did not change. The difference, according to the discussion, was in the number and extent of the encroachments. On the other hand, the professionals' focus group did not view these balcony extensions negatively. The design guide is supportive of the contribution that balconies and other façade articulation can make to design outcome but does not specifically address the matter of where they should be located in relation to the boundary. Encroachments are an opportunity for landowners to locate parts of their development on public land if there are reasons why they cannot be fully contained within their site or to accommodate some form of public amenity provided by the landowner, such as the sidewalk canopy. Approval of encroachments is dealt with outside the statutory planning process. Interestingly, where a design outcome relies on articulations extending across into public space there is always a risk the design detail would not be allowed by the encroachment evaluation process.

The literature reminds that people attach meanings to buildings and other parts of the built and natural environment when forming an aesthetic response. This fact has been confirmed, particularly through the focus group discussions where participants provided examples of positive and negative associational meanings when discussing their building preferences. Controlling or guiding design for the meanings people will attach to new developments could however, be problematic, given that meanings are personal, linking to a person's experience as well as to wider cultural or social constructs. Meanings also attach over time and can change over time. While links between certain design attributes and the meanings people relate to them are well known, the design control literature is largely silent on this issue. It may be appropriate then to undertake further research to enable better understanding of how, if at all, design control could be improved by accounting for this dimension of environmental aesthetics.

6. Conclusions

This article has discussed regulatory design control from the perspective of public opinion. After establishing the relevance of the appearance of the public realm to the lives of individuals and to society, the paper proceeded to discuss how local government controls the design of new development proposals. Although design control has a long history, only in the past half century has its use become widespread. Design control is not unproblematic and several of the tensions that arise have been identified, particularly tensions around freedom of expression and whose opinions are best to guide decisions about aesthetics.

The question as to whether design control is an effective means by which to ensure well-liked urban streetscapes, has become a research issue. The two earlier studies, both of limited scope, that were conducted in Wellington and that were discussed here did attempt to address this question. However, a significant shortcoming of both is that neither engaged with the public. The research reported in this article did consult with the public in an effort to gauge their opinions on the design quality of individual buildings, of relationships between building and of entire collections or streetscapes. Key factors affecting environmental aesthetic judgment include building height, visual interest and façade composition, the extent to which parts of a development encroach into public space and the meanings people associate with buildings and their details. Addressing the matter of height, it was noted that height is usually prescribed over wide areas with little regard for how this might affect individual relationships between buildings. Land value is clearly tied up with its development potential and allowable height is one of the key factors affecting value. Planning in the city of Bergen, Norway was cited to provide an alternative means for controlling building height. Finally, it was suggested that new research is necessary to understand how associational meanings might be referenced in design control processes.

Author Contributions

The research and writing of the paper have been substantially carried out by Morten Gjerde. The research reported in the paper is part of a larger project being conducted by Morten Gjerde as part of his PhD in urban design. Professor Brenda Vale is supervising the research and contributed directly to the Introduction section of the paper.

Conflicts of Interest

The authors declare no conflict of interest.

References

1. Punter, J. From the ill-mannered to the iconic: Design regulation in central Sydney 1947–2002. *Town Plan. Rev.* **2004**, *75*, 405–445.
2. Bentley, I. *Urban Transformations: Power, People and Urban Design*; Routledge: London, UK, 1999; p. 298.
3. Dawson, L. Architecture in the land of the long white cloud suffers from suburban creep and a lack of identity. *Archit. Rev.* **2010**, *227*, 33.
4. Pevsner, N. New Zealand. *Archit. Rev.* **1959**, *126*, 205–206.
5. Pevsner, N.; Games, S. *Pevsner: The Complete Broadcast Talks: Architecture and Art on Radio and Television, 1945–1977*; Ashgate Publishing Company: Burlington, VT, USA, 2014.
6. Madanipour, A. *Design of Urban Space: An Inquiry into a Socio-Spatial Process*; John Wiley & Sons: Chichester, UK, 1996.
7. Carmona, M.; Magalhaes, C.D.; Edwards, M. Stakeholder views on value and urban design. *J. Urban Des.* **2002**, *7*, 145–169.
8. Cuthbert, A.R. *The Form of Cities, Political Economy and Urban Design*; Blackwell Publishing: Oxford, UK, 2006; p. 290.
9. Wellington City Council. *Wellington District Plan Chapter 12: Central Area*; Wellington City Council: Wellington, New Zealand, 2013; pp. 1–64.
10. Wellington City Council. *Central Area Urban Design Guide*; Wellington City Council: Wellington, New Zealand, 2012; pp. 1–25.
11. Carmona, M.; Heath, T.; Oc, T.; Tiesdell, S. *Public Places Urban Spaces*; Architectural Press/Elsevier: Oxford, UK, 2010; p. 368.
12. Delafons, J. Democracy and Design. In *Design Review: Challenging Urban Aesthetic Control*; Scheer, B.C., Preiser, W.F.E., Eds.; Chapman & Hall: New York, NY, USA, 1994; pp. 13–19.
13. Ellin, N. *Integral Urbanism*; Routledge: New York, NY, USA, 2006; p. 173.
14. Punter, J. Developing urban design as public policy: Best practice principles for design review and development management. *J. Urban Des.* **2007**, *12*, 167–202.
15. Nasar, J.; Grannis, P. Design review reviewed. *J. Am. Plan. Assoc.* **1999**, *65*, 424–433.

16. Scheer, B.C. Introduction: The Debate on Design Review. In *Design Review: Challenging Urban Aesthetic Control*; Scheer, B.C., Preiser, W.F.E., Eds.; Chapman & Hall: New York, NY, USA, 1994; pp. 1–10.
17. Lai, R.T. Can the Process of Architectural Design Review Withstand Constitutional Scrutiny? In *Design Review: Challenging Urban Aesthetic Control*; Scheer, B.C., Preiser, W.F.E., Eds.; Chapman & Hall: New York, NY, USA, 1994; pp. 31–41.
18. George, R.V.; Campbell, M.C. Balancing different interests in aesthetic controls. *J. Plan. Educ. Res.* **2000**, *20*, 163–175.
19. McIndoe, G. *Mt. Victoria and Thorndon Residential Character Implementation Audit*; Wellington City Council: Wellington, New Zealand, 2003; p. 20.
20. Higgins, M. Urban design and the planning system in Aotearoa-New Zealand: Disjuncture or convergence? *Urban Des. Int.* **2010**, *15*, 1–21.
21. Groat, L.N. Carbuncles, Columns and Pyramids: Lay and Expert Evaluations of Contextual Design Strategies. In *Design Review: Challenging Urban Aesthetic Control*; Scheer, B.C., Preiser, W.F.E., Eds.; Chapman & Hall: New York, NY, USA, 1994; pp. 156–164.
22. Stamps, A.E. Demographic effects in environmental aesthetics: A meta-analysis. *J. Plan. Lit.* **1999**, *14*, 155–175.
23. Stamps, A.E. *Psychology and the Aesthetics of the Built Environment*; Kluwer Academic Publishers Group: Dordrecht, The Netherlands, 2000; p. 320.
24. Nasar, J.L. *Environmental Aesthetics: Theory, Research, and Applications*; Nasar, J.L., Ed.; Cambridge University Press: Cambridge, UK, 1988; p. 529.
25. Gjerde, M. Visual evaluation of urban streetscapes: How do public preferences reconcile with those held by experts. *Urban Des. Int.* **2011**, *16*, 153–161.
26. Bergen Kommune. *Kommunedelplan Sentrum*; Bergen Kommune: Bergen, Norway, 2002; p. 63. (In Norwegian)

Time-History Seismic Analysis of Masonry Buildings: A Comparison between Two Non-Linear Modelling Approaches

Michele Betti *, Luciano Galano and Andrea Vignoli

Department of Civil and Environmental Engineering (DICeA), University of Florence, Via di Santa Marta, 3-I-50139 Firenze, Italy; E-Mails: luciano@dicea.unifi.it (L.G.); avignoli@dicea.unifi.it (A.V.)

* Author to whom correspondence should be addressed; E-Mail: mbetti@dicea.unifi.it

Academic Editor: Ali Memari

Abstract: The paper presents a comparison between two numerical modelling approaches employed to investigate the seismic behavior of unreinforced masonry buildings with flexible diaphragms. The comparison is performed analyzing a two-story prototype tested on a shaking table at the CNR-ENEA research center of Casaccia (Italy). The first numerical model was built by using the finite element (FE) technique, while the second one was built by a simplified macro-element (ME) approach. Both models were employed to perform non-linear dynamic analyses, integrating the equations of motion by step-by-step procedures. The shaking table tests were simulated to analyze the behavior of the prototype from the initial elastic state until the development of extensive damage. The main results of the analyses are discussed and critically compared in terms of engineering parameters, such as accelerations, displacements and base shears. The effectiveness of both models within the investigated typology of buildings is then evaluated in depth.

Keywords: unreinforced masonry buildings; finite element (FE) models; macro-element (ME) models; time-history non-linear analysis; seismic loading; flexible diaphragms

---

1. Introduction

The seismic behavior of old masonry buildings can substantially differ from that of modern masonry constructions designed and built to meet aseismic design provisions. During an earthquake, both out-of-plane and in-plane seismic responses of the masonry walls are simultaneously activated. In the absence of a box-like behavior, the structural response is characterized by local failures with a widespread out-of-plane overturning of portions of walls. This has been frequently observed during past and recent earthquakes [1]. A satisfactory seismic performance of the structure, by means of the development of a global failure mode, can be achieved only if the out-of-plane collapses are prevented. In this case, the in-plane strength and deformation capacity of the walls are fully exploited, and they rule the global dynamic response of the building. In this respect, it is widely recognized that adequate masonry quality, proper connections between masonry walls and between walls and floors are effective measures able to prevent the local failures [2–4]. Among them, a significant role in coupling the response of the different masonry walls is played by the in-plane stiffness of the diaphragms (typically timber floors).

These few remarks point out the attention that must be paid by researchers when they analyze the seismic behavior of existing masonry buildings through global numerical models. Although in some cases simplified partial models could be used, a global model can better analyze the seismic response of the structure provided that a proper modelling of walls and floors is performed.

Dealing with the analysis methods, the largest part of them can be divided into three main categories: (i) the modal analysis, usually based on linear models, combined with the concept of structural behavior factor q to account for the energy dissipation effects; (ii) the non-linear static analysis (pushover); and (iii) the non-linear time-history analysis [5–7]. The first method is commonly adopted in the professional context, to verify the compliance of the performance level with the one required by the technical standards. Anyway, it is not fully appropriate for a reliable prediction of damages occurring in the building under earthquake loading. The pushover analysis is a procedure used to obtain the capacity diagram, which represents a basic datum to predict the strength and the ductility of the structure. In recent times, the method has become quite widespread also in engineering practice. Nevertheless, since this approach is based on a non-linear static procedure where the external loads are monotonically increased up to failure, it neglects many significant aspects of the structural response, like the damage produced by reversal loads. In addition, in the case of in-plane flexible diaphragms, even when proper connections between walls and floors prevent the local out-of-plane mechanisms, the masonry façades can exhibit an independent behavior, thus highlighting critical issues in a reliable employment of the pushover approach [3,8]. The time-history non-linear analysis method foresees a direct step-by-step integration of the equations of motion in the time domain. Provided that both a reliable non-linear numerical model of the building and a proper modelling of the seismic input are assured, this method can cover the drawbacks of the first two procedures, being able to follow the full seismic loading process (from the initial state, through the weakly non-linear behavior under service loading, up to the strongly non-linear behavior leading to collapse).

In this respect, several authors [2,3,9] have discussed the need to perform non-linear time-history analyses to reproduce the energy dissipation processes that develop in masonry buildings during the seismic event. At the same time, they showed that the use of three-dimensional finite element (FE) models requires a great amount of computational resources, which may be justified only in the case of historical and/or monumental buildings [9–11]. Analyzing traditional buildings, the need emerges for simplified, but effective (and reliable) models for the assessment of the global structural response through non-linear time-history analysis methods. Nevertheless, given the great variety of geometries, construction typologies and materials used to date, an analysis tool suitable for the analysis of all of the typologies of masonry buildings is not available. In the last few decades, for instance, an increasing development of simplified masonry modelling based on the macro-element [12] and the equivalent frame approaches [13–15] has taken place. These methods require a moderate computational effort, if compared with the traditional finite element technique. However, despite the apparent simplicity, each analysis tool requires high expertise to effectively understand its range of employability.

The above-mentioned remarks explain the difficulties that arise during the seismic assessment of existing masonry buildings. Within this framework, the paper reports the comparison between two numerical approaches employed to investigate the seismic response of traditional unreinforced masonry (URM) buildings. As a reference case study, an unreinforced masonry building (tested by a shaking table at the CNR-ENEA research center of Casaccia near Roma, in Italy) was analyzed by developing two different three-dimensional (3D) models. The first model refers to the FE technique and was built with the commercial code ANSYS [16]. The second one was based on the macro-element approach through the implementation of a non-linear algorithm in the MATLAB tool [17]. A simple identification procedure was used to define the most important masonry constitutive parameters required by the two models. To this aim, the available results by the laboratory tests performed at the CNR-ENEA center on units, mortar and diagonal compression tests on masonry wallettes were used. After identification, the two models were employed to perform linear and non-linear analyses in the time domain, by using as input the acceleration histories applied at the base of the prototype during the shaking table tests. The results obtained with the two models are compared and discussed analyzing floor accelerations, floor displacements, base shears, stiffness decay and failure modes. The comparison between experimental and numerical results was presented in a previous paper [18], to which the interested reader is referred.

2. The Masonry Building Prototype

The building assumed as a case study is one of the two masonry prototypes built at the CNR-ENEA research center of Casaccia, near Roma (Italy). The two buildings, tested through an extensive experimental investigation on a shaking table, were built within a research project partially funded by the Italian Ministry for Research [19]. The project aimed at assessing the seismic performance of scaled structures made with the typical materials and the constructive techniques employed in old masonry buildings in Central Italy. The first prototype (RM) was strengthened with the CAM system (active confinement of masonry), then it was tested to verify the effectiveness of the reinforcement; the second one (URM) was tested until collapse without any strengthening [20,21]. The prototype herein analyzed is the unreinforced one.

2.1. Geometry of the Prototype

The prototype is a 1:1.5-scaled two-story building (Figure 1), aimed at reproducing the typical ancient masonry buildings of Central Italy. The plan layout is made of a single cell with outer dimensions of 3.5 m × 3 m (Figure 2), while the inter-story heights measure approximately 2.2 m. A further masonry ring 0.15 m high is present at the top of the building, so the total height of the prototype is 4.55 m. The masonry walls have a constant thickness of 0.25 m and are composed of two vertical faces without any transversal connection (Figure 3). The horizontal structures are made with flexible timber floors. Each floor was built using five wooden joists (with a section of 0.10 m × 0.18 m) and wooden boards with a thickness of about 20 mm nailed to the wooden beams.

Figure 1. View of the unreinforced masonry (URM) prototype.

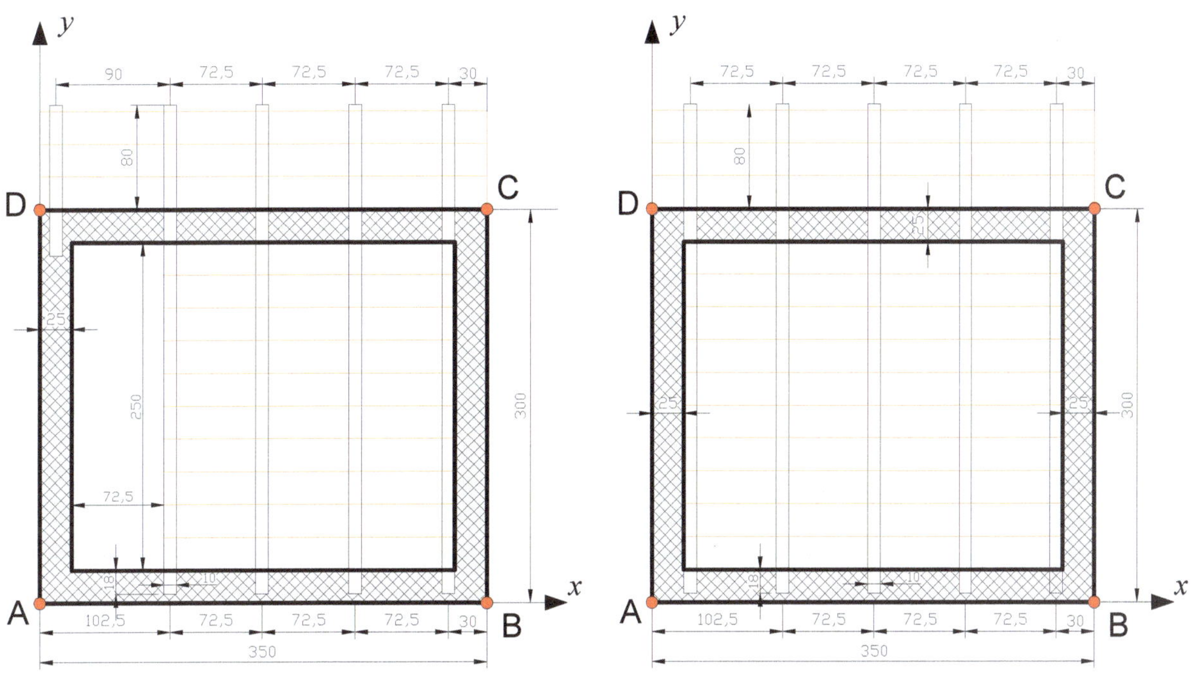

(a) (b)

Figure 2. Plan dimensions (cm) of the prototype: (**a**) first floor and (**b**) second floor.

Figure 3. Masonry texture and section of the walls of the prototype.

The total weight of the masonry is about 169.71 kN. The prototype was constructed on a reinforced concrete (RC) beam having a height of 0.4 m and a weight of 28.25 kN. The weight of the wooden floors is about 4.08 kN, and in addition, to meet the similitude scaling laws, a distributed mass of 2.5 t was placed on each floor. Hence, the total weight of the prototype over the shaking table, including the RC beam, is about 251.14 kN.

The masonry walls were constructed with calcareous tuff stones and lime-cement mortar with weight ratios of the components of 10.5 (sand), 11 (lime) to 1 (cement). The masonry units have an irregular shape and have dimensions varying from medium to large (Figure 3). The walls have a double vertical layer section with *opus incertum* texture. The average dimensions of the mortar joints were built intentionally very large both in the face and in the section. No connection between the two layers was provided. Furthermore, again in order to reproduce a poor constructive system, the walls were not connected together at the corners, and no concrete beams were realized to connect the wooden floors with the walls. A timber lintel was provided over each opening.

2.2. Mechanical Properties of the Masonry

Both destructive and non-destructive tests were preliminarily carried out on the units in order to characterize the masonry. Several samples of calcareous tuff stones were tested to evaluate the average compressive strength (f_{bc}) and the specific mass (γ_m). The obtained values were f_{bc} = 8.5 N/mm^2 and γ_m = 1700 kg/m^3. Laboratory tests on mortar provided the average compressive strength (f_{mc}) and the flexural strength (f_{mf}) equal to 0.72 and 0.14 N/mm^2, respectively. Further tests on the mortar joints were performed by the PNT-G penetrometric method [22] to assess the mortar quality; results provided an average value f_{mc} = 0.54 N/mm^2. To evaluate the compressive strength of the masonry assemblages (f_{wc}) starting from the properties of units and mortar, several relationships have been proposed [23]. Generally, these formulas are reliable only for masonries with regular texture; thus in the present case, it is only possible to obtain a feasible range of variability of this parameter. Considering f_{bc} = 8.5 N/mm^2 and f_{mc} = 0.72 N/mm^2, it is reasonable to assume that f_{wc} ranges from 2.0 to 3.5 N/mm^2 [24].

In addition to the tests on units and mortar, the masonry was characterized through three diagonal compression tests carried out on masonry wallettes. These were built with the same technique, materials and workers employed to realize the building prototypes. The wallettes were square panels with a side equal to 0.9 m and a thickness of 0.25 m. Firstly, a compressive load was applied on two opposite sides of each panel, then the diagonal load P_d was applied under displacement control up to collapse. The three panels, named MT1, MT2 and MT3, were tested imposing different values of the pre-stress (0.1, 0.15 and 0.20 N/mm^2, respectively). Results in terms of maximum diagonal load were 37.1, 27.8 and 51.5 kN for the three tests. Considering the classical interpretation of the diagonal compression test, the tensile strength of the masonry ranges from 0.06 to 0.07 N/mm^2 (the interested reader can refer to [18,24] for the formulation employed to estimate the compressive and the tensile strength of the masonry).

2.3. Experimental Failure Mode

To test the URM prototype, a 4 m × 4 m shaking table with six degrees-of-freedom (DOFs) was employed. The motion was impressed by eight hydraulic MTS dynamic jacks. The tests were carried out through the application of a normalized natural seismic input scaled at increasing intensity. The reference input was the accelerogram of Colfiorito (Perugia, Italy), recorded during the Umbria-Marche earthquake of 26 September 1997 (Richter magnitude of 6.1, epicenter in Annifo-Colfiorito, with peak ground acceleration (PGA) = 2.117 m/s^2 for the north-south (NS) component, 2.473 m/s^2 for east-west (EW) and 1.805 m/s^2 for up-down (UD); Figure 4). To be consistent with the reduced scale of the structure, the original records were scaled along the time axis by a factor equal to the square root of the geometric scale of the model: $\sqrt{3/2}$=1.225 [20]. The NS, the EW and the UD components were applied simultaneously (NS and EW in y and x directions, respectively; see Figure 2).

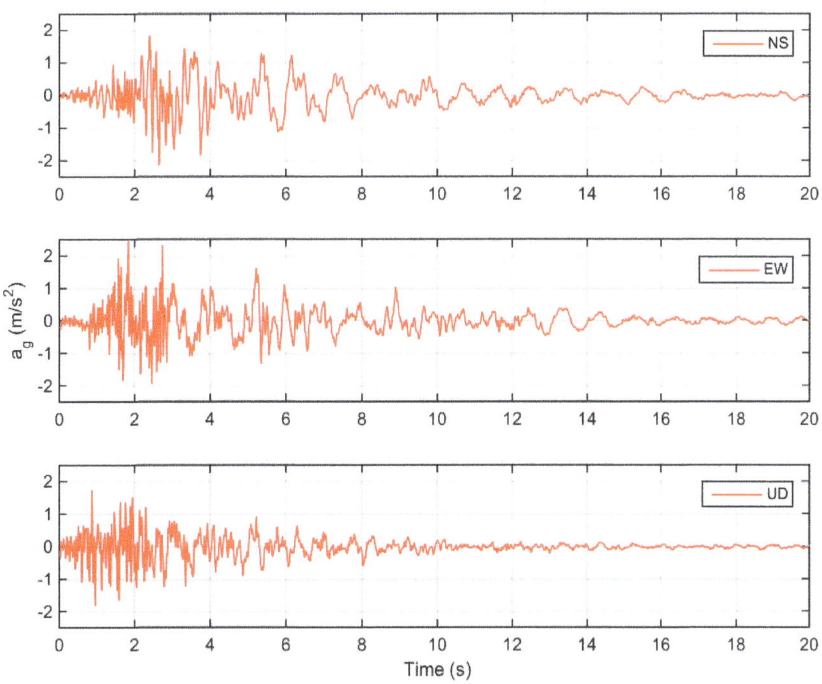

Figure 4. Colfiorito north-south (NS), east-west (EW) and up-down (UD) ground acceleration time histories.

During the tests, due to a system pump malfunction, the accelerations recorded on the shaking table were slightly altered with respect to the target input (nominal PGA in the following) [21,25,26]. The numerical analyses discussed next, although performed assuming as input the effective accelerations recorded on the shaking table, are referenced by employing their nominal PGA input value. Keeping in mind this distinction, the tests were carried out by applying the ground accelerations with increasing PGA (nominal increment of 0.05 g), starting from a low intensity (nominal PGA of 0.05 g) up to collapse (nominal PGA of 0.50 g).

During the first two shaking table tests (nominal PGA of 0.05 and 0.10 g), no damage was observed; the development of a vertical crack, approximately 1 cm thick, was first observed in the upper part of the corner between the walls CD and AD during the test with nominal PGA = 0.15 g (Figure 5a). The crack developed starting from the upper level, due to the thrust of the wooden floor. Out-of-plane flexural deformations were observed on the second level of the wall AB.

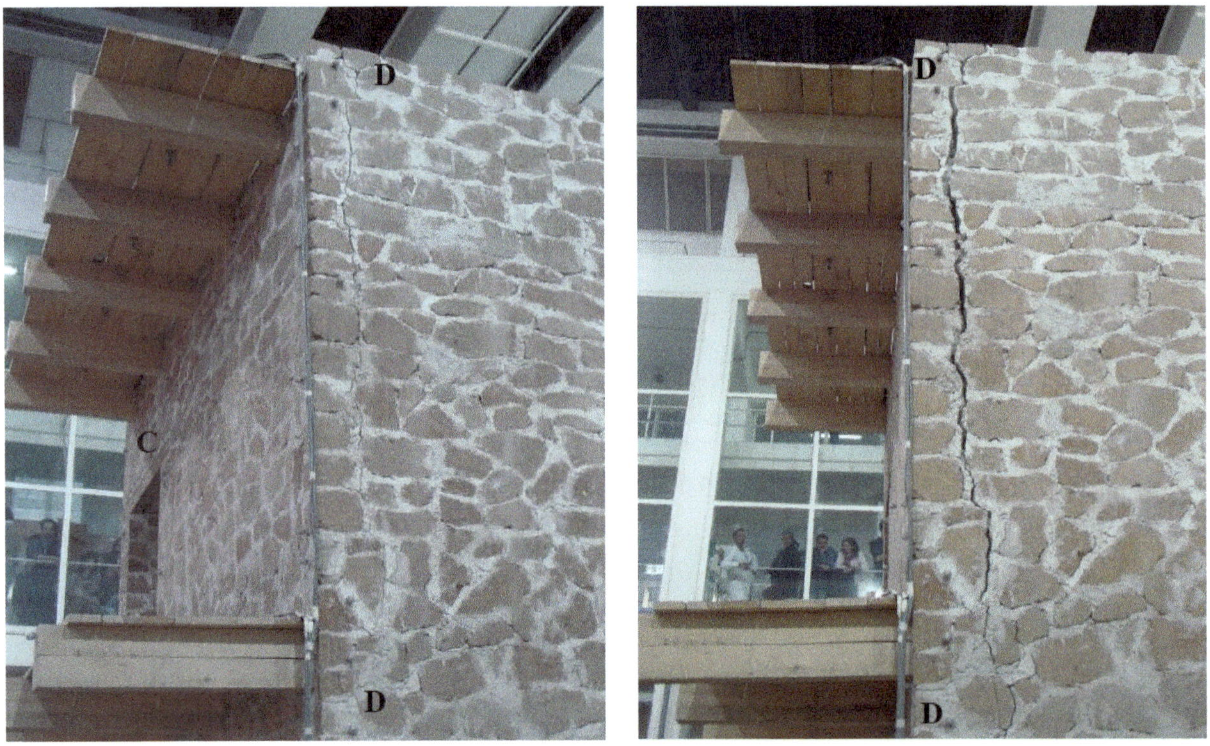

(a) (b)

Figure 5. Vertical crack in the corner between the façades AD and CD after the tests with nominal PGA of (**a**) 0.15 g and (**b**) 0.25 g.

During the test corresponding to a nominal PGA of 0.20 g, the effects observed in the previous test were accentuated, adding a strong flexural deformation on the wall BC (Figure 6a).

In the dynamic test with a nominal PGA of 0.25 g the crack on the corner between the walls CD and AD increased and widened from the roof top down to the first level (Figure 5b). At the end of the test, the crack thickness was about 3–4 cm, and fragments of mortar and stone dropped down due to the walls pounding after detachment of the corner. Diagonal cracks developed on the wall AB from the top of the façade reaching the corners in the openings. The vertical load carrying capacity of the structure was not yet compromised.

In the subsequent dynamic test (nominal PGA of 0.30 g), the corner between the walls CD and AD was completely detached, and a major crack was observed between the walls CD and BC. In addition, a cracking pattern was visible on the wall BC that extended above the timber lintel over the second level opening.

The test with nominal PGA of 0.35 g showed the worsening of previous cracks: the wall AD was completely detached from the orthogonal one (Figure 7a), while the upper part of the façade AB was subjected to considerable out-of-plane displacements with subsequent expulsion of some stone units. At the end of this test, not only the spandrels, but also the masonry piers appeared significantly damaged, specifically on the second level.

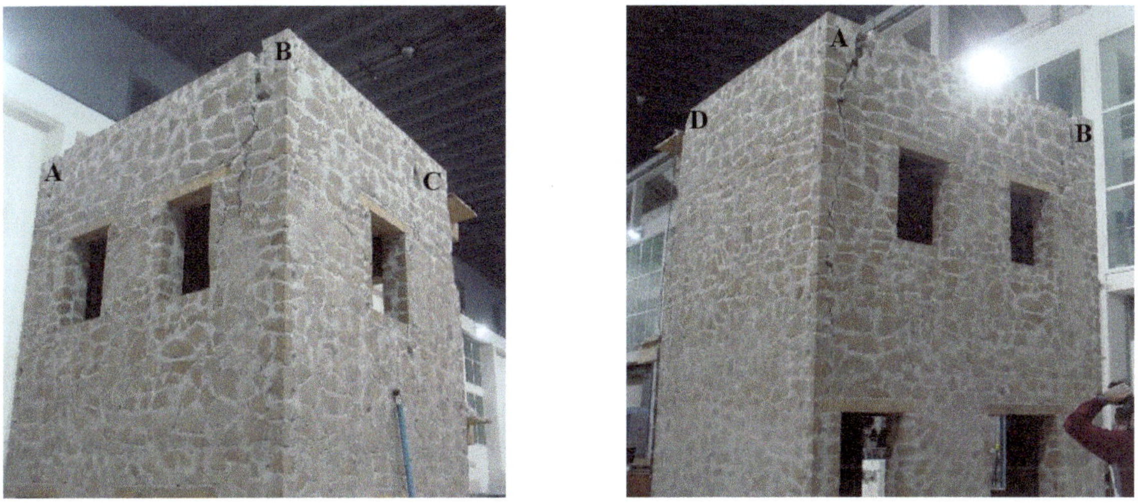

Figure 6. Diagonal cracks in the façade AB ((**a**) PGA = 0.20 g) and vertical crack in the façade AB ((**b**) PGA = 0.25 g).

Figure 7. Façade CD cracks after failures of the corners of the second level ((**a**) PGA = 0.35 g) and failure of the lintels ((**b**) PGA = 0.45 g).

The test with nominal PGA of 0.40 g was characterized by a strong global damage of the structure, with the collapse of the two corners of the wall CD and the collapse of the masonry pier supporting the lintel over the second floor opening. In addition, the already cracked corner of the wall AB partially collapsed. At this stage, the timber floors were completely detached from the load-bearing walls, compromising the box-behavior.

Two additional tests were then performed to assess the building residual capacity. The first test (nominal PGA = 0.45 g) showed the simultaneous collapse of all of the piers of the wall AB at the second level, as well as the lintel over the opening of the wall CD. The second test (nominal PGA = 0.5 g) was performed to achieve total structural collapse.

Figures 5–7 show the prototype during the dynamic tests.

3. Numerical Models of the Prototype

To simulate the dynamic behavior of the unreinforced masonry prototype, two different numerical approaches were employed. The first 3D numerical model was built by the finite element (FE) technique, employing the commercial code ANSYS [16]. The non-linear behavior of the masonry was reproduced through a smeared crack approach, and the mechanical parameters were evaluated on the basis of the experimental results by means of an identification procedure. The second model was built by the macro-element (ME) approach implementing non-linear laws of the masonry in the MATLAB tool. The parameters required by the macro-element model were assessed comparing its modal behavior with the one obtained by the FE model.

3.1. The Finite Element Model

The macro-modelling technique was employed, as it is fit for large size models. The units and the mortar joints were hence assumed to be smeared, and a homogenous isotropic material represented the masonry assemblage. Eight-node isoparametric solid elements (Solid65) were used to model both the masonry walls and the RC basement. The wooden joists were modelled by using one-dimensional two-node isoparametric elements (Beam44) while isoparametric two-dimensional elements with four nodes (Shell63) were adopted for the wood planking. Further zero-dimensional one-node elements (Mass21) were inserted on each floor to consider the additional masses. Specific weights of masonry and wood were assumed based on the experimental data (γ_w = 14.2 kN/m^3 and γ_l = 5.9 kN/m^3, respectively). The final numerical model (Figure 8) consisted of 11,081 nodes, 6824 3D Solid45 elements, 94 2D Shell63 elements, 291 1D Beam44 elements and 25 0D Mass21 elements, corresponding to 30,933 degrees of freedom (DOFs).

The constitutive behavior of the masonry was reproduced assuming an elastoplastic law with tension cut-off. To this aim, the Drucker–Prager (DP) plasticity model was combined with the Willam–Warnke (WW) failure criterion. The DP yield surface is a smooth version of the Mohr–Coulomb yield surface, defined by two parameters: the cohesion c and the internal friction angle φ. They are introduced in such a way that the circular cone of DP corresponds to the outer vertex of the hexagonal Mohr–Coulomb yield surface (Figure 9) [27]. The combination of the plasticity model with the failure criterion allows for an elastic-brittle behavior in case of biaxial tensile stresses or biaxial tensile-compressive stresses with a low compression level. On the contrary, the material behaves as elastoplastic in the case of biaxial compressive stresses or biaxial tensile-compressive stresses with a high compression level. Overall, the material behaves as an isotropic medium with plastic deformation, cracking and crushing capabilities [27,28].

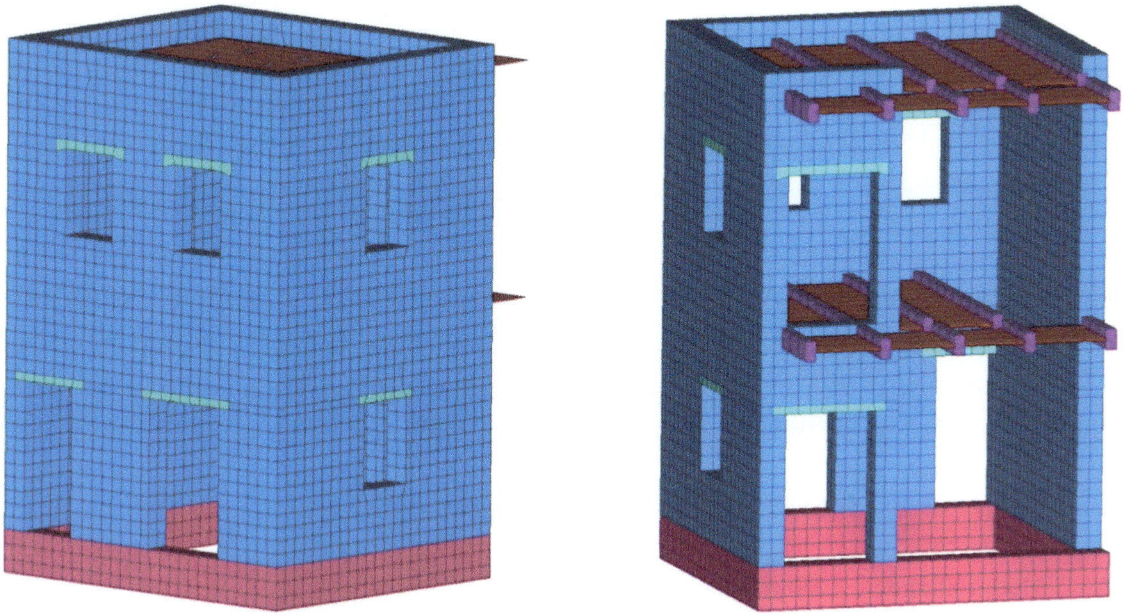

(**a**) (**b**)

Figure 8. Finite element (FE) model of the URM prototype: (**a**) external and (**b**) internal views.

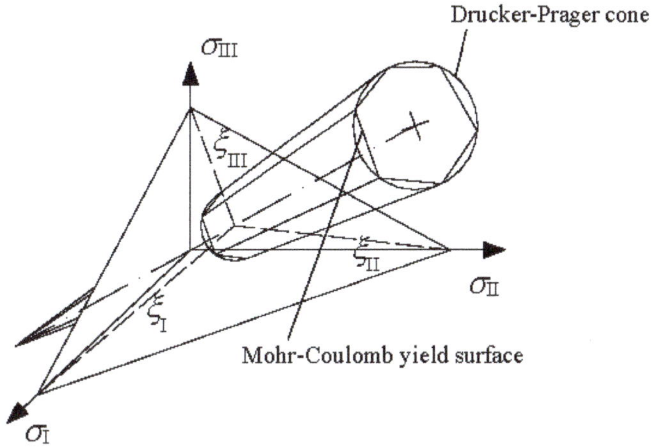

Figure 9. Drucker–Prager yield surface in the Haigh–Westergaard stress space.

The constitutive parameters required by the DP and the WW models were identified by the diagonal compression test MT1. An FE model of the square panel was built by 3D Solid65 elements, and was employed to reproduce the experimental force-vertical displacement curve. The parameters were identified through direct comparison between the experimental curve and the numerical one. The non-linear parameters are reported in Table 1, whereas the linear parameters are as follows: longitudinal modulus of elasticity E_w = 700 N/mm^2, tangential modulus of elasticity G_w = 280 N/mm^2 and Poisson ratio ν = 0.25. The identified tensile strength of the masonry (f_{wt}) is within the range estimated by using the classical interpretation of MT1 and MT3 tests [24,29].

Table 1. Finite element (FE) model parameters (f_{wc} = masonry compressive strength; f_{wt} = masonry tensile strength; β_c = shear transfer coefficient for closed cracks; β_t = shear transfer coefficient for open cracks; c = cohesion; δ = dilatancy angle; φ = internal friction angle).

Drucker–Prager Plasticity Criterion		Willam–Warnke Failure Surface	
c (N/mm^2)	0.07	f_{wc} (N/mm^2)	2.50
δ (°)	20	f_{wt} (N/mm^2)	0.065
φ (°)	40	β_c (-)	0.75
-	-	β_t (-)	0.15

The FE model of the whole building was preliminarily employed to evaluate the modal shapes by means of a modal analysis. Results show that the first modal shape is a pure translation in the x direction with a frequency of 9.03 Hz; the second modal shape is a pure translation in the y direction with a frequency of 9.37 Hz; and the third one is a torsional mode with a frequency of 15.32 Hz (Table 2).

Table 2. Modal shapes and frequencies.

Modal Shape	Direction	FE Model	ME Model
		Frequency (Hz)	Frequency (Hz)
1	x	9.03	9.07
2	y	9.37	9.37
3	Torsional	15.32	12.13

3.2. The Macro-Element Model

The second model was built by the macro-element (ME) approach. This model, originally proposed by Galano and Selleri ([17]), performs a 3D step-by-step dynamic analysis of regular masonry buildings subjected to a base ground acceleration (components in two orthogonal directions can be applied simultaneously). Each masonry wall is represented by piers (with vertical axis and height equal to the inter-story) and spandrels (with horizontal axis); however, only the piers are considered as deformable and resisting elements, whereas the spandrels have infinite stiffness and strength. The diaphragms are assumed rigid in their plane; hence, the model has three DOFs for each floor, and the masses must be concentrated or distributed over these floors. The bending failure mode and the shear failure mode with diagonal cracking are considered for each pier subjected to shear load in its plane, and a simplified trilinear diagram of shear V *vs.* drift δ is used for each failure mode, as illustrated in Figure 10. This figure also shows a sample of the hysteretic response under reversal loads with progressive damage depending on the position of the focal point F. A similar behavior is assumed for the piers subjected to out-of-plane flexure. A detailed explanation of the main features of the numerical procedure is reported in [17]. As usual, in the shear-type models, the piers behave as masonry panels constrained by double bending conditions; this hypothesis is unrealistic, especially for existing buildings with timber floors. Hence, rotational elastic springs between piers and floors were introduced in the model to reduce the shear stiffness of the panels. The four façades of the macro-element model of the prototype are depicted in Figure 11 with the indication of the piers (13 elements) and the spandrels (12 elements, denoted by "×").

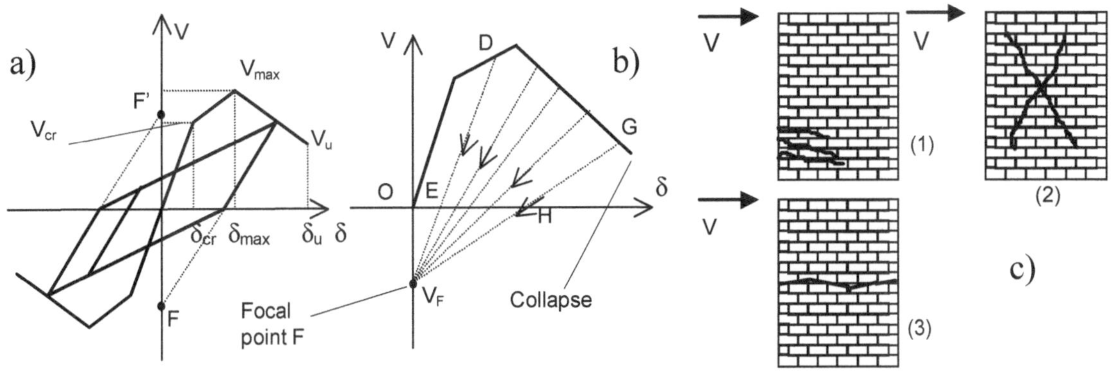

Figure 10. (**a**,**b**) Hysteretic model for piers and (**c**) failure modes: (1) bending failure; (2) diagonal cracking failure; and (3) sliding shear failure.

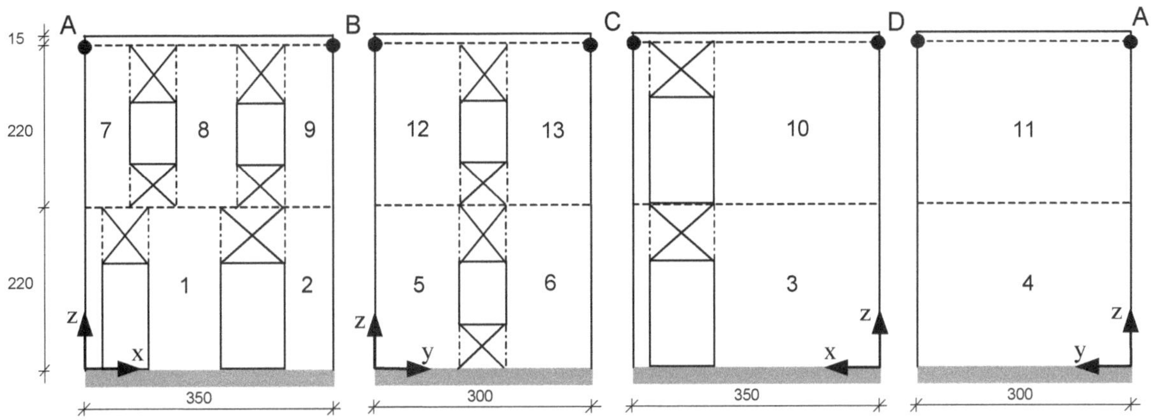

Figure 11. Macro-element (ME) model: definition and numbering of the piers.

Values of the parameters used in the ME model are: E_w, G_w, γ_w and γ_l as in the finite element model, f_{wc} = 2.5 N/mm^2, f_{wt} = 0.065 N/mm^2, ultimate drifts δ_u in the range from 7.5 mm up to 12.5 mm (from 0.34% up to 0.57% of the inter-story height), depending on the pier slenderness and the failure mode, and V_u and V_F equal to 0.25 V_{max} (Figure 10). In addition, the model requires defining the distribution of the masses and the values of the elastic stiffness of the rotational springs between piers and floors.

These parameters were selected by comparison with the modal results obtained by the FE model. A fraction of 4/11 of the pier masses of the first level was attributed to the first floor (the remaining part of 7/11 rests on the foundation), and a fraction of 7/11 of the pier masses of the second level was attributed to the first floor (the remaining part of 4/11 was placed on the top floor). The additional masses were applied on the corresponding floor. Rotational springs were introduced only for piers denoted by 3, 4, 10 and 11, which are the most relevant to define the dynamic response of the model. Table 3 collects the total masses and the coordinates of the center of mass in each floor, x_G and y_G.

The agreement between the two models both for modal shapes and frequencies was acceptable (Table 2). Further confirmations of the accord between the two models in the elastic range were obtained performing step-by-step dynamic analyses under base input seismic accelerations. Acceptable differences (less than 5%) were obtained, for instance, for the displacements of the second floor.

Table 3. Macro-element (ME) model parameters.

Floor	Mass (t)	x_G (m)	y_G (m)
1	11.344	1.76	1.56
2	6.946	1.75	1.63
Total	18.290	-	-

4. Linear and Non-Linear Time-History Results

4.1. Seismic Input and Numerical Analyses

The dynamic responses of the two models were first compared in the linear field by considering the input corresponding to the six shaking table tests with increasing nominal PGA of 0.05, 0.10, 0.15, 0.20, 0.25 and 0.30 g. Taking into account the previously mentioned differences between the target original input signals (nominal PGA) and the effective output signals recorded on the shaking table, non-proportional responses were expected (and justified) increasing the nominal input PGA.

The input of the above six tests, and that of the two additional tests with nominal PGA of 0.35 and 0.40 g, were subsequently considered in the hypothesis of non-linear behavior. The base accelerations were selected in order to reproduce the behavior of the building starting from the elastic response up to the range going from moderate to severe damage (the FE model collapsed during the test with nominal PGA of 0.40 g; collapse is here referred to as non-convergence of the numerical solution).

Only the first 10 s in the reduced time scale of the base acceleration records were used as input, because the shock has its maximum ground acceleration intensity within the first 5 s, and it is not significant after 10 s. The responses of the FE model were computed by the Newmark integration method with a time step of 0.004 s, whereas the α-Hilbert procedure [17] with a time step of 0.001 s was employed to calculate the responses of the macro-element model. These choices provided sufficient accuracy of the solutions and acceptable computational effort. Damping was considered by the classical Rayleigh formulation, and a 4% damping ratio on the first two modal shapes was selected as suggested in the literature [30].

4.2. Response of the Linear Elastic Models

Displacements and absolute accelerations at the floor levels are two of the most significant global kinematic parameters that synthesize the dynamic response of a building. In the macro-element model, each floor moves in its plane as a rigid diaphragm, while in the FE model, the timber floors are modelled as flexible diaphragms. Consequently, to perform a significant comparison, average response parameters were considered.

Denoting with $s_i(t)$ the responses (accelerations or displacements) of the four corners A, B, C and D (Figure 2), the average response $\bar{s}(t)$ was directly calculated as follows:

$$\bar{s}(t) = \frac{1}{4} \sum_{i=A,\ldots D} s_i(t) \tag{1}$$

To offer a synthetic representation of each average response over the whole time-history, a representative value, called "efficacy", was calculated by the following expression:

$$s_{eff} = \sqrt{\frac{1}{T_{max}} \int_0^{T_{max}} [\bar{s}(t)]^2 dt} \tag{2}$$

where T_{max} denotes the time length of the signal.

The agreement between the numerical responses of the two models was estimated by comparing the percentage errors defined as follows:

$$E_s = 100 \cdot \frac{\left(s_{eff}^{ME} - s_{eff}^{FE}\right)}{s_{eff}^{FE}} \tag{3}$$

According to the definition of Equation (3), it is implicitly assumed that the response of the FE model is considered as the reference in the comparisons.

The average errors obtained simulating the first six tests for displacements (E_d), absolute accelerations (E_a) and dominant frequencies of response displacement signals (E_{frd}), together with the variation coefficients, are summarized in Table 4.

Considering the significant differences between the ME and the FE models, it is possible to observe a satisfactory agreement in the amplitude of the displacements (max. error equal to 20.6%) and a perfect agreement in the frequencies. In addition, it is observed that the ME model always provides slightly higher displacements values, with more pronounced differences for the first floor in the *y* direction.

Table 4. Average (Mean) and coefficient of variation (Cv) of percentage errors (linear analyses).

Floor	Displacements E_d		Accelerations E_a		Frequencies E_{frd}	
	Mean (%)	Cv	Mean (%)	Cv	Mean (%)	Cv
F1*x*	11.1	0.11	19.9	0.74	0	0
F1*y*	20.6	0.02	56.6	0.04	0	0
F2*x*	3.0	0.61	−6.4	1.24	0	0
F2*y*	4.1	0.05	14.1	0.10	0	0

More pronounced differences were instead obtained in terms of accelerations (max. error equal to 56.6% for the first floor in the *y* direction). Again, the errors are always positive with the exception of the second floor in the *x* direction. In this case, the errors obtained in the simulations with input PGA of 0.05 and 0.10 g are slightly positive; the others are negative, hence resulting in a very high variation coefficient (1.24). As an example, the efficacy displacements and accelerations, evaluated according to the definition introduced by Equation (2), of the second floor *vs.* the target input PGA are plotted in Figure 12 (the figure shows also the same values obtained with the non-linear models, as discussed next).

The other two main response parameters to be considered for engineering purposes are the shear forces (base shear and shear at the base of the second level, in each direction) and the stiffness of the building. For linear models, it is appropriate to evaluate the maximum absolute value of the shear as follows:

$$V_{max} = \max|V(t)| \tag{4}$$

and, consequently, the percentage error:

$$E_V = 100 \cdot \frac{(V_{max}^{ME} - V_{max}^{FE})}{V_{max}^{FE}} \tag{5}$$

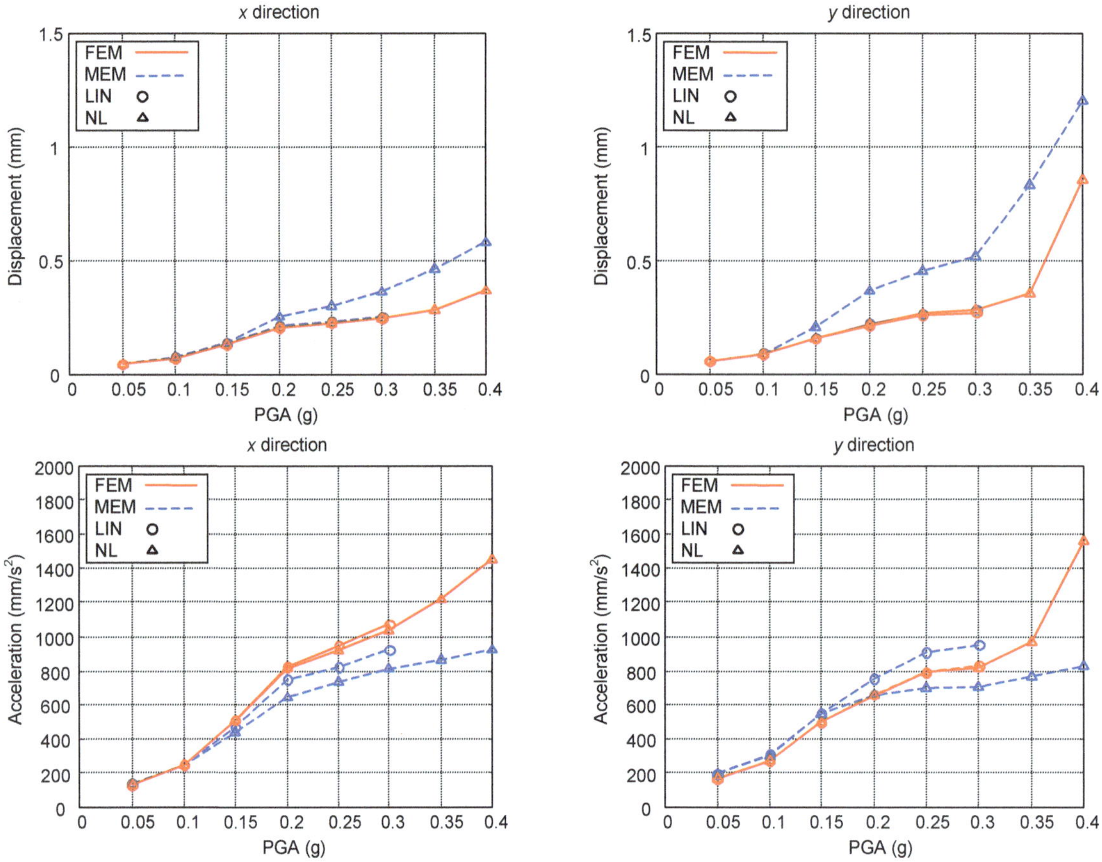

Figure 12. Efficacy values of displacements and absolute accelerations *vs.* PGA (second floor).

The calculation of the stiffness of the building by means of the results of a time-history analysis is rather complicated, even in the case of linear behavior, because in multi-degrees-of-freedom (MDOF) models, the dynamic response is composed of a set of many displacements and forces, each of them defined in the time domain. In the present case, the stiffness was computed as secant stiffness at the origin.

Starting from the cyclic response shear V *vs.* inter-story drift d of a level in a direction, the envelope diagram was considered, and then, in this diagram, the secant stiffness $K_s = V/d$ corresponding to the absolute maximum value of d was taken. This parameter is then representative of the stiffness of that level in the instant of maximum amplification of the response.

Errors were calculated according to the following equation:

$$E_K = 100 \cdot \frac{(K_S^{ME} - K_S^{FE})}{K_S^{FE}} \tag{6}$$

Table 5 summarizes the averages of the errors E_V and E_K. They are always negative, *i.e.*, the ME model provides lower values than the FE model. In particular, the stiffness of the second level in the y direction is noticeably higher for the FE model, while in the x direction, a very high coefficient of variation (0.92) is obtained. In turn, these differences explain the errors obtained in maximum shear forces (up to 30.5% for the second level in the y direction). The differences in the mass distribution between the two models can partially explain this disagreement.

Figure 13 plots the base shears and the stiffness of the building *vs.* the target input PGA (the figure shows also the same values obtained with the non-linear models, as discussed next).

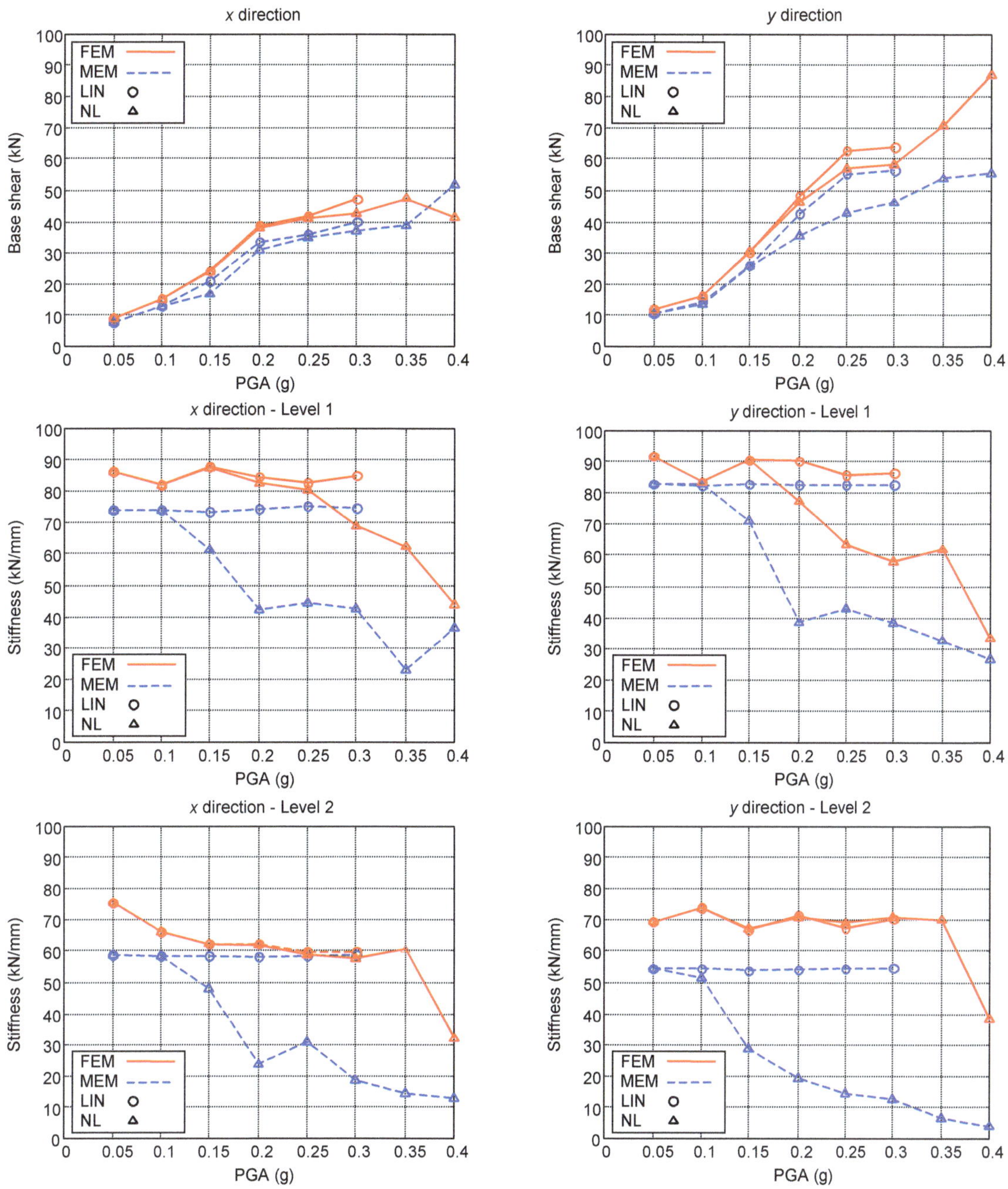

Figure 13. Maximum base shears and secant stiffness *vs.* PGA.

On the whole, the macro-element model provides dynamic responses affected by acceptable errors with respect to FEM. This suggests that in all structural problems in which the intensity of the dynamic loading is low and an almost elastic structural response is expected, the use of a simplified model, such as the one employed here, is the most acceptable choice. This is true both for the ease of use (the construction of the numerical model) and for the small computational times (each time-history analysis performed with the FE model required about 7 h against the about 5 min required by the ME model). As an example, the ME model can be an effective tool for analyzing the response of secondary elements supported by the floors of a masonry building under low to moderate earthquakes (museums, history archives, *etc.*). On the other hand, given the errors obtained for the base shears, the use of the FE model is more appropriate for structural analyses aimed at investigating the seismic safety.

Table 5. Average (Mean) and coefficient of variation (Cv) of percentage errors (linear analyses).

Level	Shear Forces E_V		Stiffness E_K	
	Mean (%)	Cv	Mean (%)	Cv
F1x	−13.7	0.08	−12.3	0.22
F1y	−11.9	0.08	−6.0	0.56
F2x	−15.3	0.44	−8.4	0.92
F2y	−30.5	0.04	−22.2	0.12

4.3. Response of the Non-Linear Models

The discussion of the results obtained with the non-linear models is performed still considering the dynamic responses under the increasing accelerations applied at the base. As a preliminary step, the progress of the damage obtained by both models is reported below.

Figure 14 illustrates the damage progress in the masonry walls predicted by the ME model from a qualitative point of view. According to the constitutive equations, if the deformation in a pier reaches the value δ_{cr} or δ_{max} (Figure 10), the pier is assumed to be in the "cracking state" or in the "plastic state", respectively. Collapse occurs if the deformation reaches the limit value δ_u. The dead loads of the prototype are rather low, so the ME model predicts flexural failures for all of the masonry piers with the exception of Pier 3 (Figure 11), where a shear failure is always predicted.

During the first two shocks (nominal PGA = 0.05 and 0.10 g), no appreciable damage is detected, and the model response is almost linear elastic. The first damages appear during the third shock (nominal PGA = 0.15 g), when in-plane cracking occurs in Piers 5, 12 and 13 in the y direction and in Piers 2 and 9 in the x direction. This indicates that the first damages are concentrated along Corner B between the walls AB and CB. In the subsequent two shocks (PGA = 0.20 and 0.25 g), the largest parts of the piers crack, especially those of the second level: {1, 2, 7, 8, 9, 12, 13} in the x direction, and {5, 6, 8, 9, 11, 12, 13} in the y direction. Some piers crack with out-of-plane deformation. In addition, the deformation of Pier 12 goes beyond δ_{max} in the plastic state.

During the shock with nominal PGA = 0.30 g, the same trend is observed, and also, Pier 13 reaches the plastic state. This indicates the beginning of a flexural failure mode on the second level of the building in the y direction. With the shock with PGA = 0.35 g, also Piers 6 and 9 reach the plastic state. Finally, in the last shock (PGA = 0.40 g), Piers 7 and 8 together with 2, 5 and 11 reach the plastic state. In all of the simulations, the deformations never reach the ultimate value δ_u, denoting that the ME model predicts additional capacity reserve.

Overall, the ME model predicts the collapse of the second level under a shock with a nominal PGA greater than 0.40 g, with main deformations in the y direction. This can be explained taking into account the warping of the timber floors. They are sustained by the walls parallel to the x direction, and the orthogonal walls have low compressive stress and low shear strength. The piers collapse in flexure in their plane; hence, the model predicts the failure of the second level with a II mode mechanism. This disagrees with the failure mode of the prototype detected during the experimental tests, characterized by the development of main cracks along the corners up to the overturning of the walls of the second level. Due to the assumption of rigid diaphragms, this collapse configuration cannot be correctly reproduced by the ME model.

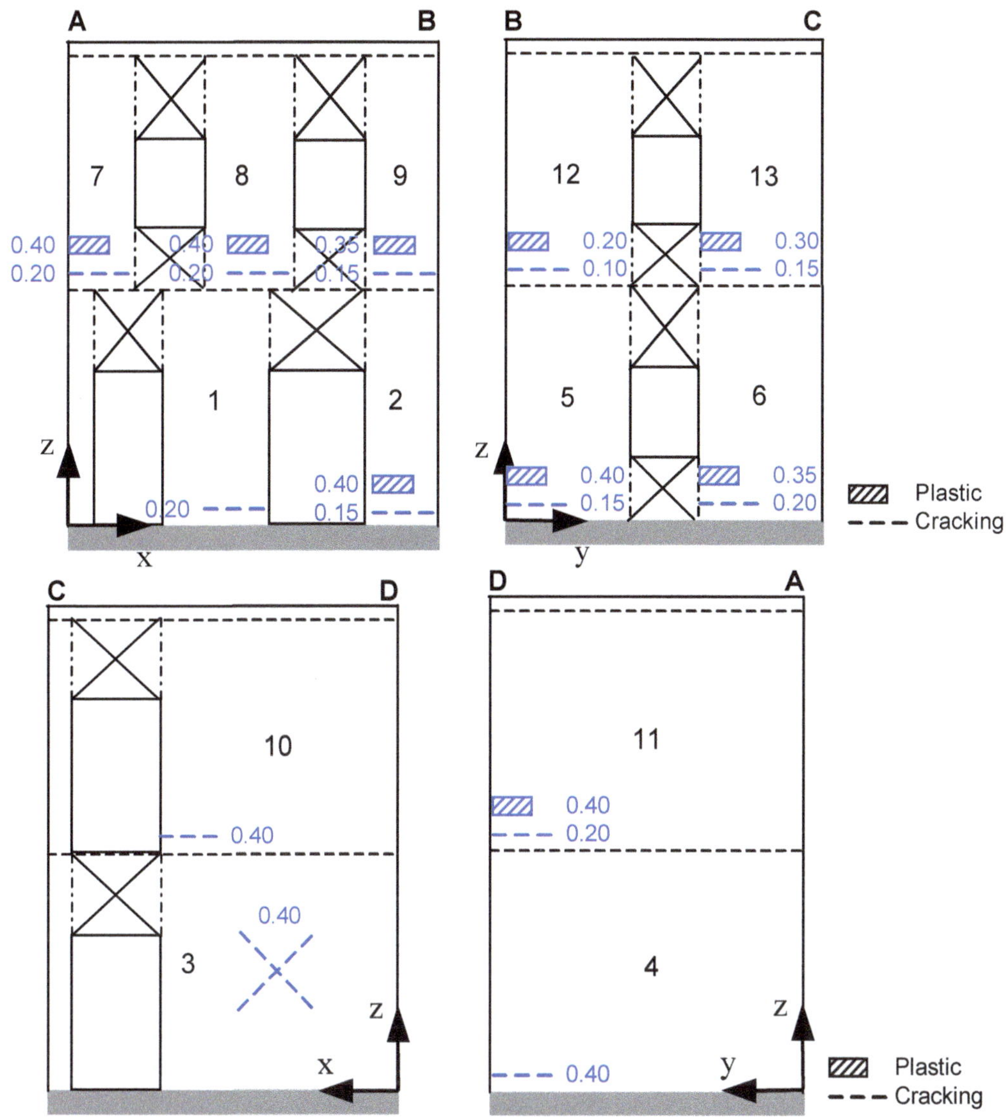

Figure 14. Damage progress obtained by the ME model.

The FE model provides a linear response with the shocks with nominal PGA of 0.05 and 0.10 g, and an almost linear response under the shock with nominal PGA of 0.15 g. The first noticeable damages appear during the shock with nominal PGA equal to 0.20 g, when a main vertical crack develops along Corner D starting from the level of the first floor (as in the experiments; Figure 5). Other horizontal cracks appear at the base of the building (Façade AB) in the contact zone with the RC basement. In the subsequent analysis, the previous cracking pattern becomes wider, and a further main diagonal crack develops on the first level (Façade BC) starting from the corner of the window. The analysis with nominal PGA = 0.30 g confirms the previous trend, and in addition, a second main diagonal crack appears on the façade BC in which the typical cracking pattern with the "×" profile is recognized. At the same time, another vertical crack develops along Corner A. In the analysis with nominal PGA = 0.35 g, a diagonal crack develops on the façade DA, on both levels, and the vertical crack along Corners A and D increases. During the shock with PGA = 0.40 g, at 5.87 s, the numerical analysis stops due to the non-convergence of the solution (numerical collapse of the model). Displacements reach high values, and the cracking pattern confirms the previous trend.

The progress of the cracks, for all of the analyses, is summarized in Figure 15.

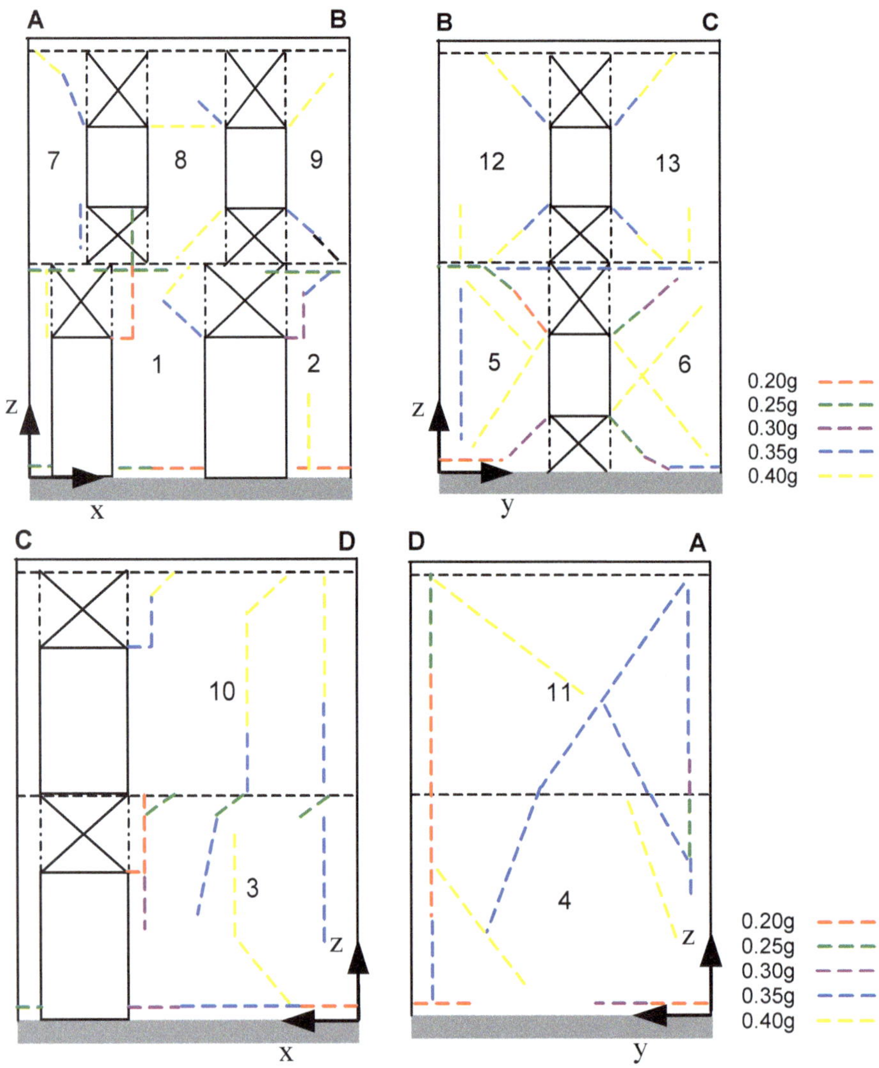

Figure 15. Damage progress obtained by the FE model.

As a general remark, the FE model substantially matches the experimental collapse mode, reproducing the larger part of the damages. In addition, the phenomenon of opening of the corners is reasonably well reproduced.

A further comparison of the dynamic responses obtained through the two approaches is made analyzing the same parameters discussed in the previous section. Stiffness decay is shown in Figure 13. The ME model predicts a noticeable decay of the secant stiffness in the analyses with nominal PGA from 0.10 to 0.20 g, in both directions and for both levels of the building. The ME model response, in fact, is very sensitive to the cracking of the first pier (each pier is modelled as a single macro-element). In the subsequent shocks the stiffness continues to decrease, although with less evidence. In particular, stiffness decay is predominant at the second level in the *y* direction, where a stiffness reduction equal to about ten-times the initial elastic value is observed.

The evolution predicted by the FE model is different. The stiffness of the second level remains almost the same up to the shock with nominal PGA = 0.35 g, when a sudden decrease occurs. The stiffness of the base level gradually decreases starting from the shocks with nominal PGA = 0.30 g (*x* direction) and 0.15 g (*y* direction), until it reaches the same values predicted by the ME model. The premature stiffness decay predicted by the ME model with respect to the FE model is outlined also by the maximum base shears (Figure 13). Figure 16 plots the shear forces *vs.* the corresponding drifts, comparing the responses obtained with the non-linear analyses with nominal PGA = 0.30 g. The differences between the two models are clearly visible.

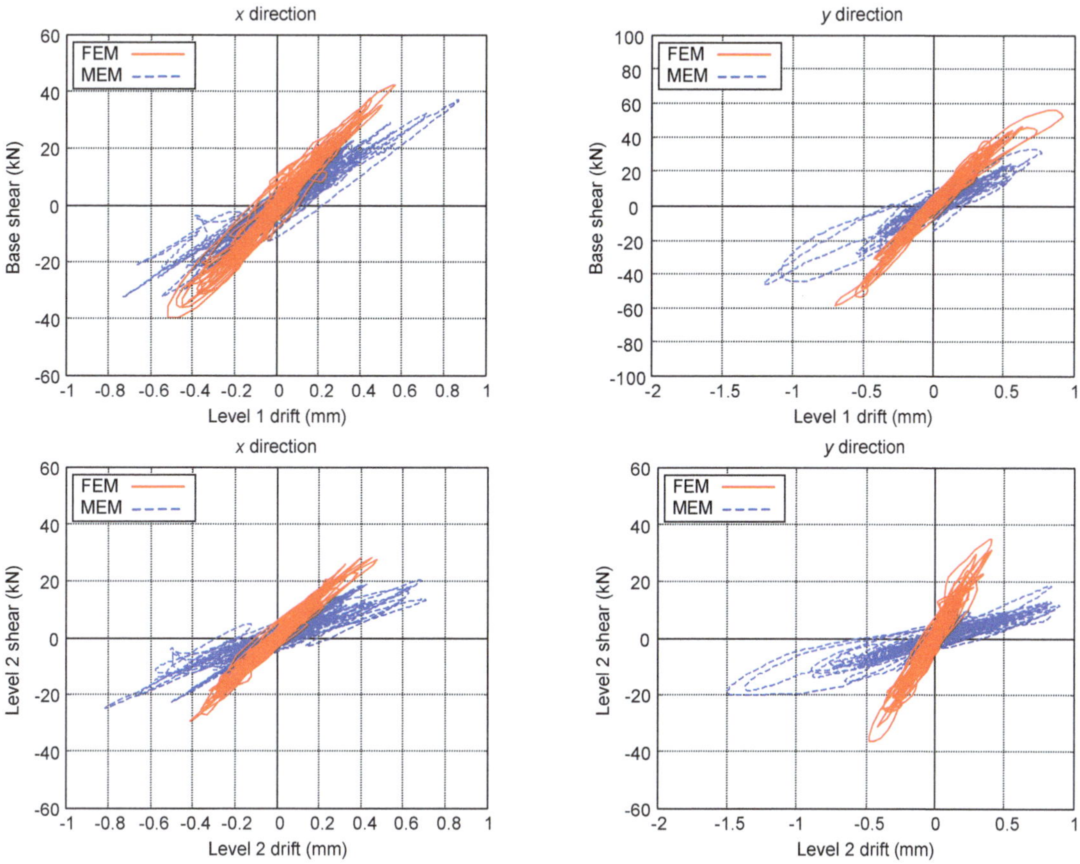

Figure 16. Shear forces *vs.* corresponding drifts for the analyses with nominal PGA = 0.30 g.

Responses in terms of displacements and accelerations are less significant for non-linear analyses. As reported before, the ME model provides higher values of displacements than the FE model. The errors E_d are significant: for example, at the second level, errors are detected up to 58.8% and 85% in the *x* and *y* directions, respectively (Table 6). On the contrary, for the shocks with PGA greater than 0.20 g, the absolute accelerations of the second level predicted by the ME model are lower than those obtained by the FE model (with errors E_a up to 36.4% and 47% in the two directions; Table 6).

Table 6. Percentage errors (non-linear analyses).

Floor	Displacements E_d (%)			Accelerations E_a (%)		
	0.2 g	0.3 g	0.4 g	0.2 g	0.3 g	0.4 g
F1*x*	22.7	41.4	29.0	3.0	7.2	−26.9
F1*y*	63.4	54.9	−22.4	32.3	25.1	−27.2
F2*x*	23.0	50.7	58.8	−21.0	−21.1	−36.4
F2*y*	70.4	85.0	41.0	0.6	−13.7	−47.0

The errors E_V and E_K are summarized in Table 7. These differences are confirmed by the spectral analysis of the signals. The dominant frequencies obtained with the FE model are higher than those obtained by the ME model, with differences increasing as the PGA increases. Analyzing the displacements, the errors on the dominant frequency, E_{frd}, range from 20% to 50%, while for accelerations the same errors vary from 20% to 80%.

As an example, Figure 17 shows the displacements amplitude spectra of the second floor for the simulations with nominal PGA = 0.30 g (linear and non-linear analyses, *x* and *y* directions). The spectra were obtained by a standard fast Fourier transform (FFT) procedure, operating on subsequent temporal windows with an overlap of 50%. The good agreement obtained with the linear models and the differences predicted by the non-linear ones are evident.

Table 7. Percentage errors (non-linear analyses).

Level	Shear Forces E_V (%)			Stiffness E_K (%)		
	0.2 g	0.3 g	0.4 g	0.2 g	0.3 g	0.4 g
F1*x*	−18.3	−12.9	24.2	−48.7	−38.0	−17.1
F1*y*	−22.5	−20.6	−36.0	−50.0	−34.1	−19.2
F2*x*	−28.4	−14.5	2.4	−61.8	−67.3	−59.4
F2*y*	−33.1	−45.1	−56.9	−72.6	−82.3	−89.7

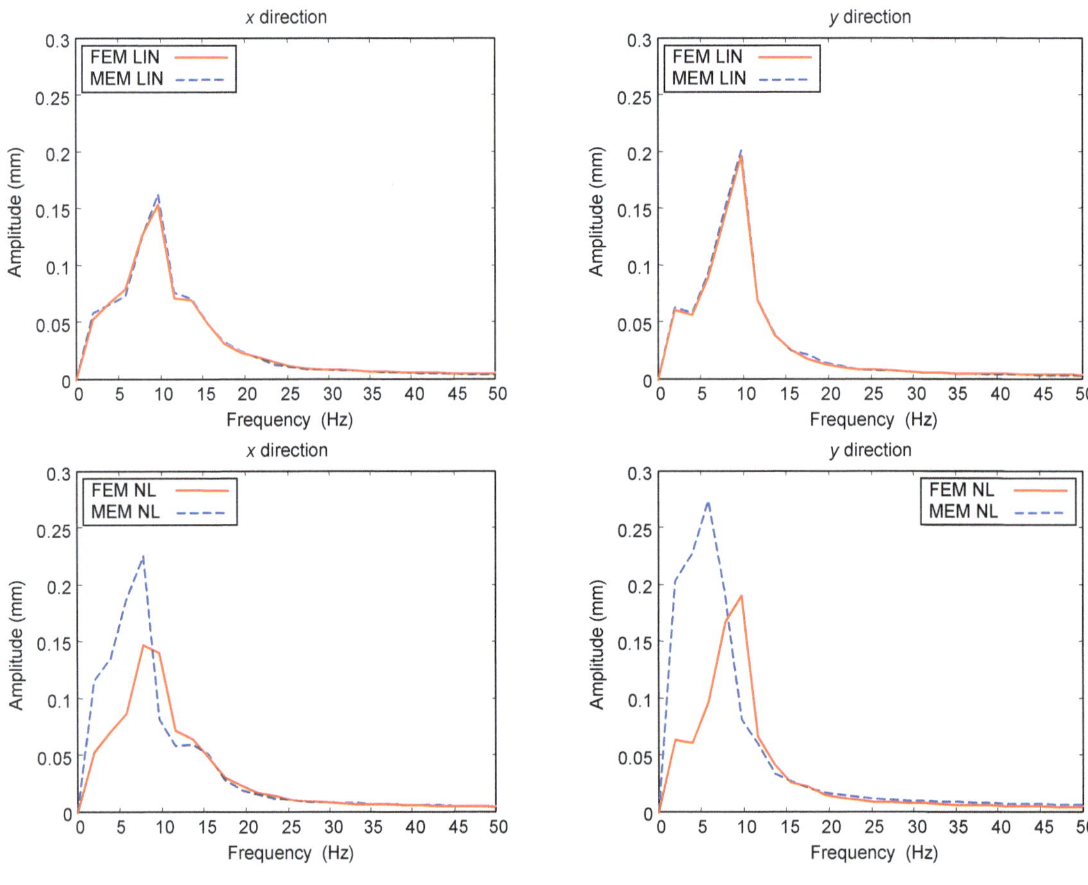

Figure 17. Displacement amplitude spectra of the second floor for linear and non-linear analyses with nominal PGA = 0.30 g.

In general, even if both models predict a collapse PGA value equal to or greater than 0.40 g, they provide different collapse modes. This is substantially due to the differences in the modelling hypotheses between ME and FE approaches. Smeared cracking produces many and distributed cracks that result in a small, but constant stiffness decrease as the PGA increases (as observed during the experimental tests). When the damage has spread over the entire building, a sudden loss of strength and stiffness occurs.

On the one hand, the comparison between the two codes shows that the FE model is capable of reproducing with great confidence the experimental results (progress, location and extent of the damaged portions until collapse). The macro-element model, due to the intrinsic hypothesis of rigid floors, is capable of predicting the collapse load, but not providing a satisfactory reconstruction of the actual collapse mechanism (wall overturning due to the flexibility of the wooden floors). On the other hand, the results highlight that each model has a range of validity, which needs to be carefully understood, and the use of such tools requires high expertise. With respect, for instance, to the floor modelling, it is confirmed that different assumptions on the diaphragms' stiffness significantly affect the building overall seismic response. In fact, in the case of flexible floors (like in the FE models), there is no load transfer from collapsed walls to still efficient structural elements. On the contrary, in the case of stiff floors (like in the ME models), the load transfer is overestimated, producing, in turn, an overestimation of the building seismic capacity. In this respect, the results of the macro-element model, despite its intrinsic simplifications, should be considered as an upper limit of the building capacity, which can be reached only when the out-of-plane and the local damage mechanisms are not previously activated.

5. Conclusive Remarks

The paper reported a comparison between two numerical models used to investigate the seismic behavior of existing masonry buildings. As a reference case study, an unreinforced masonry prototype with flexible diaphragms tested on a shaking table until collapse was considered. The first numerical model was built by the finite element method through a macro-model technique; the second one was built by using a macro-element approach through the implementation of a non-linear algorithm in the MATLAB tool.

The models adopt different mechanical hypotheses to account for the non-linear behavior of the masonry and different techniques to reproduce the building geometry. The two procedures were employed to simulate the tests, by linear and non-linear time-history analyses. The comparison was performed considering the displacements and the absolute accelerations, as these are two of the most significant global kinematic parameters that synthesize the dynamic response of a building. In addition, the shear forces, the stiffness decay and the damage evolution were also considered, as parameters with significant engineering relevance.

The results highlighted that the FE model is capable of reproducing with good confidence the experimental damages, while the macro-element model, due to the intrinsic hypothesis of rigid floors, is capable of predicting the collapse load, but not providing a satisfactory reconstruction of the actual collapse mechanism. Errors on displacements and accelerations are acceptable in the simulations with low-to-moderate PGA intensities, *i.e.*, in the almost linear range of the dynamic response. In all of the simulations, the ME model underestimates the shear forces, since it provides a more pronounced stiffness decay with respect to the FE model.

As the availability of an effective non-linear tool for the seismic assessment of masonry structural elements is still a crucial requirement, the paper suggests that the seismic analysis of traditional masonry buildings (with flexible floors and poor connections between the walls) should be examined through a cross-numerical approach.

Acknowledgments

This research was developed in the framework of the ReLUIS-DPC project (Line 1: "Assessment and reduction of the vulnerability of masonry buildings") funded by the Italian Department of Civil Protection.

Author Contributions

All the authors contributed equally to this paper.

Conflicts of Interest

The authors declare no conflict of interest.

References

1. D'Ayala, D.; Paganoni, S. Assessment and analysis of damage in L'Aquila historic city centre after 6th April 2009. *Bull. Earthq. Eng.* **2011**, *9*, 81–104.
2. Penna, A. Seismic assessment of existing and strengthened stone-masonry buildings: Critical issues and possible strategies. *Bull. Earthq. Eng.* **2015**, *13*, 1051–1071.
3. Lourenço, P.B.; Mendes, N.; Ramos, L.F.; Oliveira, D.V. Analysis of masonry structures without box behavior. *Int. J. Archit. Herit.* **2011**, *5*, 369–382.
4. Diaferio, M.; Foti, D.; Giannoccaro, N.I.; Ivorra, S. Optimal model through identified frequencies of a masonry building structure with wooden floors. *Int. J. Mech.* **2014**, *8*, 282–288.
5. Kappos, A.J.; Penelis, G.G.; Drakopoulos, C.G. Evaluation of simplified models for lateral load analysis of unreinforced masonry buildings. *ASCE J. Struct. Eng.* **2002**, *128*, 890–897.
6. Casolo, S.; Uva, G. Nonlinear analysis of out-of-plane masonry façades: Full dynamic *vs.* pushover methods by rigid body and spring model. *Earthq. Eng. Struct. Dyn.* **2013**, *42*, 499–521.
7. Vamvatsikos, D.; Cornell, C.A. Direct estimation of the seismic demand and capacity of oscillators with multi-linear static pushovers through IDA. *Earthq. Eng. Struct. Dyn.* **2006**, *35*, 1097–1117.
8. Galasco, A.; Lagomarsino, S.; Penna, A. On the Use of Pushover Analysis for Existing Masonry Buildings. In Proceedings of the 13th European Conference on Earthquake Engineering, Geneva, Switzerland, 3–8 September 2006.
9. Casolo, S.; Uva, G. Full Dynamic and Pushover Analyses of Out-of-Plane Masonry Facades: Application and Comparisons by RBSM. In Proceedings of the 14th World Conference on Earthquake Engineering, Beijing, China, 12–17 October 2008.

10. Bartoli, G.; Betti, M. Cappella dei principi in Firenze, Italy: Experimental analyses and numerical modeling for the investigation of a local failure. *ASCE J. Perform. Constr. Facil.* **2013**, *27*, 4–26.
11. Betti, M.; Galano, L. Seismic analysis of historic masonry buildings: The vicarious palace in Pescia (Italy). *Buildings* **2012**, *2*, 63–82.
12. Caliò, I.; Marletta, M.; Pantò, B. A new discrete element model for the evaluation of the seismic behaviour of unreinforced masonry buildings. *Eng. Struct.* **2012**, *40*, 327–338.
13. Magenes, G.; Della Fontana, S. Simplified Non-Linear Seismic Analysis of Masonry Buildings. In Proceedings of the 5th International Masonry Conference, London, UK, 13 July 1998.
14. Bucchi, F.; Arangio, S.; Bontempi, F. Seismic Assessment of an Historical Masonry Building using Nonlinear Static Analysis. In Proceedings of the 14th International Conference on Civil, Structural and Environmental Engineering Computing, Sardinia, Italy, 3–6 September 2013.
15. Lagomarsino, S.; Penna, A.; Galasco, A.; Cattari, S. TREMURI program: An equivalent frame model for the nonlinear seismic analysis of masonry buildings. *Eng. Struct.* **2013**, *56*, 1787–1799.
16. *ANSYS* v.11.0; ANSYS Manual Set; ANSYS Inc.: Canonsburg, PA, USA, 1998.
17. Galano, L.; Selleri, F. Analisi sismica degli edifici in muratura: Un modello semplificato a macroelementi. *Ing. Sismica* **2004**, *21*, 28–45. (In Italian)
18. Betti, M.; Galano, L.; Vignoli, A. Comparative analysis on the seismic behaviour of unreinforced masonry buildings with flexible diaphragms. *Eng. Struct.* **2014**, *61*, 195–208.
19. De Canio, G.; Dolce, M.; Goretti, A.; Marnetto, R. Progetto TREMA–Tecnologie per la Riduzione degli Effetti sismici sui Manufatti Architettonici in muratura e in ca. MURST Legge n. 449/1997, D.M. 10 Maggio 2000. (In Italian)
20. Dolce, M.; Ponzo, F.C.; di Croce, M.; Moroni, C.; Giordano, F.; Nigro, D.; Marnetto, R. Experimental Assessment of the CAM and DIS-CAM Systems for the Seismic Upgrading of Monumental Masonry Buildings. In Proceedings of the 1st International Conference on Protection of Historical Constructions, Rome, Italy, 26–28 October 2009.
21. Dolce, M.; Ponzo, F.C.; Moroni, C.; Nigro, D.; Goretti, A.; Giordano, F.; de Canio, G.; Marnetto, R. 3D Dynamic Tests on 2/3 Scale Masonry Buildings Retrofitted with Different Systems. In Proceedings of the 14th World Conference on Earthquake Engineering, Beijing, China, 12–17 October 2008.
22. De Vekey, R.C.; Sassu, M. Comparison on Non-destructive *In-Situ* Mechanical Tests on Masonry Mortars: The PNT-G Method and the Helix Method. In Proceedings of the 11th International Brick/Block Masonry Conference, Shanghai, China, 14–16 October 1997.
23. Sahlin, V. *Structural Masonry*; Prentice-Hall: Englewood Cliffs, NJ, USA, 1971.
24. Galano, L.; Vignoli, A. Resistenza a taglio di murature di pietrame: Analisi comparata della rappresentatività di tipiche prove di taglio *in situ*. In Atti del X Congresso Nazionale ANIDIS L'Ingegneria Sismica in Italia, Potenza-Matera, Italia, 9–13 settembre 2001. (In Italian)
25. Magliulo, G.; Petrone, C.; Capozzi, V.; Maddaloni, G.; Lopez, P.; Manfredi, G. Seismic performance evaluation of plasterboard partitions via shake table tests. *Bull. Earthq. Eng.* **2014**, *12*, 1657–1677.

26. Giamundo, V.; Lignola, G.P.; Maddaloni, G.; Balsamo, A.; Prota, A.; Manfredi, G. Experimental investigation of the seismic performances of IMG reinforcement on curved masonry elements. *Comp. Part B Eng.* **2015**, *70*, 53–63.
27. Avossa, A.M.; Malangone, P. Seismic performance assessment of masonry structures with a modified "concrete" model. *Bull. Earthq. Eng.* **2015**, doi:10.1007/s10518-015-9726-1.
28. Betti, M.; Vignoli, A. Numerical assessment of the static and seismic behaviour of the basilica of Santa Maria all'Impruneta (Italy). *Constr. Build. Mater.* **2011**, *25*, 4308–4324.
29. Calderini, C.; Cattari, S.; Lagomarsino, S. The use of the diagonal compression test to identify the shear mechanical parameters of masonry. *Constr. Build. Mater.* **2010**, *24*, 677–685.
30. Elmenshawi, A.; Sorour, M.; Mufti, A.; Jaeger, L.G.; Shrive, N. Damping mechanisms and damping ratios in vibrating unreinforced stone masonry. *Eng. Struct.* **2010**, *32*, 3269–3278.

13

IDEAhaus: A Modular Approach to Climate Resilient UK Housing

Greg Keeffe [1,*] and Ian McHugh [2]

[1] Architecture at Queens, School of Planning, Architecture and Civil Engineering, Queens University Belfast, Belfast BT9 5AG, UK

[2] Green Triangle Studio, 134 Sandy Lane, Manchester M21 8TZ, UK; E-Mail: green.triangle@btinternet.com

* Author to whom correspondence should be addressed; E-Mail: g.keeffe@qub.ac.uk

External Editor: Brian Ford

Abstract: This paper describes the result of a project to develop climate adaptation design strategies funded by the UK's Technology Strategy Board. The aim of the project was to look at the threats and opportunities presented by industrialized and house-building techniques in the light of predicted future increases in flooding and overheating due to anthropogenic climate change. The paper shows that the thermal performance of houses built to the current UK Building Regulations is not adequate to cope with changing weather patterns, and in light of this, develops a detailed design for a new house: one that is industrially produced and climatically resilient, but affordable. This detailed concept *IDEAhaus* of a modular house is not only flood-proof to a water depth of 750 mm, but also is designed to utilize passive cooling, which dramatically reduces the amount of overheating, both now and in the future.

Keywords: climate change adaptation; thermal comfort; passive cooling; flood resilience; industrialised housing; mass customization

---

1. Introduction

Recent research has shown that current UK housing, even that designed to the current Building Regulations is poorly suited to deal with the changing climate the UK is likely to experience over the next 50 years. This paper describes the design of a novel modular house, which aims to answer some of these short-comings and provide a cost-effective and reproducible solution to living within a changing climate.

Anthropogenic climate change is now an undeniable phenomenon [1] and already in the UK, models of future climates have been developed, allowing architects and designers to predict the thermal performance of designs using predicted weather data for a range of climate scenarios over the next 70 years [2]. The change in climate is not only to hotter drier summer conditions, but also to milder much wetter winters. These two changes create a new difficult scenario for the architects of housing where passive cooling needs to be factored into the design, along with flood resilience.

In addition, in the UK there is a great need for new affordable mass housing and indeed greater industrialization of the process that could bring better quality, speed and predictability to its delivery. However, in the UK, factory-made housing has not been able to provide the variety and flexibility necessary to respond adequately enough to differing site contexts and programmatic requirements. Therefore, rather than engage with modular solutions, UK house-builders have looked to use timber-framed solutions, to make a cost effective, high build quality solutions. This is not without its drawbacks: indeed much of the recently built, highly insulated, air tight, timber-framed housing is already suffering summer overheating. In addition it is also highly susceptible to flood damage. These are both risks that will be exacerbated in the future with climate change.

The Design for Future Climate projects are part of the UK Technology Strategy Board's vision to support UK industry to develop innovation in new technologies. Climate change resilience is a key area of that work, where solutions are seen as holistic and design based, rather than technological add-ons. The IDEAhaus project as part of the Design for Future Climate, focused on two main objectives: firstly to assess the performance of current newly-designed housing within future UK climate scenarios, and secondly to develop a design for a new house that was affordable and could be easily replicated, which could adapt in a better and more passive way to the changing future climate context.

As the case study, the project took a proposed social housing development in Liverpool UK. This development consisted mainly of semi-detached and terraced low-rise houses, which were typically three bedroomed. Once the energy and thermal performance of these houses had been modelled utilizing UK Met Office climate projections through the century up to 2080 [3], it became clear, that although these houses had been designed to current UK Building Regulations, the house type was unsuitable for the future climate change scenarios expected. The thermal analysis showed significant risks of overheating not only in the future, but also today, and also that the cooling demand for the house would exceed heating demand within 20–30 years. The houses had been designed to meet a scenario where heating in winter was the key driver, and had not considered issues of cooling.

Taking this performance data, the project developed designs for a building system for low-rise housing which would be more resilient to flood damage and better at resisting overheating through passive cooling techniques, such as shading, and the use of thermal mass and natural ventilation strategies. The system was to be based around a limited number of components which could be assembled to provide different sized homes, with a modular service/circulation core and a range of cladding options. The design illustrated how spatial flexibility, in-built flood protection, combined with thermal improvements and future adaptability could be used to develop an affordable but resilient house type that could cope with future extreme weather scenarios.

2. Future Climate in the UK

Climate change forecasting is an uncertain science. In the UK, the Department of Environment, Food and Rural Affairs have produced a range of Climate Projection scenarios known as UKCP09 [4]. These cover a range of years (2030, 2050, 2080), emissions scenarios (low medium high) and probability (33, 50, 66 and 90 percentiles). The emission scenarios are based on future levels of carbon emissions, however it seems unlikely at present, that anything other than a high emissions scenario will be likely. The probabilities here are the likelihood of a certain climate being not exceeded, so the 90th percentile model is the most extreme with only a 1 in 10 chance being exceeded. Thus, this 90th percentile is a likely occurrence in one year in that decade. From these scenarios, the Prometheus Project at the University of Exeter [5] have projected climate as a year-long set of results that mimic the CIBSE dsy (design summer year) and try (test reference year) data. The scenarios are in the form of hourly data, in Energy Plus format (epw) and can be used in most energy modeling software.

In general the anticipated pattern of change in the UK is toward hotter drier summers, milder wetter winters, stronger winds and more frequent, more extreme events such as heat-waves, storms and flooding. Climate change means that buildings will have to be, firstly more adaptable to cope with the changing climatic conditions, and secondly more resilient to cope with extreme events. Both these functions, will also have to happen within a changing energy and materials scenario, as oil diminishes and materials deplete. In addition, these two functional requirements, of adaption and resilience, are also not necessarily commensurate and create a complex design context, which can be difficult to negotiate.

The initial case-study was situated in Toxteth, Liverpool, UK. The current climate in Liverpool can be considered "mild maritime" with a January average of 6 °C and a July average of 13.5 °C. By plotting temperature and humidity data for the test years on a psychrometric chart, a bioclimatic analysis of the predicted future climate could be envisaged. Shown in Figure 1 are the charts for 2010, 2030, 2050, and 2080, for one year in ten using a high emissions scenario (the most likely). This shows a gradual change in summer from a mild maritime climate to a more Mediterranean one. This is significant as there is a drift of temperatures during this period, across the comfort zone, developing from a heating dominated operation, to a more cooling orientated one. In addition to this, quantitative analysis of the data showed there is also a predicted increase in extreme wet weather in winter, with over 20% more rain during the winter months.

The current data (2010) from the Chartered Institute of Building Services Engineers' (CIBSE) Design Summer Year (DSY) demonstrates clearly the current position: there should be very little need for any environmental control in summer, other than solar control. However, as we move into the future, the situation begins to change: by 2030 there is a need for a coordinated cooling strategy—say increased thermal mass. Continuing into the mid-century period, analysis of the DSY data showed that by 2050 even with a medium emissions scenario, there was a 50% chance (*i.e.*, one year in two) that the summer would be similar in temperature and humidity to that of a Northern Mediterranean city, such as Lyon, France. By 2080 the predicted weather had altered again: the data for 2080 with a high emission Scenario with a 90% chance of not being exceeded, gave a very different summer climate, one more similar to that currently experienced in Rome, Italy. In this scenario the increase in average summer temperatures would be an astounding 9 °C and the maximum temperature some 11 °C higher than experienced today.

Figure 1. Changing climate: bioclimatic charts: 2010–2080, Liverpool, UK.

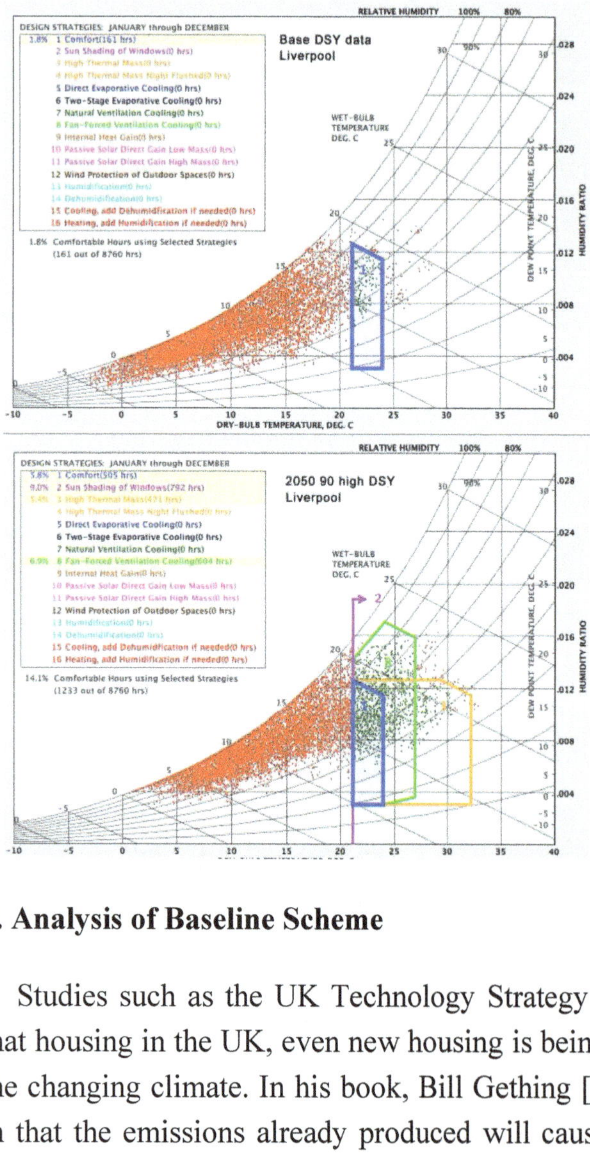

3. Analysis of Baseline Scheme

Studies such as the UK Technology Strategy Board's Design for Future Climate [6] have shown that housing in the UK, even new housing is being designed and built with little or no consideration of the changing climate. In his book, Bill Gething [7] claims that we are now beyond climate mitigation in that the emissions already produced will cause unfettered climate change throughout the century, and thus we must now embark on a program of adaptation. This adaptation will have to engage with both new build and retrofit schemes, as much of the built environment of 2050 is already in existence. Retrofit ideas are already well documented, some involve fabric measures such as the addition of shading devices, or passive cooling techniques, whilst others involve the adaptation of the landscape in which the properties sit, to suit warmer and wetter conditions [8].

The IDEAhaus concerned itself with the new build market, particularly that for low cost housing suitable for rental. In this market in the UK, housing design is highly regulated, particularly with regards to spatial standards, life-cycle costings, and energy use. The energy use of the house is modeled using an approved energy model known as the Standard Assessment Procedure (SAP) developed by the Building Research Establishment [9]. This model, originally based on a modified degree-day type calculation only really concerns itself with heating (and lighting), thus houses are adapted specifically for cold, not warm weather, that is for heating and not cooling.

By considering a housing scheme already designed as a baseline, the project aimed first to critique current practice and assess the performance of housing that met the criteria with respect to the climate that the houses will face in the foreseeable future. From this critique, a new modular house was designed with a customizable façade, which was designed with the future climate and weather in mind. One that was robust and energy efficient, but also comfortable for the occupants.

The baseline scheme for the project was an existing design for Plus Dane Housing Association that was proposed to be built in the near future. The site was located in Toxteth, Central Liverpool, UK (see Figure 2). This scheme was used to develop the base condition for comparative thermal and energy modeling, for different constructions within the same site layout. The layout of housing on the site adopted a diagonal solar orientation with houses facing SW, SE, NW and NE in square urban block arrangements.

Figure 2. Housing layout, Toxteth, Liverpool, UK.

A typical 3-bedroom 5-person, 2-story, semi-detached/end terraced house (Figure 3) with traditional elevations was selected. The construction specification was for a high performance closed panel timber frame system. The fabric, designed to current UK Building Regulations (REF), was therefore highly insulated (wall, floor and roof *U*-values at approximately 0.1 W/m^2K) and reasonably airtight ($5\ m^3/h/m^2$) but not so airtight that it required a whole house Mechanical Ventilation with Heat Recovery (MVHR) system: the house was naturally ventilated in both summer and winter.

Figure 3. Baseline 3-bed 5-person-house type.

Initially, detailed thermal modeling using 2010 climatic data was carried out with IES software for each orientation of the properties on site. However the different orientations of the house type made very little difference to the thermal modeling results with regard to internal temperatures and energy use. The most likely reason for this being that the front and rear elevations had similar proportions of glazing and that the diagonal solar aspect (SE/NW) of the layout tended to equalize the exposure to sunlight, so a property with a SE orientation was chosen for the further comparative analysis.

Table 1 illustrates overheating in a typical SE facing double bedroom in the timber-framed designs. In the UK, CIBSE guidance recommends that internal temperatures should not exceed 25 °C for >5% of annual habitable hours and 28 °C for less than 1% of the hours [10]. This guidance is not currently applied to single dwelling-houses in the UK but is used in other residential buildings. The detailed modeling used 2010 CIBSE Design Summer Year (DSY) data and compared this with a "worst case scenario" that being the hottest summer in the decade around 2080. (This equates to 2080 DSY, with a high emissions scenario, 90th percentile projected weather set from the Prometheus database.) Findings showed that overheating in bedrooms (internal temperature above 28 °C) would occur for 6.6% of the time in 2010 rising to 50.5% of the time in 2080. If only the summer months of July and August are considered, this would equate to 28.4% of the time in 2010 and 74.6% in 2080. Peak internal temperatures experienced in the bedroom are 35 °C in 2010 and 39 °C in 2080. The conclusion inferred from this analysis is that the current house type is inadequate with respect to summer over-heating, as the internal overheating is already well above the CIBSE guidelines now, and becomes severe by 2080.

Table 1. Summary of internal overheating modeling (IES) on baseline house-type for SE facing bedroom in 2010 and 2080* Hi-emissions 90th percentile.

CIBSE	Guidelines	Period	Peak Temp (°C)	Int Temp (h)	>25 °C (% Time)	Int Temp (h)	>28 °C (% Time)
Timber-framed	2010 DSY	annual	35	573	19.6	192	6.6
house		July/August	35	310	62.5	141	28.4
	2080*	annual	39	2605	89.2	1474	50.5
		July/Aug	39	496	100.0	370	74.6
Masonry	2010 DSY	annual	29	396	13.6	81	2.8
house		July/Aug	29	286	57.7	42	8.5
brick/block	2080*	annual	33	1391	47.6	934	32.0
		July/Aug	33	496	100.0	262	52.8

Next, a similar analysis was performed on a similar house, that was designed utilizing what would be considered a "traditional" UK building method, namely a masonry construction of brick external skin and medium weight concrete block inner, with a wet plaster finish. This change of construction shows a significant reduction in overheating, with the house being above 28 °C for 2.8% of the time above in 2010, and 32.0% in 2080. Results for July and August show overheating for 8.5% of the time in 2010 and for 52.8% in 2080. Peak internal temperatures are also significantly reduced at 29 °C and 33 °C respectively, but these are still high.

Further energy modeling of the timber framed house was carried out using Sefaira Concept software [11]. This produced an energy consumption analysis based on the UK Government's SAP analysis [12].

This analysis shows the energy footprint in kWh for the heating, hot water, lighting and power for the house (see Figure 4). It is clear that the space heating demand (red) reduces considerably to a minimal level by 2080. Hot water, lighting and appliances (yellow, orange and green) are constant. The modeling was then repeated with the addition of air-conditioning for comfort cooling; switching on at 25 °C (Figure 5).

This analysis shows that reduced demand for heating (yellow) that occurs in 2080 is counteracted by an increasing cooling demand (blue), and thus the overall energy demand for the house would actually increase over time. Further IES analysis shows that energy demand for cooling in the house could overtake its heating demand before 2040 in a 90th percentile year (Figure 6).

In fact, the "carbon crossover" rather than the energy crossing would, in practice, arrive even earlier, as cooling energy is generally more carbon intensive (with cooling being produced using electricity) than space heating (from gas central heating) in the typical UK situation. The team therefore concluded that if we are to design for future climate as well as the present, then summer performance will become more crucial and will need to be accommodated in today's designs. In fact, the importance of winter performance and heat loss as a design strategy will diminish and more attention in the future will need to be paid to overheating and reducing the demand for cooling and summer energy use.

Figure 4. Energy analysis for 2010 and 2080 high emissions scenario 90th percentile energy analysis.

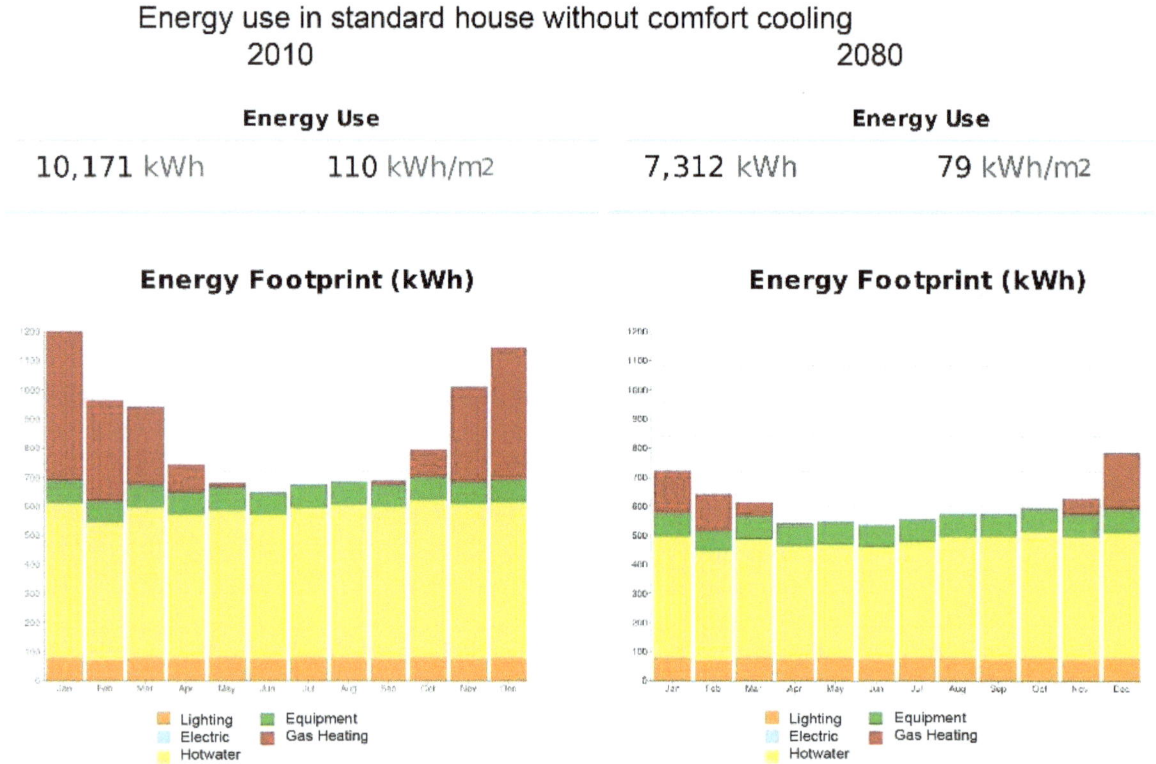

Figure 5. Energy analysis with air-conditioning at 25 °C set point for 2010 and 2080 high emissions scenario 90th percentile.

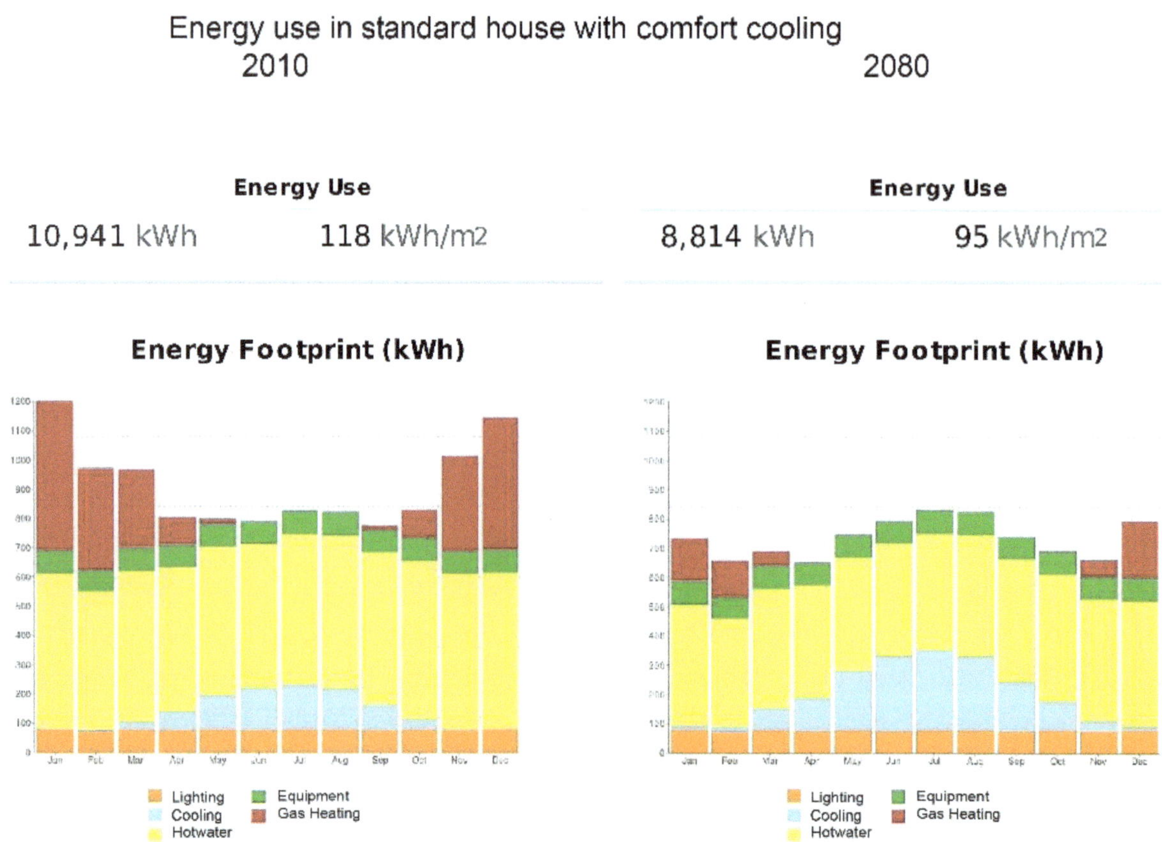

Figure 6. Timber frame—Space Heating *versus* Cooling energy demand in the house using high emission scenario 90th percentile projections.

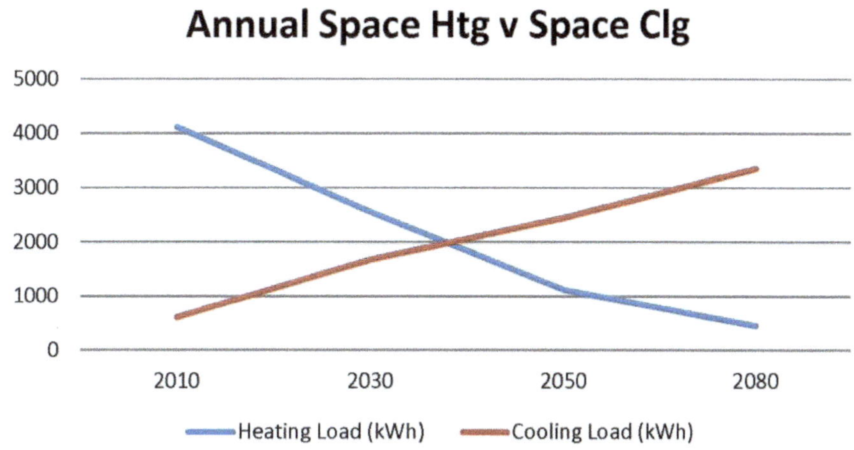

4. Climate Change Risk Assessment

A risk analysis of the effects of climate change on the housing development was performed. This emphasized two issues, firstly the likelihood of the risk, the number of impacts from the risk and the severity of these. Over 100 risks were identified, but only three had a likelihood and severity that made them serious enough to consider now for the future. These future serious risks were mainly concerned with extreme weather events or extended heat waves: the house of the future would have to cope high wind-speeds, high rainfall and flooding, and summertime overheating.

4.1. Flooding

Future flood return intervals are difficult to assess since they are based on historic data. In the UK, flood risk is predicted by creating "occurrence in a time period" data, so sites are given a clearance for a "once in 20 year flood", or a "once in 50 year flood", or a "once in 100 year flood" [13]. Climate change is affecting this analysis, as a site with a 1 in 30 year flood risk in 2010 will have a 1 in 16 year risk of flooding by 2080 (see Figure 7), and a current 100-year flood event is projected to become a 50-year event by 2080 within a widening probability band [14]. This can be seen clearly in Figure 7.

Figure 7. UK Met Office graphs showing changes in return period for winter rainfall events in Liverpool 2012–2080.

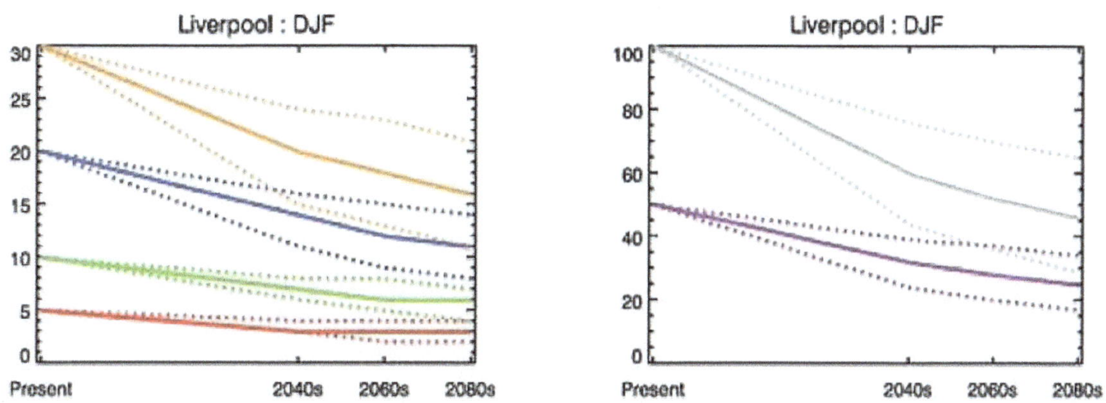

However, there is no new projected 100-year flood data for 2080. What is certain, however, is that the flood risk certainty is diminishing, and it is likely that most of the urban space in the UK will be at a much increased flood risk in the future. Flooding in the UK can be both fluvial and pluvial, but typically water depths are likely to be below 750 mm and the duration is likely to be no more than 2 weeks.

4.2. Wind

Data on possible extreme wind events is difficult to obtain, and the hourly data from Prometheus does not accommodate storms, as it is based on a model year, which by its nature has non-extreme weather. The Prometheus model seems to predict general wind increases but not extreme gusts. It is likely that hurricane-force winds will be more common, and Liverpool has already seen gusts of over 150 km/h in the past five years [15]. Timber frame performs well in wind and simple improvements such as changing the size and frequency of fastenings for roof tiles can be easily made that have little effect on the overall design and cost, but greatly improve resilience to wind. These new safety factors should easily be incorporated into all housing.

4.3. Overheating

Prolonged overheating presents a danger to health. Death rates amongst elderly people rise sharply above roughly 28 °C. In France for example, in 2003, 15,000 extra deaths of elderly people occurred due to an extended heat wave that lasted for only three-weeks, during which the daytime temperatures were only 4–5 °C above normal [16]. However, it appears that our own physiological response to the heat stress, rather than the maximum temperature experienced, is what causes death: thus an equivalent temperature in the UK could be more life threatening the same in Greece. This behavioral adaptation may affect the exact trigger temperature, but what is clear that once the temperature exceeds for some period a temperature of around 28 °C, mortality rates start to rise [17].

The other effect of overheating is that it raises the likelihood of carbon intensive cooling systems such as air conditioning being retrofitted to the property. Most of these systems use electrical means for chilling, which makes them very inefficient and costly to run, which in turn will increase household payments for occupiers, use precious resources, and indeed create a localized heat island effect which could exacerbate the problem.

5. IDEAhaus Concept

The final stage in the project was the design development of an idealized future housing product which could be mass produced, yet solve these future climatic issues. The design needed to be industrialized and cost effective, yet flexible in design and meet current spatial and energy standards without adaptation. The house would also have to be resilient to the risks inherent in the UK's future climate and resource scenario: it would therefore need to include passive design strategies for dealing with issues of flooding and overheating. Branded the IDEAhaus, this would be: Industrialized, Delightful, Efficient, and Adaptable (see Figures 8 and 9).

Figure 8. Exploded view highlighting component kit.

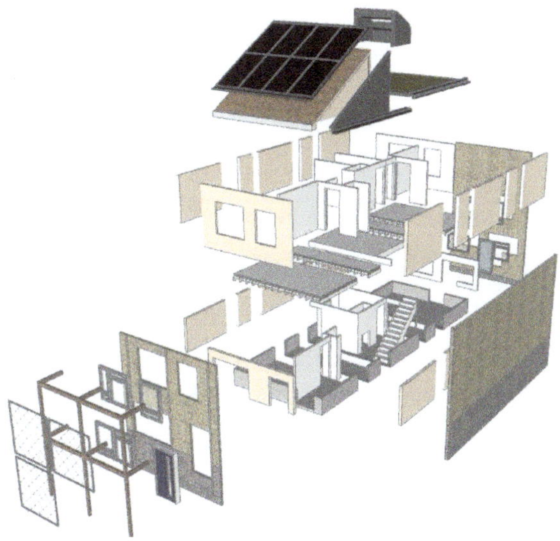

Figure 9. Elevational options—brick, timber, render/panel.

5.1. Industrialized

The house would need to compete in the current market, in terms of price-point, quality and amenity. This lends itself to the development of an industrialized, product-based system:

- **Standardization**—mass production of regular core components for the superstructure;
- **Manufacturing quality**—enhanced quality achieved by production under factory conditions;
- **Predictable cost and delivery**—through repetitive design, specification and construction method;
- **Economies of scale**—through bulk purchasing power and availability of stock items.

5.2. Delightful

The house would need to be valued as a place to live, with excellent architectural features that enhance livability:

- **Spacious**—designs based on detailed furniture layouts and activity spaces, good ceiling heights and central light-well;

- **Individualized**—different possible room layouts, fenestration, cladding and finishing options;
- **Comfortable**—through use of thermal mass, good ventilation, shading options and radiant heating systems;
- **Quality**—high quality products with finishing options.

5.3. Efficient

The resource scenario for the rest of the century is uncertain to say the least. Peak oil production was reached in 2009 and we are now moving to a low-carbon economy [18]. Increasing differentials between rich and poor in the UK are already producing areas of "fuel poverty" where occupiers cannot afford to heat their homes adequately. It is likely that the necessity of mechanical cooling will exacerbate the problem. Fuel and therefore materials are likely to become more rather than less expensive in real terms. The house if it is to be successful as a model for social housing will need to very low energy and resource efficient.

- **Passive design**—highly insulated fabric with thermal mass, good controllable natural ventilation and shading options (see Figure 10);
- **Renewable energy options**—ability to incorporate renewable energy systems;
- **Low impact materials**—sustainably sourced materials, engineered to minimize waste;
- **Fast construction**—predictable design time and quick to erect watertight shell construction.

Figure 10. Long section.

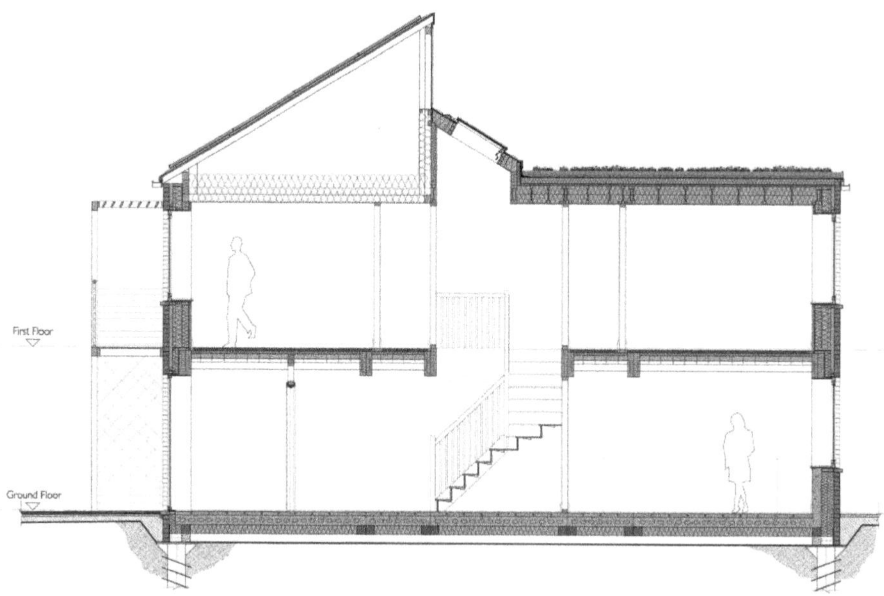

5.4. Adaptable

The market for social housing is difficult, due to the cost of available land in the UK, which is one of the most densely populated countries in Europe. Typically, social housing is forced to the margins, occupying space that has low value for development, in that it is awkward to occupy, in danger of flooding, or it is brownfield, with a complex history of industrial occupation. In order to maximize density on these sorts of sites, the product must be spatially adaptable to occupy the site effectively. In addition, the range of occupiers for this sort of house is wide in the UK, and there is a need for 2, 3 4, or 5 bedroomed properties, often on 2 or 3 floors. Lifetime Homes is a UK standard that promotes flexibility for house designs so that they are adaptable for the needs of aged or disabled persons.

- **Flexible layout**—designed to the UK's Lifetime Homes generous space and accessibility standards (see Figure 11);
- **Climate resilient**—flood resilient and overheating resistant construction;
- **Additive features**—construction allows for exo-structure options and vertical extension;
- **Upgradable performance**—allowing for replaceable cladding, solar panels and services.

Figure 11. Four bed six person house floor plans.

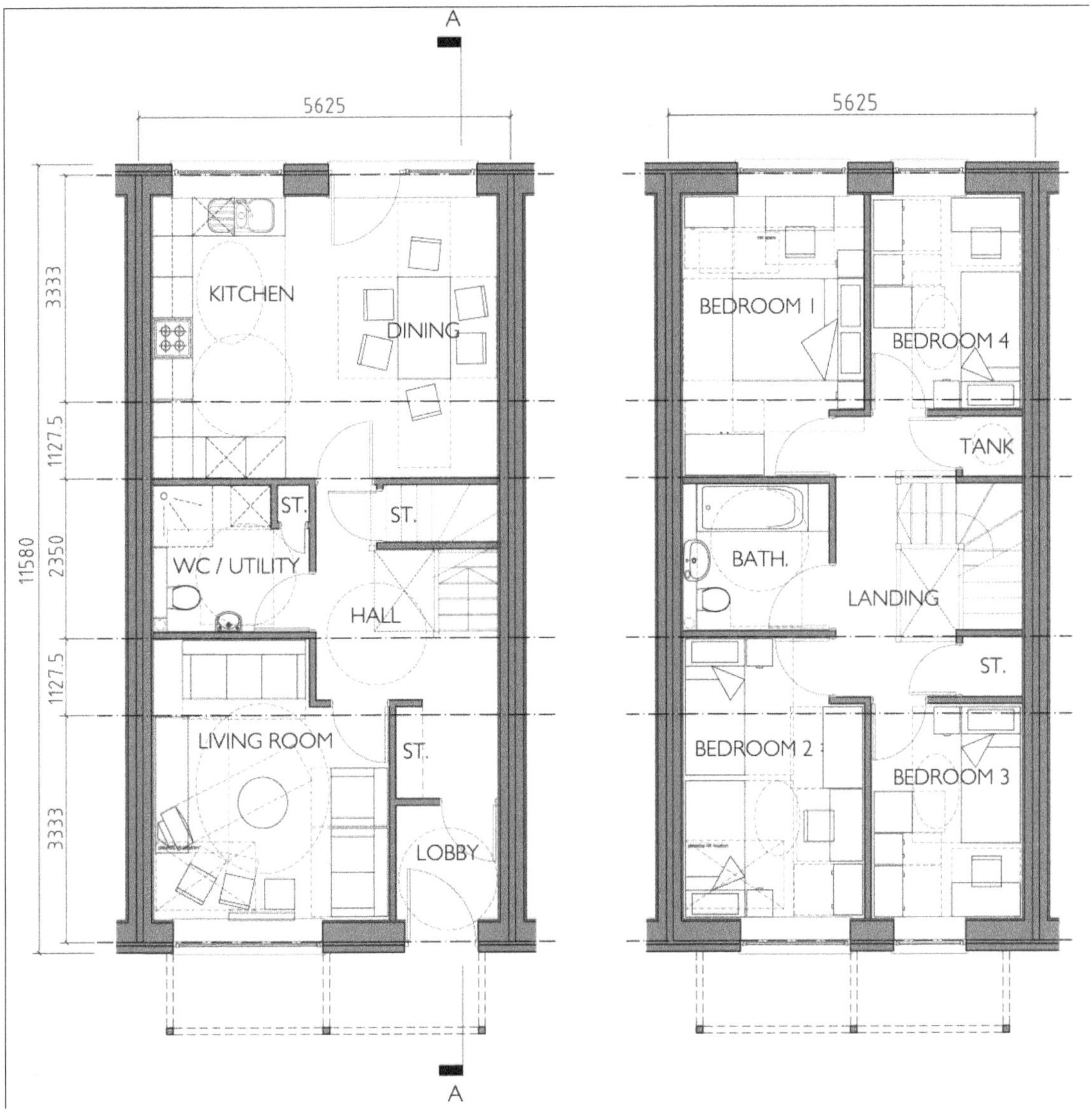

6. IDEAhaus Detailed Construction

The idea of mass customization is that of "*producing goods and services to meet individual customer's needs with near mass production efficiency*" [19]. This has been described as combining the customization of bespoke tailoring with the efficiency of the production line. Following the philosophy of mass customization, the construction of the IDEAHaus is considered as a three-part system consisting of Core construction, Additive components, Adaptable services.

The house construction is made of five elements, three that constitute the core:

- **Floor and foundations:** the development of a flood resilient ground system including foundations, ground floor and lower walls [Core];
- **Volumetric central services area:** this factory assembled unit contains stairs, bathrooms and services [Core];
- **Cassette based upper systems:** This is a series of insulated and structural panels that complete the interior of the house [Core];
- **Customizable external cladding system:** This allows for aesthetic contextual adaption, or possibly climatic adaptation through retrofit [Additive];
- **Utility components:** This is a series of components that enhance the utility of the house, such as photovoltaics or solar domestic hot water systems [Adaptive].

6.1. Core Construction

This part of the system is the spatial construct that defines the house as a living space. This consists of the foundations and ground floor, the wall, upper floor and roof constructions and the service pods for kitchen and bathroom.

6.1.1. Foundations

Helical steel screw piles are proposed to suit virtually any site conditions (e.g., urban housing on filled brownfield sites) with minimal disruption and preparation. They can reduce site excavation and minimize cost of landfill taxes. They allow large shading trees to be located closer to buildings without root damage to foundations (see Figure 12).

Figure 12. Ground floor to external wall detail, highlighting flood resilient floor slab upstand.

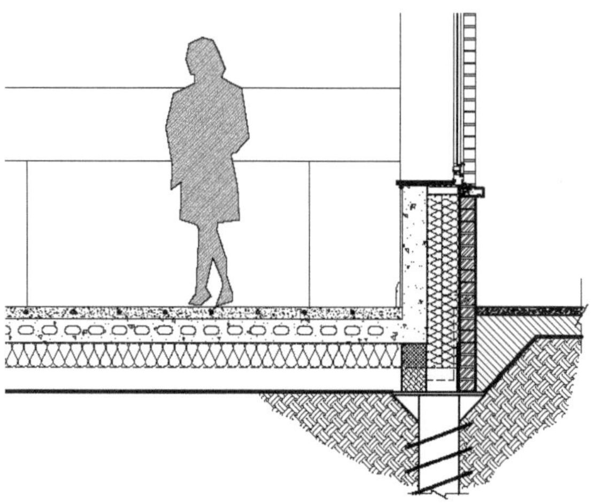

6.1.2. Ground Floor

The floor is made of a series of large precast concrete units with flood-resilient, up-stand edges, with a bonded damp membrane and closed cell insulation to the outer faces giving an effective damp-proof course some 750 mm above floor level (to repel flood water) and a high thermal mass. The units span between pile caps on insulated blocks with a reinforced up-stand edge beam. Units are designed to a standard house width of 5.6 m to suit 2, 3 and 4 bedroom house types. The modular floor allows a standard position for front and rear doors within two large 3.3 m length elements. In the center of the plan a 2.4 m floor unit is designed to suit a WC/utility and stairwell and if necessary 1.1 m infill units can be used to extend the floor depth of the house to suit the number of bedrooms required.

6.1.3. Wall Cassettes

The walls were manufactured with pre-insulated timber frame wall cassettes with 120 mm pre-cast "Hemcrete" insulation and 200 mm of hemp fiber insulation quilt [20]. The hemcrete product provides excellent thermal mass and has phase-change properties which enhance its cooling performance. The cassette has a breathable construction that enhances humidity control in the house. Window openings can be individually designed and the proposals shown are set out to suit standard brick dimensions, for ease of external cladding.

6.1.4. Upper Floor Cassettes

The upper floors are manufactured using open panel cassettes. Over the main living spaces these comprise exposed engineered timber edge beams and joists, which are then in-filled on site with hollow clay blocks based on the Ibstock "Coolvault" system [21]. This provides exposed thermal mass in the house, seen as a self-finished vaulted ceiling to ground floor rooms and a timber boarded finish above (see Figure 13). The central area around bathrooms and stairs are closed panel with plasterboard ceilings to allow space for services distribution.

Figure 13. Party wall/upper floor detail.

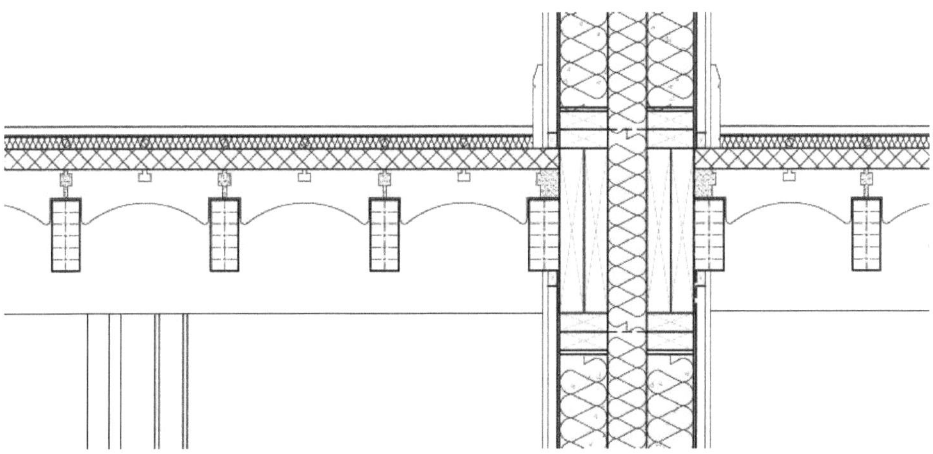

6.1.5. Central Volumetrics

The highly serviced central area with bathrooms, stairs and main heating system is standardized for all the house types. This would be manufactured off-site and be of volumetric construction. These could even be stock items. Finishes and fittings could be completed to standard or customized to individual order. The upper volume has a pre-assembled roof cassette to match the main flat roofs. This can be rotated to suit solar orientation.

6.1.6. Roof

The roof can be rotated during construction to maximize solar collection. The roof is proposed with a 30 degree pitch for optimum solar collection potential. This can be pre-assembled (on or off-site) in trussed rafter and purlins spanning between party walls. It is constructed with a plywood boarded finish to provide racking and allow for different cladding options, such as photovoltaics.

6.2. Additive Components

These components offer the first part of mass customization. Changes to external components can affect both the aesthetics of the house as well as its thermal performance. Loggias or shades can be added to the external walls or green roof systems above to help with environmental control. The choice of external cladding can also help with planning issues re conservation (see Figure 14).

Figure 14. Eaves detail.

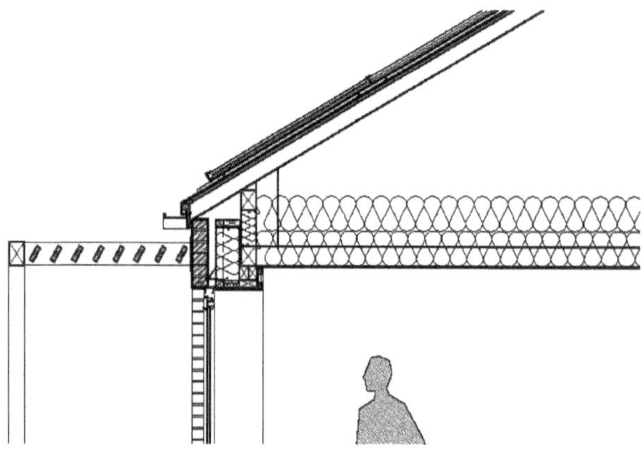

6.2.1. External Cladding

The proposal shows cavity wall brickwork cladding and zinc clad roofs with solar panels over the pitched roof, however, other finishes, such as brick or a rain-screen are equally viable.

6.2.2. Green Roof/Roof Garden

The North-facing flat roofs lend themselves to a green roof/garden finish to aid in bio-diversity, rainwater attenuation and cooling micro-climate through evapo-transpiration.

6.2.3. Fit Out

The layouts shown are based on highly specified UK social housing standards. This gives the flexibility to vary room sizes and shapes or adapt to a more open plan depending on the overall size of house.

6.2.4. Exo-Structure and Components

A grid of thermally broken fixing points is built into the façade for an optional 1.2 m deep timber framed exo-structure with a range of porches, shading devices, balconies, trelliswork, *etc*. (Figure 15).

Figure 15. Exo-structure variations.

6.2.5. Extra floors

The structural system will support additional floors with a second staircase added for vertical extension. Roof cassettes can be demounted and reused.

6.3. Adaptive Services

These components change to performance of the house from an operational viewpoint. There is a choice of renewable systems that can be added at the design stage or retrofitted at a later date.

6.3.1. PV-Thermal

Hybrid PV-T panels which produce both electricity and hot water are proposed to the south facing roof combining solar hot water collectors under photovoltaic cells. PV-T's can give a 40% greater energy yield for equivalent areas of roof than separate panel systems. A large hot water tank is provided at first floor level.

6.3.2. Under-Floor Heating and Cooling

Both the ground floor and the first floor are shown with under-floor heating pipework for comfortable radiant heat at low temperatures which also allows the use of a ground to water heat pump. This pipework could also be used for summer cooling if necessary, to disperse heat from the structure. There is space in the service module for a gas-fired boiler if specified to provide heating and hot water in combination with the renewable energy sources.

6.3.3. Ventilation

The design has focused on a natural ventilation strategy rather than whole house MVHR. Window patterns open top and bottom to enhance single sided ventilation airflow in rooms and the rooflight increases the options for stack effect ventilation during warm periods. Windows can be securely restrained re high winds and insect blinds can be added in the reveals. Opening sizes shown allow a night-time purge ventilation rate of 6 air changes/hour at a modest air speed of 0.5 m/s. Individual extract fans with heat recovery are proposed for kitchens and bathrooms.

6.3.4. Services Distribution

The external and party walls are dry-lined above 750 mm to allow a service zone for cables and pipework this protects services from flood damage, and the plasterboard from the same. Wiring for the ground floor lighting runs in the top of the "coolvault" units and drops through the ceiling where required.

7. Performance Modeling of the IDEAhaus Design

Thermal modeling with particular focus on overheating was carried out on the IDEAhaus using IES, again utilizing hourly weather data from Prometheus for today and 2080. The results are summarized in Table 2.

Table 2. Summary of internal overheating modeling (IES) on IDEAhaus SE facing bedroom, for 2010 and 2080* Hi-emissions 90th percentile dataset.

House	Climatic dataset	Period	Peak Internal Temp	Temp above	25 °C	Temp above	28 °C
			(°C)	(h)	(% Time)	(h)	(% Time)
CIBSE Guidelines					5		1
IDEAhaus	2010 DSY	annual	28	41	1.4	5	0.2
		July/August		34	6.9	5	1.0
	2080*	annual	34	990	33.9	352	12.1
		July/August		496	100.0	200	40.3

The performance of the house, without active cooling appears to be excellent. The findings show the IDEAhaus proposal reduces the overheating problem significantly more than even the brick/block option. ISO 7730 recommends several ways of assessing thermal comfort in occupied buildings. The researchers chose method A as described in Annex H, long-term evaluation of the general thermal comfort conditions: "*Calculate the number or percentage of hours during the hours the building is occupied, the PMV or the operative temperature is outside a specified range*" [22]. The reason this more simple method was chosen, is that it is the most reasonable due to the compound error that is always present in future casting utilizing data that has been derived from theoretical climate modeling. In addition, it is difficult to assess future comfort exactly, as there will likely be a physical and behavioral adaption over time to the changing climate, which makes people's reaction in the future difficult to assess [23].

The results thus show, annual habitable hours with internal temperatures above 28 °C are 0.2% of the time in 2010 and 12.1% in 2080. In the summer months of July and August, overheating was also reduced to just 1.0% of the time in 2010 and 40.3% in 2080. This reduction is a 76% improvement over the existing timber framed design for 2080. This considerable improvement is related not only to the availability of thermal mass in the house, not only in the floor and walls, but also in the ceiling, compared to the masonry and timber framed designs. In addition, the breathable Hemcrete has phase-change properties due to its moisture content, that also helps prevent overheating [24]. The improved ventilation utilizing the openable roof lights above the stairwell, also helps cool the house, by increasing ventilation rate using the considerable stack effect of the high-level opening.

IES analysis of the IDEAhaus heating and cooling demand shows a lower demand than the timber frame baseline house which equates to a halving of the energy requirement for heating and cooling in 2080. The adoption of the IDEAhaus construction also changes the crossover point between heating and cooling demand to around 2050 (about 10 years later than the standard house). This is shown in Figure 16. The building was designed to be more resilient to climate change with respect to comfort, rather than economics. The team realize however that the cost–effectiveness of the solution needs to be understood. This not only includes the additional cost of manufacture, but also the future energy savings described above. A further assessment of the financial investment will be needed, to take the house into the production and deployment phase. It is intended to extend this work in the prototyping stage to produce an accurate cost model of the building after value engineering, which will allow the team to further assess the IDEAhaus as a cost-effective solution through a life cycle cost assessment.

Figure 16. IDEAhaus—Space Heating *versus* Cooling energy demand—Hi-emission, 90th percentile projections.

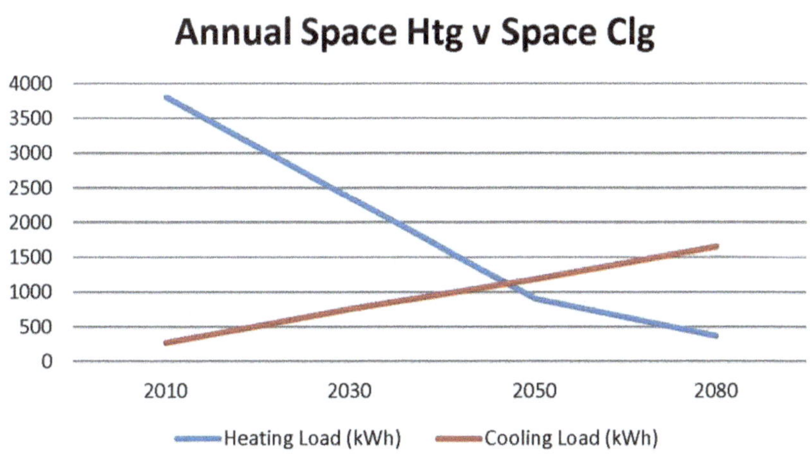

8. Conclusions

Future climate will challenge contemporary house design in the UK. Firstly there is a need to recognize that overheating is already an issue in highly insulated, lightweight construction and cooling for comfort is already becoming an issue. This will become a serious issue within 20 years, when more energy will be spent cooling a timber-framed house than heating it. Secondly that extreme winter weather will become the norm in the near future and new houses need to be designed to withstand higher wind speeds, possibly up to Category Five Hurricane, and be able to resist possible flood damage.

8.1. Passive Cooling

There is a need for cooling demand to be recognized in housing design in the UK. Appropriate passive cooling strategies can provide more comfortable and energy efficient houses. In the UK, although climate change will be extreme by the end of the century, passive cooling techniques involving strategic use of thermal mass and night time ventilation, will mitigate much of the change. With careful design, such as in the IDEAhaus, thermal mass can be incorporated into lightweight structures, and improve the performance of the dwelling.

8.2. Flooding

Flooding is a serious and increased risk during this century in the UK [25] Timber frame structures are very susceptible to damage from floods, but can be made flood resistant if adequately protected. The IDEAhaus utilizes a concrete basin-type ground floor with closed cell insulation and doorway protection to give protection against water ingress to a depth of 750 mm.

8.3. Modular Construction

The use of a hybrid open system of modular construction gives new flexibility that suits the demands of the UK housing market (see Figure 17). The use of mass-customization and customizable façade cladding can provide an industrialized but attractive choice of products, and this, an adaptable construction, can allow for change over time, which can mitigate for future uncertainty.

The IDEAhaus is one blue-print for a more adaptable UK house, than can withstand the rapid climate change expected in the twenty-first century.

Figure 17. Typical IDEAhaus.

Acknowledgments

The authors would like to acknowledge the following for their contributions to and support of the project: Triangle Architects (Matt Hargreaves, Peter Fisk, David Ward, Mark Trayhorn), Queens University Belfast School of Architecture (Morgan Grennan), Leeds Metropolitan Univeristy School of Architecture (Lucy Andersson), The Energy Council (Matthew Adams), Sutcliffe Consulting Engineers (Simon Brady, Ian MacIver), Markhams (Mike Hornsby, Helen Riley), Plus Dane Housing Group (Inger Leach, Steve Elliot), Technology Strategy Board (Julie Meikle, Mark Wray), Homes and Communities Agency (Allan Foster), Hemcrete Projects Ltd (Ian MacCarthy).

Author Contributions

The co-authors both contributed equally to this research project and the writing and review of this article.

Conflicts of Interest

The authors declare no conflict of interest.

References

1. The Guardian. Global Warming: Undeniable Evidence; 2010. Available online: http://www.theguardian.com/commentisfree/2010/feb/01/climate-change-university-east-anglia (accessed on 14 October 2013).
2. UK Met Office. Climate Projections up to 2080. Available: online: http://www.metoffice.gov.uk/services/climate-services/uk/ukcp (accessed on 14 October 2013).
3. Prometheus Project. University of Exeter, Exeter, UK. Available online: http://emps.exeter.ac.uk/research/energy-environment/cee/projects/prometheus/ (accessed on 14 October 2013).
4. Department of Environment, Food, and Rural Affairs. UK Climate Change Projections. Available online: http://ukclimateprojections.defra.gov.uk/content/view/868/531/ (accessed on 1 April 2013).
5. Eames, M.; Kershaw, T.; Coley, D. On the creation of future probabilistic design weather years from UKCP09. *Build. Serv. Eng. Res. Technol.* **2011**, *32*, 127.
6. Design for Future Climate. Available online: https://connect.innovateuk.org/web/design-for-future-climate/projects-outputs (accessed on 16 February 2014).
7. Gething, B.; Puckett, K. *Design for Climate Change*; RIBA Publishing: London, UK, 2012.
8. Williams, K.; Joynt, J.L.R.; Payne, C.; Hopkins, D.; Smith, I. The conditions for, and challenges of, adapting England's suburbs for climate change. *Build. Environ.* **2012**, *55*, 131–140.
9. *Investigation into Overheating in Homes*; Department for Communities and Local Government (DCLG): London, UK, 2012.
10. Chartered Institute of Building Services Engineers (CIBSE). *How to Manage Overheating in Buildings (CIBSE Knowledge Series 16): A Practical Guide to Improving Summertime Comfort in Buildings*; CIBSE: London, UK, 2012.

11. Sefaira Concept Software. Available online: http://www.sefaira.com/products/sefaira-concept/#sefaira-concept (accessed on 21 June 2012).
12. UK Government. *Conservation of Fuel and Power 2010 Approved Document Part L1a*; UK Building Regulations HMSO: London, UK, 2010.
13. Lane, S.N. Climate change and the summer 2007 floods in the UK. *Geography* **2008**, *93*, 91–97.
14. Sanderson, M. *Changes in the Frequency of Extreme Rainfall Events for Selected Towns and Cities*; UK Met Office: London, UK. 2010.
15. Macdonald, N. *Liverpool Hit by High Winds*. Available online: http://www.liverpoolecho.co.uk/news/liverpool-news/live-merseyside-hit-high-winds-6375437 (accessed on 5 December 2013).
16. Vandentorren, S.; Bretin, P.; Zeghnoun, A.; Mandereau-Bruno, L.; Croisier, A.; Cochet, C.; Ribéron, J.; Siberan, I.; Declercq, B.; Ledrans, M. August 2003 heat wave in France: Risk factors for death of elderly people living at home. *Eur. J. Public Health* **2006**, *16*, 583–591.
17. Keatinge, W.R. Death in heatwaves. *Br. Med. J.* **2003**, *327*, 512.
18. Bardi, U. Peak oil: The four stages of a new idea. *Energy* **2009**, *34*, 323–326.
19. Tseng, M.M.; Jiao, J. Mass Customization. In *Handbook of Industrial Engineering, Technology and Operation Management*, 3rd ed.; Wiley: New York, NY, USA, 2001.
20. Lime Technology Hemcrete Insulation. Available online: http://www.limetechnology.co.uk/hemcrete.htm (accessed on 13 July 2012).
21. Ibstock. Coolvault. Available online: http://www.ibstock.com/ (accessed on 8 March 2012).
22. *ISO 7730: 2005 Ergonomics of the Thermal Environment*; British Standards Institute: London, UK, 2005.
23. Nicol, F.; Humphreys, M.; Roaf, S. *Adaptive Thermal Comfort: Principles and Practice*; Routledge: London, UK, 2012.
24. Jankovic, L. A Method for Reducing Simulation Performance Gap Using Fourier Filtering; In Proceedings of the Building Simulation 2013 Conference, Chambery, France, 25–30 August 2013.
25. Evans, E.; Hall, J.; Penning-Rowsell, E.; Sayers, P.; Thorne, C.; Watkinson, A. Future flood risk management in the UK. *Water Manag.* **2006**, *159*, 53–61.

Permissions

All chapters in this book were first published in Buildings, by MDPI; hereby published with permission under the Creative Commons Attribution License or equivalent. Every chapter published in this book has been scrutinized by our experts. Their significance has been extensively debated. The topics covered herein carry significant findings which will fuel the growth of the discipline. They may even be implemented as practical applications or may be referred to as a beginning point for another development.

The contributors of this book come from diverse backgrounds, making this book a truly international effort. This book will bring forth new frontiers with its revolutionizing research information and detailed analysis of the nascent developments around the world.

We would like to thank all the contributing authors for lending their expertise to make the book truly unique. They have played a crucial role in the development of this book. Without their invaluable contributions this book wouldn't have been possible. They have made vital efforts to compile up to date information on the varied aspects of this subject to make this book a valuable addition to the collection of many professionals and students.

This book was conceptualized with the vision of imparting up-to-date information and advanced data in this field. To ensure the same, a matchless editorial board was set up. Every individual on the board went through rigorous rounds of assessment to prove their worth. After which they invested a large part of their time researching and compiling the most relevant data for our readers.

The editorial board has been involved in producing this book since its inception. They have spent rigorous hours researching and exploring the diverse topics which have resulted in the successful publishing of this book. They have passed on their knowledge of decades through this book. To expedite this challenging task, the publisher supported the team at every step. A small team of assistant editors was also appointed to further simplify the editing procedure and attain best results for the readers.

Apart from the editorial board, the designing team has also invested a significant amount of their time in understanding the subject and creating the most relevant covers. They scrutinized every image to scout for the most suitable representation of the subject and create an appropriate cover for the book.

The publishing team has been an ardent support to the editorial, designing and production team. Their endless efforts to recruit the best for this project, has resulted in the accomplishment of this book. They are a veteran in the field of academics and their pool of knowledge is as vast as their experience in printing. Their expertise and guidance has proved useful at every step. Their uncompromising quality standards have made this book an exceptional effort. Their encouragement from time to time has been an inspiration for everyone.

The publisher and the editorial board hope that this book will prove to be a valuable piece of knowledge for researchers, students, practitioners and scholars across the globe.

List of Contributors

Andrew Kauffman
Barnhart Crane and Rigging Company, Monroe, MI 48161, USA

Ali M. Memari
Department of Architectural Engineering and Department of Civil and Environmental Engineering, The Pennsylvania State University, University Park, PA 16802, USA

John Napier
Department of Architecture, University of Lincoln, Brayford Pool, Lincoln LN6 7TS, UK

Mark Bassett
BRANZ Ltd., P.O. Box 50908, Porirua City 5240, New Zealand

Greg Overton
BRANZ Ltd., P.O. Box 50908, Porirua City 5240, New Zealand

Toni L. Griffin
J. Max Bond Center on Design for the Just City, Spitzer School of Architecture, The City College of New York, 141 Convent Avenue, New York, NY 10031, USA

Dan Cramer
Grassroots Solutions, 2828 University Avenue SE, Suite #150, Minneapolis, MN 55414, USA

Megan Powers
Grassroots Solutions, 2828 University Avenue SE, Suite #150, Minneapolis, MN 55414, USA

Zulfikar A. Adamu
School of Civil and Building Engineering, Loughborough University, LE11 3TU Leicestershire, UK

Andrew Price
School of Civil and Building Engineering, Loughborough University, LE11 3TU Leicestershire, UK

Ricardo Antunes
Department of Civil and Environmental Engineering, University of Auckland, 20 Symonds Street, Auckland 1142, New Zealand

Vicente Gonzalez
Department of Civil and Environmental Engineering, University of Auckland, 20 Symonds Street, Auckland 1142, New Zealand

Roger Birchmore
Department of Construction, Unitec Institute of Technology, P.O. Box 92025 Auckland, New Zealand

Andy Pivac
Department of Building Technology, Unitec Institute of Technology, P.O. Box 92025 Auckland, New Zealand

Robert Tait
Department of Building Technology, Unitec Institute of Technology, P.O. Box 92025 Auckland, New Zealand

Bruno Palazzo
Department of Civil Engineering, University of Salerno, Via Giovanni Paolo II, 84084 Fisciano, Italy

Paolo Castaldo
Department of Civil Engineering, University of Salerno, Via Giovanni Paolo II, 84084 Fisciano, Italy

Ivana Marino
Department of Civil Engineering, University of Salerno, Via Giovanni Paolo II, 84084 Fisciano, Italy

Natee Singhaputtangkul
School of Engineering and Technology, Central Queensland University, Sydney, NSW 2000, Australia

Sui Pheng Low
Department of Building, School of Design and Environment, National University of Singapore, 4 Architecture Drive, Singapore 117566, Singapore

Mohammadali Noktehdan
Department of Civil and Environmental Engineering, The University of Auckland,

Mehdi Shahbazpour
Department of Mechanical Engineering, The University of Auckland, Private Bag 92019, Auckland 1142, New Zealand

Suzanne Wilkinson
Department of Civil and Environmental Engineering, The University of Auckland,

Morten Gjerde
School of Architecture, Victoria University of Wellington, P.O. Box 600, Wellington 6140, New Zealand

Brenda Vale
School of Architecture, Victoria University of Wellington, P.O. Box 600, Wellington 6140, New Zealand

Michele Betti
Department of Civil and Environmental Engineering (DICeA), University of Florence, Via di Santa Marta, 3-I-50139 Firenze, Italy

Luciano Galano
Department of Civil and Environmental Engineering (DICeA), University of Florence, Via di Santa Marta, 3-I-50139 Firenze, Italy

Andrea Vignoli
Department of Civil and Environmental Engineering (DICeA), University of Florence, Via di Santa Marta, 3-I-50139 Firenze, Italy

Greg Keeffe
Architecture at Queens, School of Planning, Architecture and Civil Engineering, Queens University Belfast, Belfast BT9 5AG, UK

Ian McHugh
Green Triangle Studio, 134 Sandy Lane, Manchester M21 8TZ, UK